Van Nostrand Reinhold Soil Science Series

Editor: Charles W. Finkl, Jnr., Florida Atlantic University

SOIL CLASSIFICATION / *Charles W. Finkl, Jnr.*
CHEMISTRY OF IRRIGATED SOILS / *Rachel Levy*
SOIL SALINITY: Two Decades of Research in Irrigated Agriculture /
    H. Frenkel and A. Meiri
ANDOSOLS / *Kim H. Tan*
PODZOLS / *Peter Buurman*
SOIL NUTRIENT AVAILABILITY: Chemistry and Concepts / *Y. K. Soon*
ADSORPTION PHENOMENA / *Robert D. Harter*
SOIL EROSION AND ITS CONTROL / *R. P. C. Morgan*
SOIL MICROMORPHOLOGY / *Georges Stoops and Hari Eswaran*

SOIL MICROMORPHOLOGY

Edited by

GEORGES STOOPS
State University of Ghent, Belgium

and

HARI ESWARAN
Soil Management Support Services
Washington, D.C.

A Hutchinson Ross Publication

VAN NOSTRAND REINHOLD COMPANY
New York

Copyright © 1986 by **Van Nostrand Reinhold Company Inc.**
Van Nostrand Reinhold Soil Science Series
Library of Congress Catalog Card Number: 85-11628
ISBN: 0-442-28058-0

All rights reserved. No part of this work covered by the copyrights
hereon may be reproduced or used in any form or by any means—
graphic, electronic, or mechanical, including photocopying,
recording, taping, or information storage and retrieval systems—
without permission of the publisher.

Manufactured in the United States of America.

Published by Van Nostrand Reinhold Company Inc.
115 Fifth Avenue
New York, New York 10003

Van Nostrand Reinhold Company Limited
Molly Millars Lane
Wokingham, Berkshire RG11 2PY, England

Van Nostrand Reinhold
480 Latrobe Street
Melbourne, Victoria 3000, Australia

Macmillan of Canada
Division of Gage Publishing Limited
164 Commander Boulevard
Agincourt, Ontario M1S 3C7, Canada

15 14 13 12 11 10 9 8 7 6 5 4 3 2 1

Library of Congress Cataloging in Publication Data
Main entry under title:
Soil micromorphology.
  (Van Nostrand Reinhold soil science series)
  "A Hutchinson Ross publication."
  Bibliography: p.
  Includes indexes.
  1. Soil micromorphology—Addresses, essays, lectures.   I. Stoops, Georges.
II. Eswaran, Hari.   III. Series.
S593.2.S65   1986   631.4'3   85-11628
ISBN 0-442-28058-0

CONTENTS

Series Editor's Foreword	ix
Preface	xiii
Contents by Author	xvii
Introduction	1

PART I: ROLE AND SCOPE

Editors' Comments on Papers 1, 2, and 3   10

  1   KUBIËNA, W. L.:   The Principle of Micropedology   12
        Micropedology, Collegiate Press Inc., Ames, Iowa, pp. 5-8 (1938)

  2   OSMOND, D. A.:   Micropedology   15
        Soils and Fertilizers **21:**1-6 (1958)

  3   BULLOCK, P.:   The Changing Face of Soil Micromorphology   21
        Soil Micromorphology, vol. I. Techniques and Applications,
        P. Bullock and C. P. Murphy, eds., AB Academic Publishers, Berkhamsted,
        pp. 1-18 (1983)

PART II: GENERAL CONCEPTS OF FABRIC

Editors' Comments on Papers 4, 5, and 6   40

  4A   KUBIËNA, W. L.:   Introduction   43
        Micropedology, Collegiate Press Inc., Ames, Iowa, pp. 125-128 (1938)

  4B   KUBIËNA, W. L.:   Elementary Fabric   47
        Micropedology, Collegiate Press Inc., Ames, Iowa, pp. 129-153
        and 171-173 (1938)

  5   BREWER, R. and SLEEMAN, J. R.:   Soil Structure and Fabric. Their
        Definition and Description   75
        Jour. Soil Sci. **11:**172-185 (1960)

  6   ALTEMÜLLER, H.-J.:   Gedanken zum Aufbau des Bodens und seiner
        begrifflichen Erfassung   89
        Zeitschr. Kulturtechnik **3:**328-334, 335, 336 (1962)

PART III: SPATIAL RELATIONSHIPS OF BASIC SOIL COMPONENTS

Editors' Comments on Papers 7, 8, and 9   98

  7   STOOPS, G. and A. JONGERIUS: Proposal for a Micromorphological
        Classification of Soil Materials. I. A Classification of the
        Related Distributions of Fine and Coarse Particles   102
        Geoderma **13:**189-199 (1975)

Contents

8 BREWER, R., and S. PAWLUK: Investigations of Some Soils Developed in Hummocks of the Canadian Sub-Arctic and Southern-Arctic Regions. 1. Morphology and Micromorphology 113
Canadian Jour. Soil Sci. **55**:304-311 (1975)

9 ESWARAN, H., and Cl. BAÑOS: Related Distribution Patterns in Soils and Their Significance 120
An. Edafologia Agrobiologia **35**:33-45 (1976)

PART IV: ORGANIC MATTER

Editors' Comments on Papers 10 Through 13 134

10 JONGERIUS, A., and J. SCHELLING: Micromorphology of organic matter formed under the influence of soil organisms, especially soil fauna 138
Internat. Congr. Soil Sci. 7th Trans. Madison, Wisc. **3**:702-710 (1960)

11 BABEL, U.: Humuschemische Untersuchung eines Buchen-Rohhumus mittels mikroskopischer Methoden 146
Mitt. Vereins f. Forstl. Standortkund. u. Forstpflanzenzüchtung **15**:33-38 (1965)

12 BARRATT, B.C.: A Revised Classification and Nomenclature of Microscopic Soil Materials with Particular Reference to Organic Components 153
Geoderma **2**:257-261, 264-271 (1969)

13 DE CONINCK, F., D. RIGHI, J. MAUCORPS, and ROBIN, A. M.: Origin and Micromorphological Nomenclature of Organic Matter in Sandy Spodosols 166
Soil Microscopy, G. K. Rutherford, ed., The Limestone Press, Kingston, pp. 263-273, 279-280 (1974)

PART V: MICROSTRUCTURE

Editors' Comments on Papers 14 and 15 180

14 BECKMANN, W., and E. GEYGER: Entwurf einer Ordnung der natürlichen Hohlraum-, Aggregat- und Strukturformen im Boden 184
Die mikromorphometrische Bodenanalyse, W.L. Kubiëna, ed., Ferdinand Enke Verlag, Stuttgart, pp. 163-188 (1967)

15 DUMANSKI, J., and R. J. ST. ARNAUD: A Micropedological Study of Eluvial Soil Horizons 210
Canadian Jour. Soil Sci. **46**:287-292 (1966)

PART VI: CLAY REARRANGEMENTS

Editors' Comments on Papers 16 Through 20 220

16 MINASHINA, N. G.: Optically Oriented Clays in Soils 225
Soviet Soil Sci. **4**:424-430 (1958)

17 STEPHEN, I.: Clay Orientation in Soils 232
Science Progress **48**(190):322-331 (1960)

18	**BUOL, S. W., and F. D. HOLE:** Clay Skin Genesis in Wisconsin Soils *Soil Sci. Soc. America Proc.* **25**:377-379 (1961)	242
19	**NETTLETON, W. D., K. W. FLACH, and B. R. BRASHER:** Argillic Horizons without Clay Skins *Soil Sci. Soc. America Proc.* **33**:121-125 (1969)	245
20	**FEDOROFF, N.:** Classification of Accumulations of Translocated Particles *Soil Microscopy,* G. K. Rutherford, ed., The Limestone Press, Kingston, Ontario, pp. 695-700, 711, 712 (1974)	250

PART VII: AMORPHOUS AND CRYSTALLINE NEOFORMATIONS

Editors' Comments on Papers 21 Through 25		258
21	**KUBIËNA, W. L.:** Die taxonomische Bedeutung der Art und Ausbildung von Eisenoxyhydratmineralien in Tropenböden *Zeitschr. für Pflanzenern., Düng., Bodenkunde* **98**:205-214 (1963)	262
22	**PARFENOVA, E. I., and E. A. YARILOVA:** Characteristic Features of Certain USSR Soils in Thin Sections *Mineralogical Investigations in Soil Science,* Israel Programs for Scientific Translations, Jerusalem, 1965, pp. 78-96	271
23	**FLACH, K. W., J. G. CADY, and W. D. NETTLETON:** Pedogenic Alteration of Highly Weathered Parent Materials *Internat. Congr. Soil Sci. 9th Trans.* **4**:343-351 (1968)	290
24	**TURSINA, T. V., I. A. YAMNOVA, and S. A. SHOBA:** Combined Stage-by-Stage Morphological, Mineralogical and Chemical Study of the Composition and Organization of Saline Soils *Soviet Soil Sci.* 81-94 (1980)	299
25	**MIEDEMA, R., A. G. JONGMAN, and S. SLAGER:** Micromorphological Observations of Pyrite and Its Oxidation Products in Four Holocene Alluvial Soils in the Netherlands *Soil Microscopy,* G. K. Rutherford ed., The Limestone Press, Kingston, Ontario, pp. 772-794	313

Author Citation Index	337
Subject Index	341
About the Editors	345

SERIES EDITOR'S FOREWORD

The Van Nostrand Reinhold Soil Science Series attempts to provide cogent summaries of the field by reproducing classical and modern papers, ones that provide keys to understanding of critical turning points in the development of the discipline. Scientific literature today is so vast and widely dispersed, especially in a multifaceted discipline like soil science, that much valuable information becomes ignored by default. Many pioneering works are now coveted by libraries, and retrieval from the archives is not easy. In fact, many important papers published in the ephemeral literature are no longer available to serious or committed researchers through interlibrary loan. Other professionals devoted to teaching or burdened with administrative duties must be hard pressed to keep up with comprehensive arrays of technical literature spread through scores of journals. Most of us can, at best, skim only a few select journals to make copies of tables of contents, abstracts and summaries, and reviews in order to remain abreast of specialized and often limited aspects of the robust field of soil science as a whole.

This series in soil science, developed as a practical solution to this problem, reprints key papers and investigative landmarks that relate to a common theme. The papers are reproduced in facsimile, either in their entirety or in significant part, so readers can follow major original events in the field, not peruse paraphrased or abbreviated versions of others. Some foreign works have been especially translated for use in the series. Occasionally short, foreign language articles are reproduced from French or German journals.

Essays by the volume editor provide running commentaries that introduce readers to highlights in the field, provide critical evaluation of the significance of the various papers, and discuss the development of selected topics or subject areas. It is hoped that the volume editor's comments will ease the transition for the seasoned investigator who wishes to step into a new field of research as well as provide students and professors with a compact working library of most important scientific advances in soil science.

Areas of specialization in soil science are divided by the International Society of Soil Science into seven divisions or "commissions." The first six commissions cover soil physics, chemistry, mineralogy, biology, fertility, and technology. Because the scope of the field is so great, we concentrate initially on topics traditionally devoted to the seventh commission: soil morphology, genesis, classification, and geography. The series thus begins

Series Editor's Foreword

with volumes dealing with the major soils of the world: their recognition, characteristics, formation, distribution, and classification. Other volumes concentrate on topics in agronomy, soil-plant relationships, soil engineering topics, or melds of pure science with soil systems. The Van Nostrand Reinhold Soil Science Series plows deeply through the field, picking significant but timely topics on an eclectic basis.

Each volume in the series is edited by a specialist or authority in the area covered by the book. The volume editor's efforts reflect a concerted worldwide search, review, selection, and distillation of the primary literature contained in journals and monographs and in industrial and governmental reports. Individual volumes thus represent an information-selection and repackaging program of value to libraries, students, and professionals.

The books contain a preface, introduction, and highlight commentaries by the volume editor. Many volumes contain rare papers that are hard to locate and obtain, as well as landmark papers published in English for the first time. All volumes contain author citation and subject indexes of the contained papers, usually twenty to fifty key papers in a given subject area.

This volume deals with soil micromorphology, a technical subject that has come of age in a mere fifty years. For many disciplines it is often difficult to clearly define *the* founding father because contributions come from different quarters at about the same time. This is not the case for soil micromorphology because its sequential development is accurately recorded in the literature. W. L. Kubiëna, an Austrian scientist, was the first person to use magnifying instruments to systematically study soil materials. By 1931 he had recorded some of his first observations, albeit in some rather obscure scientific journals. Since that time, however, there have been several international meetings and working groups that have dealt with aspects of the subject. Laboratories specifically equipped for micromorphological research have been set up at research institutes and universities in several countries; these establishments collectively encourage research on a broad basis. The development of a new method for preparing thin sections using polyester resins in the early 1960s revolutionized this young science. The introduction of this method for vacuum impregnation of soil materials made it possible to more easily prepare large numbers of quality thin sections, especially large sections (12×18 cm) that support soil micromorphometry, the quantitative study of distribution patterns. Thin sections and polished blocks are normally examined under a petrological microscope using different types of illumination, for example ordinary transmitted and reflected light, plane polarized light, crossed polarized light, and circularly polarized light. The advent of scanning electron microscopes and microprobes in the 1970s brought micromorphology into the realm of submicroscopy, now an extremely important technique that is essential to the detailed study of mineral alterations and neoformations. More recent techniques for examining thin sections and polished blocks include electron analyses (e.g. transmission and scanning electron microscopy), X-ray analysis, fluorescence, staining, and autoradiographs, among others.

Data derived from micromorphological research is finding increased

Series Editor's Foreword

application in diverse fields, viz. agriculture, archaeology, geotechnical engineering, geomorphology, paleoclimatology, pedology and paleopedology, soil microbiology, and soil microzoology. Because the kinds of information that can be obtained from thin sections of soils and unconsolidated materials is quite diverse, they may be used for analytical or interpretative purposes and to predict behavior under different environmental conditions. The usefulness of micromorphology, although not yet fully appreciated, is apparent from the rapid expansion of the field in micropedological investigations and in other fields as well.

The evolution of micropedological concepts, terminologies, and classification systems is carefully traced in this volume by facsimile reproduction of the original classic articles. This collection of benchmark papers highlights important stepping stones in the development of soil micromorphological research. Areas that are specifically examined include concepts of soil fabric, the spatial relations of basic soil components, soil organic matter, microstructure, rearrangements of clay and clay-sized particles, and mineral neoformations. Researchers pioneering in this field noted that the soil is full of unfathomable mysteries and that even with a lifetime of investigation of a single soil, some facets of this extremely complicated microcosm would still be unknown. Because the methodologies of micromorphology have advanced at pace with improvements in techniques for examining the nature and properties of small spaces, it behooves micropedologists and other interested investigators to reflect on the past achievements of their predecessors. This monograph thus surveys to advantage some important turning points in the development of the field of soil micromorphology, a rapidly growing endeavor.

CHARLES W. FINKL, JNR.

PREFACE

It is now about half a century since the first papers dealing with micropedological observations were published (e.g., Kubiëna, 1931). Since then this discipline has undergone quite an impressive evolution: concepts and terms were created, used, and replaced by newer ones, many techniques were developed and then sometimes forgotten, and a wide range of applications were invented.

Micromorphology contributed enormously to a better understanding of soil genesis, and many aspects of the formation of some soils could never have been explained without the study of thin sections. It proved to be a very useful tool for soil classification; as such, micromorphology was used in Soil Taxonomy (USDA) in identifying some diagnostic horizons.

Since the 1960s this discipline has expanded rapidly worldwide. Both the range of topics treated and the quantity of studies and publications have increased rapidly. As a result, it seemed impossible to select a series of papers that cover the whole field and still form a unit. Therefore we thought it more appropriate to cover only one subject. The most urgent need seemed to be a volume illustrating the evolution of the concepts, terminologies, and classification systems formerly or currently used, or now developing in soil microphology. We are aware that we have omitted very important and interesting fields of micromorphological research, such as soil physics (quantitative determinations of soil structure and/or porosity), soil genesis, geomorphology, paleopedology, and archeology—but for each of these fields we could have selected enough important papers to fill another volume.

We chose this topic for two reasons:
1. Our years of teaching experience have shown us that students might encounter difficulties in interpreting micromorphological studies that mention older or less common concepts and terminology. Although the *Glossary of Soil Micromorphology* (Jongerius and Rutherford, 1979) has partially solved this problem, a study of the original, generally well-illustrated papers is generally preferable—and even necessary in the case of complex concepts.
2. A knowledge of the variety of micromorphological concepts and an understanding of their evolution is without doubt essential to gaining a better insight into the field and new developments. This is certainly a must for scientists involved in the theoretical development of this discipline.

As this volume deals to a large extent with terminology, the question of the importance of a specific system of micromorphological terminology

Preface

may arise. Some soil scientists claim that a specific system of terminology and classification is superfluous, and that micromorphological descriptions can just as easily be made using the everyday language. We, along with many other micromorphologists, do not agree with this point of view, for several reasons: (a) A clear, precise, and well-defined standardized terminology is necessary for the appropriate communication of scientific observations, and is especially indispensable for communication between scientists who speak different languages. Many scientists whose mother tongue is English or French do not realize (or do not want to realize) the difficulty of precisely expressing observations in a general, everyday foreign language, or even of understanding such descriptions. (b) Standardized terminology and classification are essential to the development of data-storing techniques and worldwide comparison of micromorphological characteristics of soil materials. (c) A clear terminology and classification system is also an important didactic tool. Indeed, many years of experience with students at different levels have shown us that students at first remember only those things for which we have a specific name. A comprehensive classification, moreover, obliges them to analyze and record every feature of everything they observe. In addition, it is also a good memory aid, even for a trained micromorphologist.

It will probably strike the reader that we have given a relatively large amount of space to papers discussing the orientation of clay. This was done because the concepts illustrated there (clay orientation, illuviation, mechanical destruction) belong to the most intensively studied domains of micromorphology.

The papers reprinted in this volume have been selected from a wide variety of publications. We have limited ourselves to two types of papers: papers giving the original definitions of concepts and terms and papers that show the evolution of the concepts in this discipline. Although the final selection has been aided by some practical factors, such as the language of the papers, copyright clearances, and so on, our main criterion was the paper's real contribution to the field of soil micromorphology, as reflected, for instance, in recent papers and handbooks (e.g., FitzPatrick, 1984; Bullock et al., in press).

In addition, we have tried to demonstrate the historical development of some concepts and terms. Although we included some of the older papers mainly for their historical value, their scientific level is generally so high that they still bear a message even for the modern researcher.

Moreover, some of the older papers are practically unavailable to scientists working in relative new institutions where the libraries don't have earlier volumes of journals and congress transactions and some important papers were published in local bulletins with a very limited circulation.

We acknowledge the critical comments of Prof. Dr. A. Bronger, Dr. P. Bullock, Dr. M. Kooistra, Dr. A. McKeague, Drs. H. Mücher, and Dr. E. Van Ranst.

GEORGES STOOPS
HARI ESWARAN

REFERENCES

Bullock, P., N. Fedoroff, A. Jongerius, G. Stoops, and T. Tursina, *Handbook for Soil Thin Section Description,* Waine Research Publications, Wolverhampton, (in press).

FitzPatrick, E. A., 1984, *Micromorphology of Soils,* Chapman & Hall, New York and London, 433p.

Jongerius, A., and G. K. Rutherford, eds., 1979, *Glossary of Soil Micromorphology,* Centre for Agricultural Publishing and Documentation, Wageningen, 138p.

Kubiëna, W., 1931, Mikropedologische Studien, *Arc. Pflanzenbau* **5**(4):613-648.

CONTENTS BY AUTHOR

Altemüller, H.-J., 89
Babel, U., 146
Baños, Cl., 120
Barrat, B. C., 153
Beckmann, W., 184
Brasher, B. R., 245
Brewer, R., 75, 113
Bullock, P., 21
Buol, S. W., 242
Cady, J. G., 290
De Coninck, F., 166
Dumanski, J., 210
Eswaran, H., 120
Fedoroff, N., 250
Flach, K. W., 245, 290
Geyger, E., 184
Hole, F. D., 242
Jongerius, A., 102, 138
Jongman, A. G., 313
Kubiëna, W. L., 12, 43, 47, 262

Maucorps, J., 166
Miedema, R., 313
Minashina, N. G., 225
Nettleton, W. D., 245, 290
Osmond, D. A., 15
Parfenova, E. I., 271
Pawluk, S., 113
Righi, D., 166
Robin, A. M., 166
Schelling, J., 138
Shoba, S. A., 299
Slager, S., 313
Sleeman, J. R., 75
St. Arnaud, R. J., 210
Stephen, I., 232
Stoops, G., 102
Tursina, T. V., 299
Yamnova, I. A., 299
Yarilova, E. A., 271

SOIL MICROMORPHOLOGY

INTRODUCTION

WHAT IS MICROPEDOLOGY?

Micropedology is a modern method of studying undisturbed soil samples with the aid of microscopic and ultramicroscopic techniques to identify the different constituents and to determine their mutual relations in space and—as far as is possible—in time. Its aim is to search for the processes responsible for the formation not only of the soil in general but also of specific features, both natural (clay skins, nodules) and artificial (irrigation crusts, surface slacking). Consequently, it has become an important tool for investigations into the genesis, classification, and management of soils.

Micropedology is based on two principles: the undisturbed sample and functional investigation. The first principle dictates that investigations be carried out on undisturbed (and mostly naturally oriented) soil samples. Micropedology thus contrasts with mineralogical, chemical, and physical analyses, which require that the samples be mixed, ground, solubilized, or fractionated (Paper 1). The second principle, functional investigations, is self-explanatory: all observations should be directed to reaching an understanding of the function of each soil constituent, and of their interrelationship.

The research domain of micropedology encompasses all microscopic observations of undisturbed soil samples, including studies of thin sections, micromanipulations, microchemical and microphysical methods, and even the more recently developed ultramicroscopic techniques. Of all the study areas, the most developed is soil fabric analysis, also called micromorphology, and its quantitative counterpart, soil micromorphometry.

In contrast to most other soil scientists, micropedologists are in general more interested in specific features, observed in the profile or in thin sections, than in the average composition of the material. Their interpretation of these features sometimes clearly reveals the genesis of the soil.

Introduction

HISTORICAL REVIEW

Observations of soil materials with a more or less strong hand lens, in the field or in the laboratory, probably have been made since the early days of soil science. The first person who used magnifying instruments in a systematic way to study the soil was the Austrian scientist W. L. Kubiëna, who is considered the spiritual father of micropedology. By 1931 Kubiëna had published his first observations in some smaller papers, but his work did not achieve an international reputation until 1938, when his manual *Micropedology* was published. Kubiëna devoted an important part of this book to the explanation and illustration of his morphoanalytical approach to the soil microfabric. The term *morphoanalytical* refers to analysis of the fabric according to pure morphological criteria. All soil components, and their mutual relationships are considered. The genetic interpretation of the data obtained follows only as a second step. Kubiëna introduced completely new terminology to name the different fabric types he described. His book, then, can be considered the basis for the further development of micromorphology.

Kubiëna distinguished two main soil constituents: fine material (of colloidal size), which he called the "plasma" of the soil that is supposedly more or less mobile, and the coarser, immobile material (mostly the sand and the silt fraction), which he called the "skeleton" of the soil. He further distinguished eight main types of "elementary fabrics," that is, mutual distribution patterns of plasma and skeleton (Paper 4). In addition, he delineated higher levels of fabric based on the structure (aggregates, cleavage blocks, coherent soils) and specific features, such as crystallizations of calcite, gypsum, iron segregation, and so forth.

Although World War II brought a temporary halt to most scientific research on soils, Kubiëna was able to continue his investigation in Austria and Spain. In 1948 he published *Entwicklungslehre des Bodens,* in which he explained his ideas about soil genesis, based partly on his micromorphological observations. He devoted part of the book to the presentation of his new morphogenetic approach to the soil microfabric.

Whereas the morphoanalytical approach calls for an analysis of the complete fabric, the morphogenetic approach considers only some characteristic features, and then uses them to determine the position of the material in question within the presumed evolutional sequence of soils. Kubiëna (1948, 1953) believed in a hierarchical genetic classification of soils, whereby some types may evolve into others depending on internal and external conditions.

In this morphogenetic system no individual features are considered;

rather, all characteristics as a whole are related to a specific soil type, after which the microfabric is named. The Braunlehm type is considered as a central concept from which many others may be derived (Paper 21). Two main groups of microfabrics can be distinguished: the Lehm-fabric and the Erde-fabric. The former group is characterized by a high mobility of the plasma, expressed by the presence of clay orientations (Paper 16), as dense packing and a weak structure resulting in a high sensitivity to erosion. The Erde-fabric, on the other hand, is characterized by a relative immobility of the plasma, resulting in a stable structure and a high degree of porosity. Kubiëna also described transitional types. Only some of these fabric terms were used in micromorphological texts (Paper 2).

This morphogenetic approach to the study of microfabrics involves a genetic interpretation of the soil, beginning with the descriptive phase of study. The terminology is restricted to the soil types described by Kubiëna; the terms should therefore not be used for soil materials with a similar fabric but a different genetic evolution (e.g., polycyclic soils).

Until the end of the 1950s, few papers on micromorphology appeared, and only a small number of laboratories were equipped for these investigations, which were restricted more and more to the study of thin sections. Some of the most active centers of that period were Reinbek, West Germany, where several younger scientists were working under Kubiëna's leadership; Arnhem, the Netherlands (A. Jongerius), Braunschweig, West Germany (H.-J. Altemüller), Ghent, (J. Laruelle), Moscow (E. I. Parfenova and E. A. Yarilova), and Zurich (E. Frei).

In 1958 a first International Working Meeting on Soil Micromorphology, organized by H. Frese and H.-J. Altemüller in Braunschweig, contributed a great deal to the development of this young science. Most authors were using Kubiëna's terminologies and concepts, although some developed their own ideas. H.-J. Altemüller (1962) presented a new method of thin section preparation, introducing the use of polyester resins for impregnations. Although his paper is purely technical, its importance in the further development of micromorphology cannot be sufficiently emphasized. The impregnation method facilitated the preparation of large numbers of high-quality thin sections and of large thin sections (e.g., 12 × 18 cm), which in turn permitted the development of the quantitative study of pore patterns and the like.

The early 1960s saw an expansion of micromorphological studies in many countries. Several general trends became evident, for instance: micropedology became more and more restricted to morphological studies on soil thin sections (micromorphology); several laboratories became interested in the quantification of features observed in thin sections (e.g., W. Beckmann, 1962; E. Geyger, 1962; A. Jongerius, 1963; G. Reuter, 1964); and the morphogenetic approach of Kubiëna and his school was found to be unsatisfactory in several ways—as a result, R. Brewer and J. R. Sleeman in Australia developed a new morphoanalytical system for micromorphological descriptions of inorganic soil components (Paper 5). Brewer (1964) included it in his book *Fabric and Mineral Analysis of Soils*.

According to Brewer, the fabric analysis of soil material is based on a recognition of the orientation and distribution patterns of the constituents. Individual, stable mineral grains are called skeleton grains, and soluble grains and materials of colloidal size form the plasma. Orientation patterns of the clay particles within the plasma as observed between crossed polarisers, are described as different types of plasmic fabric. Specific arrangements (e.g., concentrations, accumulations, reorientations) of the constituents are recognized and called pedological features. These features are subdivided according to their position with respect to the ped surfaces, their genesis, or their morphology. Five morphological groups of pedological features are considered: cutans (e.g., clay coatings), glaebules (e.g., concretions), pedotubules (e.g., crotovina), crystallaria (e.g., pores filled with gypsum crystals), and subcutanic features (e.g., slickensides). The description of several different types of voids gives an idea of the porosity pattern of the soil.

The Second International Working Meeting on Soil Micromorphology, held in Arnhem in 1964, was attended by many scientists from all over the world, indicating that soil micromorphology had been accorded recognition as an important subfield of soil science. In fact, the second half of the 1960s witnessed an explosive growth of the field of soil micromorphology. Several new centers for study were established in Great Britain, France, Spain, and elsewhere in Europe, and researchers in Africa, South America, and Asia began to show interest in micromorphology. Some of this growth of interest was due to the influence of institutions such as the post graduate training centers at Ghent and Wageningen, and ORSTOM in Paris.

Presentations at the Third International Working Meeting on Soil Micromorphology (held in Wroclaw, Poland, in 1969) made it clear that most micromorphologists were using the Brewer system. Application of this system, and the study of specific arid and tropical soils with previous unknown features, led several scientists to attempt to complete the system by adding new terms or adapting some of Brewer's concepts. To prevent as much as possible a proliferation of terminology, an International Working Group was established to propose an international acceptable terminology and classification of micromorphological features, and to publish a multilingual glossary of terms (I.S.S.S. Bull. 35 [1969] and 37 [1970]. The glossary was published in 1979 (Jongerius and Rutherford) and a handbook is in press (Bullock et al.).

Meanwhile, some scientists tried to extend the use of the Brewer system to studies of the organic matter present in soils (Paper 12; Bal, 1973; Paper 13). The concepts of plasma and skeleton grain and their related distribution patterns gave rise to some scientific discussion and proposals for adaptation of the Brewer system (Papers 7, 8, and 9).

Although some work had been done on quantification of soil components in the early 1960s (e.g., Paper 3), only in the second half of that decade did this type of research become important. The publication of *Die mikromorfometrische Bodenanalyse,* edited by Kubiëna in 1967, can be considered a milestone in this respect. The evolution of electronic equipment in the 1970s made it possible to apply electro-optical methods to the quantification of pore spaces (Jongerius, Schoonderbeek and Jager, 1972; Murphy, Bullock and Turner, 1977; Murphy, Bullock and Biswell, 1977; Bouma et al., 1977). Thus many more scientists and agronomists became interested in applied micromorphology.

The recent evolution of micropedology is due in part to the increased availability, beginning in the 1970s, of scanning electron microscopes (SEMs) and microprobes in many laboratories. SEMs allowed a more detailed study of fine crystalline neoformations and transformations in the soil, whereas micro-analyzers were very helpful in the determination of mineral alterations and neoformations. For an excellent review of the possibilities and applications of submicroscopic techniques in micropedology, see Bisdom (1981) and Bisdom and Ducloux (1983). An example of the integrated application of SEMs in micromorphology is given in Paper 24.

A few words concerning the parallel development of micromorphology in the USSR may be useful. Several centers (e.g., Dokuchaev Institute and the Universities of Moscow and Leningrad) are known for their micromorphological investigations. Most Soviet scientists use the traditional terminologies of the petrographic school, although specific terms (*polynite* and *clay pseudomorph,* among others) have been introduced (Paper 22). Some Soviet scientists also follow Kubiëna's system, and have proposed ameliorations and additions to it.

POSSIBLE PROGRESS IN METHODOLOGY AND CONCEPTS

A micromorphological analysis involves four steps: (a) preparation of the sample, (b) observation, (c) presentation of the data, and (d) interpretation.

Introduction

Sample Preparation

It is clear that the quality of thin sections could be improved by making them thinner and cleaner, and that their fabrication could be more automatized. In addition, better dehydration techniques are needed for making samples suitable for impregnation—unless a water-compatible resin can be found.

Observation

The definition of micropedology makes it obvious that observation is an important phase in the study of a sample. Little progress can be expected in optical microscopy, except for a better application of specific techniques such as ultraviolet fluorescence (e.g., Van Vliet-Lanoe, 1980), interference microscopy, phase contrast, and so forth. In addition, the use of microchemical reactions on uncovered soil thin sections might be extended, allowing more complete mineralogical and chemical identifications. Without doubt, much progress can be made with submicroscopic techniques and electro-optical image analyzers; however, a discussion of this topic is beyond the scope of this volume.

Representation of Data

With respect to the presentation of data, considerable progress may still be expected from better description techniques using standardized terminology. Eventually we may see the data stored in computers. Comparing descriptions made by different authors from different "schools" is now very difficult, as most authors describe only some of the features they have observed. A more uniform presentation of data would make comparison (with or without the help of a computer) possible.

Interpretation

This ability to compare results would clearly ameliorate interpretation, as would the availability of a complete collection of reference material. Ideally, this reference material would comprise both standard thin sections or descriptions, and would correlate micromorphological data with other physical or chemical soil characteristics, or with specific processes and evolutionary phases. Such data can be obtained only through a systematic and standardized study of all different types of soils and soil horizons. This will in turn give rise to the need for a higher level of classification of soil fabrics, requiring the categorizing and naming of typical or diagnostic combinations of features rather than of individual features. In petrographic terms, this would mean calling a particular rock "granite" rather than merely referring to it as a mixture of alkali-feldspar, plagioclase, quartz, and mica with a subhedral granular fabric. Not only new names, but also new concepts and new rules, must be created. A first attempt in this direction, restricted to some diagnostic horizons of soil taxonomy, has been made by Fedoroff and Bullock (1978). It is obvious that such a classification would also be of great help in the study of paleosols and polycyclic soils.

Much progress has still to be made in the interpretation of the results obtained by morphometric methods. Indeed, the pedological meaning of some quantitative results seems doubtful. A study on methodology is therefore necessary, using procedures that yield a maximum of information and selecting the best parameters.

To obtain more realistic and useful interpretations, a more systematic approach is necessary. Until now, most scientists were satisfied with a static, momentary image of the soil fabric. In reality, however, the soil is dynamic, changing throughout the seasons. This means that for detailed studies of soils in contrasting climates, samples should be collected in each season. This method of sampling is especially justified for surface horizons and crusts from which the microstructure or composition, or both, is strongly influenced by changes in temperature and humidity. Litter layers and salt crusts, although extreme examples, clearly demonstrate this need. The study of samples gathered at different seasons may yield a cinematographic image of soil dynamism.

REFERENCES

Altemüller, H.-J., 1962, Verbesserung der Einbettungs- und Schleiftechnik bei der Herstellung von Bodendünnschliffen mit Vestopal, *Zeitschr. PflErnähr. Düng.* **99:**164–177.

Bal, L., 1973, *Micromorphological Analysis of Soils. Lower Levels in the Organization of Organic Materials,* Soil Survey Paper No. 6, Netherlands Soil Survey Institute, Wageningen, 175p.

Beckmann, W., 1962, Zur Mikromorphometrie von Hohlräumen und Aggregaten im Boden, *Zeitschr. PflErnähr. Düng.* **99:**129–139.

Bisdom, E. B. A., ed., 1981, *Submicroscopy of soils and weathered rocks,* 1st Workshop of the Intern. Working-group on Submicroscopy of Undisturbed Soil Materials (IWGSUSM), Wageningen, Centre for Agricultural Publishing and Documentation, 320p.

Bisdom, E. B. A., and J. Ducloux, 1983, *Submicroscopic Studies of Soils,* Elsevier Publ. Co., Amsterdam, Oxford and New York, 356p.

Bouma, J., A. Jongerius, O. Boersma, A. Jager, and D. Schoonderbeek, 1977, The function of different types of macropores during saturated flow through four swelling soil horizons, *Soil Sci. Soc. America Jour.* **41**(5): 945-950.

Brewer, R., 1964, *Fabric and Mineral Analysis of Soils,* J. Wiley & Sons, London, New York, and Sydney, 470p.

Bullock, P., N. Fedoroff, A. Jongerius, G. Stoops, and T. Tursina, *Handbook for Soil Thin Section Description,* Waine Research Publications, Wolverhampton, (in press).

Fedoroff, N., and P. Bullock, 1978, Principes et méthodologie de la description microscopique des sols, in *Micromorfologia de Suelos,* M. Delgado, ed., 5th International Working Meeting Soil Micromorphology, Proc. Granada, pp. 59-92.

Geyger, E., 1962, Zur Methodik der mikromorphometrischen Bodenuntersuchung, *Zeitschr. PflErnähr. Düng.* **99**:118-129.

Jongerius, A., 1963, Optic-volumetric measurements on some humus forms, in *Soil Organisms,* ed. J. Doeksen and J. Van der Drift, 137-148. Amsterdam.

Jongerius, A., D. Schoonderbeek, and A. Jager, 1972, The application of the Quantimet 720 in Soil Micromorphometry, *The Microscope* **20**(3): 243-254.

Jongerius, A. and G. K. Rutherford, eds., 1979, *Glossary of soil Micromorphology,* Centre for Agricultural Publishing and Documentation, Wageningen, 138p.

Kubiëna, W. L., 1938, *Micropedology,* Collegiate Press Inc., Ames, Iowa, 242p.

Kubiëna, W. L., 1948, *Entwicklungslehre des Bodens,* Springer Verlag, Vienna, 215p.

Kubiëna, W. L., 1953, *The Soils of Europe,* Thomas Murby & Co., London, 314p.

Kubiëna, W. L., 1967, *Die mikromorphometrische Bodenanalyse,* Ferdinand Enke Verlag, Stuttgart, 196p.

Murphy, C. P., P. Bullock, and R. H. Turner, 1977, The measurement and characterization of voids in soil thin sections by image analysis. Part I. Principles and techniques, *Jour. Soil Sci.* **28**:498-508.

Murphy, C. P., P. Bullock, and K. J. Biswell, 1977, The measurement and characterization of voids by image analysis. Part II. Applications, *Jour. Soil Sci.* **28**:509-518.

Reuter, G., 1964, Zur Mikromorphologie lessivierter Böden in verschiedenen Klimagebieten, in *Soil Micromorphology,* A. Jongerius ed., Elsevier, Amsterdam, pp. 213-218.

Stoops, G., 1978, *Provisional Notes on Micropedology,* Geologisch Instituut, R.U.G.

Van Vliet-Lanoe, B., 1980, Approche des conditions physico-chimiques favorisant l'autofluorescence des minéraux argileux, *Pédologie* **30**(3): 369-390.

Part I
ROLE AND SCOPE OF SOIL MICROMORPHOLOGY

Editors' Comments
on Papers 1, 2, and 3

1   **KUBIËNA**
    The Principle of Micropedology

2   **OSMOND**
    Micropedology

3   **BULLOCK**
    The Changing Face of Soil Micromorphology

Half a century ago, micropedology and soil micromorphology were still unknown terms and concepts for most soil scientists. Micropedology was first introduced to a broad international audience by the *Micropedology,* by W. L. Kubiëna, published in 1938. In the first chapter (Paper 1), Kubiëna clearly explicated its principles. For many years, this text has served as a source of inspiration for introductory lectures on micropedology.

Paper 2 shows the state of the art 20 years later. Most investigations in soil science were hindered during World War II and the following years, but sufficient progress was made in micromorphology that the relationship between different soil types and their micromorphological identity was better recognized, and new fields of application were discovered. The paper gives a complete review of the progress made in preparation techniques and of the results obtained in the fields of soil genesis and classification and experimental pedology.

During the 1960s, micromorphology enjoyed an increasing interest from soil scientists, and the number of research workers grew rapidly. Several papers published in the 1960s address the concepts and role of micromorphology. Zimmerman (1964), for example, discussed the position of this discipline within the context of soil science, and considered the methodology of micromorphology from a theoretical point of view. In addition, he analyzed the steps from sampling to final interpretation. Kubiëna (1967) discussed how to interpret micromorphological observations and the possibilities of applied quantitative micromorphology (micromorphometry). Zachariae (1967) clearly explained the application of micromorphological methods in soil zoology, including paleopedology and archaeology.

Editors' Comments on Papers 1, 2, and 3

The invention of electronic instruments had a major impact on micromorphological research in the 1960s. Formerly investigations had been practically restricted to direct optical microscopic observations, but now more and more sophisticated instruments became available: electro-optical systems allowing automatic image analyses and quantifications of soil porosity; scanning electron microscopes, allowing very high magnifications of undisturbed soil materials; and different types of microprobes for *in situ* chemical analyses of micron-sized fields. Paper 3 presents an overall review of this rapidly expanding micromorphological research.

REFERENCES

Kubiëna, W. L., 1967, Mikromorphologie und Mikromorphometrie, in *Die mikromorphometrische Bodenanalyse,* Ferdinand Enke Verlag, Stuttgart, pp. 4-18.

Zachariae, G., 1967, Der Einsatz mikromorphologischer Methoden bei bodenzoologischen Arbeiten, *Geoderma* **1:**175-196.

Zimmerman, K., 1964, Uber die Stellung der Mikromorphologie in Rahmen der Bodenkunde, in *Soil Micromorphology,* A. Jongerius, ed., Elsevier, Amsterdam, pp. 505-524.

Copyright © 1938 by Collegiate Press, Inc.
Reprinted from pages 5-8 of *Micropedology,* Collegiate Press, Inc., Ames, Iowa, 1938, 242p.

The Principle of Micropedology

W. L. Kubiëna

What is micropedology? What use has it in general pedology, and how do its principles differ from the latter?

The first question which natural science generally asks in opening a new investigation of an object in nature is: How does it appear? If we would state this question more exactly, we would have to say: How does it seem to us? How does it seem to our special human senses? Because, as the philosophers tell us, we know nothing about the things as such, we only know how they seem to us.

How does the soil appear to human beings? It looks like a more or less plastic mass, in most cases somewhat drab in color and possesses certain physical and chemical properties.

If we were of the size and if we had the manner of living of the microorganisms, our answer would be quite a different one. Perhaps we would say: The soil is a huge system of many-shaped cavities, which are built of glassy material, partially clear and colorless, partially intense green, red, yellow or brown, almost entirely transparent, seldom translucent and only infrequently opaque. In the cavities active organisms are to be found; in some only few, in others very many, according to the size, climate, and food condition of the various cavities.

For centuries science in general advanced knowing little of matter save what man could observe under ordinary conditions. Since the construction of microscopes and other instruments, man has been able to penetrate deeper, to see and to know more. But this "knowing more" developed only through accident, and still the things are perceived not as they really are, but only as they appear to us, for knowledge depends very much on the instruments and the methods which are available to us. Through the use of improved instruments, or instruments which work in another way, and better methods, it has been shown that many concepts developed through use of older instruments and methods have led us to false conclusions. We shall always continue to have new and improved instruments and better methods in the future.

Did we not use the microscope for soil investigation in the past? It was used primarily for examining soil microorganisms or for determining mineral particles by optical methods developed by petrographers. It was used for the investigation of a small group of soil constituents, isolated from the soil, and not for the investigation of the soil itself as a formation and as an entity. We were able to isolate a few ingredients of the soil and to work with them in isolated stages, but we knew very little about how these ingredients were related in the soil, in what microscopic location they were to be found, how they arrived at these places, and what role they played in their environment. Since many objects occurring in the soil spaces are so fragile that they are destroyed by the least disturbance, our knowledge of them was negligible.

Take, for instance, the city of New York. If it were put in a large glass vessel with water or hydrochloric acid, as we do with the soil, and shaken for twenty-four hours, one would not then be able to reconstruct Broadway, Fifth Avenue, or the Empire State Building, or to find out what kind of goods are found in the large warehouses on the New York harbor. The first thing to know, in order to get an idea of New York, is not so much the nature of its chemical composition as a whole, but how it looks in detail as a structural entity.

This is not to be read as a criticism of the existing methods and procedures of soil research. They will find their use in the future as they have in the past, but they need amendment. In microscopic dimensions the soil is not just a mass, but a whole world. We are able to get an idea of what we know of this world if we think of it in terms of our world translated down to microscopic dimensions. All its towns, villages, houses, church towers, trees, men, and animals would be visible only by the use of a good microscope. We could not have known very much about it if we had not lived in it ourselves. It would appear to us as does the soil, quite like a mass. Naturally then we cannot treat this soil world *en masse* if we wish to know what is actually going on within it.

What principles have we to observe in investigating the microscopic world of the soil? First of all the soil must not be disturbed in its natural arrangement. The spaces must be opened but not destroyed. How can this be done? Generally we split open the undisturbed soil sample with a strong needle or we break it open. Now we take some single soil spaces and try to translate all these unimaginable small objects into the more familiar dimensions of our own world. We have to use special microscopes for this purpose which work, not with transmitted light, but with incident light. They enable us to investigate formations in nature directly without the necessity of making special preparations. They allow an observation of things in nature *in vivo* and *in situ*.

Microscopes employing reflected vertical light have found extensive use in metal microscopy, but in metal microscopy the preparations have plane and polished surfaces. In soil microscopy we examine cavities, many of which are comparatively deep. This prevents us from the free use of the highest magnifications, and in part also from the use of the so-called vertical illuminators.

We need, furthermore, not only to see the microscopic objects and organisms in the undisturbed soil, but also to handle everything to be found in the soil spaces, to take things out, to put others in, to investigate some of the constituent substances chemically and physically, to carry out measurements in a selected region, and to arrange experiments in small cavities. The possibilities for doing all this are disclosed to us by methods which in their totality we designate as "microtechnical methods." The term "microtechnique" or "microtechnical methods" is used here in the sense of methods which have to be performed in microscopic dimensions. They are always confined to the use of the microscope, and that not only for mere observation or temporary control of the effect of manipulations, but for the actual performance of these manipulations. Methods of this character which are used in micropedology can be divided into three groups: (1) micrurgical methods, (2) microchemical methods, and (3) microphysical methods. The name of the first group is derived from micrurgy, which means the art of manipulating microscopic objects. The term is used here in a somewhat wider sense, i. e., not only for manipulations performed by using the highest magnifications, special manipulation apparatus and finest glass tools, but for every kind of manipulation performed with the aid of a microscope. Micropedological work needs, for the great part, methods of a kind of semi-micrurgy, using lower magnification, incident light, free hand manipulation, and resterilizable metal tools.

In view of the particular necessity for the use of microtechnical methods in micropedological research it is essential to know what similar methods have been developed by other sciences, how these methods may be used for soil investigations, how they may be adapted to this purpose and what other methods, not yet developed, remain to be devised as parts of the technique necessary to achieve our aim.

MICROPEDOLOGY

by

D. A. OsMOND

Micropedology treats of the investigation of soils by methods adapted to the use of small quantities of material and of the study and interpretation of details of soil morphology that are not visible to the unaided eye. Much can be seen and a different outlook on soil morphology and genesis can be obtained if the soil is studied under a dissecting or petrological microscope in the laboratory or field though the use of either out of doors is not very convenient.

The first book dealing solely with this subject was written in 1938 by W. L. Kubiena and in it the use of the microscope and microchemical methods in the study of soils is described and several examples are given of the kinds of observations made and how they are to be interpreted. This was followed ten years later by his "Entwicklungslehre des Bodens" in which the development of soils and development-sequences of soils are discussed, with particular reference to those on calcareous rocks. The importance of biological activity in soil formation is emphasised and micromorphological evidence is used in support of a genetic approach to the problems of soil development. In 1953 "The Soils of Europe" appeared in which Kubiena introduced a classification of soils giving great weight to soil genesis and making much use of micromorphological observations in recognizing and describing soil types.

Mick (1949) has remarked that "in the final analysis a study is only as good as the samples on which it is based", and in micropedological studies it is essential that the profiles investigated should be carefully chosen from sites whose history is known with some certainty and that samples taken should be good examples of the phenomena being studied. Samples of known orientation should be collected so that observations can be made on both vertical and transverse faces or sections (Kubiena, 1938). Coarse-structured soils can be sampled by carefully removing small blocks or entire peds* from the face and labelling them so that their orientation is known. Looser materials such as sandy or weakly structured soils and highly organic horizons are more conveniently collected in a sample frame (Kubiena, 1953) consisting of a shallow box, 3 x 2.5 x 1.5 in., with removable top and bottom; this can be eased into the soil with a stout knife and an undisturbed sample taken. A similar frame 1 x 1 x 1 in. has been found of more use in stony soils and has the additional advantage that, if necessary, the whole block can be removed and impregnated for sectioning.

Preparation of thin sections. During recent years attention has been focussed on the interpretation of thin sections although only a few pages of Kubiena's book (1938) are given to a description of their preparation. While of use in studies of organic and surface soils, there is a wealth of information to be obtained from sections of mineral soils and particularly of the subsurface layers. Several methods are available for the preparation of sections, but it is essential that all pore spaces should be completely filled with the impregnating medium so that the peds are firmly held in their relative positions during subsequent operations. Soils differ so much that probably no one medium is suitable for all. Natural waxes and mixtures of natural resins have been successfully used (Kubiena, 1938) but there are now available suitable transparent plastics having refractive indices near that of Canada balsam. In a recent paper by Dalrymple (1957) the names and addresses of suppliers of several suitable impregnating materials are given. While it is possible to impregnate a large porous ped, it may be necessary to use a small block or slice if the soil is dense and difficult to impregnate. After impregnation a thin slice is cut or sawn off and one side, ground flat and polished, is cemented to a flat microscope slide with Canada balsam, Kollolith or Lakeside 70. The excess thickness is removed by grinding with different grades of carborundum powder or other abrasive either dry or wetted with paraffin (kerosene) until the slice is thin enough. For most purposes a thickness of 30μ is adequate and may be checked by observing the polarization colours of suitable mineral grains, e.g. quartz or feldspar. Structures and some other features may sometimes be more readily recognized if a somewhat thicker slice is used. When finished the slide is thoroughly cleaned and a cover glass cemented on with cold Canada balsam or other cement.

---

* A ped is an individual natural soil aggregate. (Soil Survey Manual, U.S.D.A.)

Polished impregnated blocks or thin sections can be used for estimating the volume percentage pore space by Rosiwall's method and for determining its variability (Swanson & Peterson, 1940). Portions can be removed from uncovered sections for microchemical investigation although little use has been made of this approach beyond ascertaining the chemical nature of efflorescences and concretions.

Stereoscopic microscope observations. The description and identification of the materials of superficial organic horizons are not easy in the field, but can be fairly readily done in the laboratory by dissecting a frame sample, when the essential similarities of the organic profile under different vegetation covers become apparent although their morphologies differ according to the flora and fauna present. During dissection some of the fauna can be captured for later identification and it is evident that the number and species of fauna profoundly affect the rate of destruction of plant remains and the kind of humus formed (Kuhnelt, 1950; Murphy, 1953). Useful classifications of soil fauna based on their action on plant remains have been made by Jacot (1940) and Fenton (1945).

The L-layer, of recently fallen plant remains, can be readily distinguished from earlier and partly destroyed fallings whose destruction by bacterial, fungal and faunal activity becomes obvious in the F-layer in which fungi can be recognized as well as the partly eaten plant remains and droppings of soil fauna that occur as black, round or cylindrical pellets of various sizes either loose, attached to or sometimes inside the remains. With increasing depth, droppings and their remains become more obvious at the expense of the plant remains which are eaten and reduced in size. The H-layer consists almost entirely of droppings, some mineral grains and a few resistant plant fragments. The droppings are recognizable in the upper part, but under weak pressure or through dispersion by percolating solutions their shape is lost and the base of the layer is amorphous, black and greasy. Thin sections of droppings show that they may contain a few very small fragments of undigested plant remains (plant splinters) and very small mineral grains.

By studying the micromorphology of organic horizons Kubiena (1953) has classified the humus forms and has described the main groups as dy, gyttja, sapropel and peat, formed under sub-aqueous conditions and raw-humus, moder and mull found under terrestrial conditions. There are several subtypes, each with a recognizable micromorphology, related more specifically to climate, vegetation and fauna. The identification and recognition of humus forms are of importance, for certain of them are associated with definite soil forms and are useful in classification. A detailed study of humus forms and their distribution in heath and woodland soils has been made by Jongerius (1956) who subdivides moder according to the fauna from which it originates, and finds a relation between the formation and mechanical transport of moder and of dispersed colloidal humus depending on the degree of podzolization. In poorer soils there is only infiltration of dispersed humus while more mechanical mixing occurs in the less podzolized soils.

Microscopic investigation with good field work assists in elucidating the genesis and course of development of soils such as the rendzina-like soils that have been carefully worked out by Kubiena (1948). These are developed on calcareous rocks and have A- and C-horizons only. In the early stages of development the A-horizon is a loose, rendzina-moder consisting of partly destroyed plant remains, numerous droppings from the drought-resistant fauna and mineral grains of calcite or dolomite. The droppings are well preserved and, as in all moders, mineral and organic matter can be mechanically separated. These primitive soils are shallow and support a scanty vegetation, but in the course of development a similar but deeper profile occurs under a grassy vegetation suitable for rough grazing. With increasing weathering, more clay is released, the soil becomes deeper and a more varied and moisture-loving fauna can exist. By passage through larger fauna, particularly earthworms, organic and mineral matter is inseparably mixed in the humus form of mull which is almost completely devoid of recognizable plant remains. Microscopic examination of such mull rendzinas shows the A-horizon to be composed almost entirely of worm-casts forming rounded aggregates. Such a soil is well aerated, for the stable crumb peds do not fit closely, but it also retains moisture adequate for crop growth.

A similar course of development of ranker-like soils can be traced on silicate rocks (Kubiena, 1953) but owing to the lack of calcium carbonate their subsequent development differs from that of the rendzina-like soils. Which stage of development of rendzina- or ranker-like soils is prominent in a landscape will depend, among other things, on such factors as the climate and nature of the rock.

Brown earths have a more complex profile of the type A (B)* C and in all sub-types except one (Alpine sod brauenerde) the humus form is mull. Microscopic examination of peds of the (B)-horizon shows that they lack the smooth, rounded shapes of those in the A-horizon and are irregularly, often subangularly, shaped owing to shrinkage cracks developing. The soil is more or less spongy and the peds are porous or, in some cases when they are relatively dense, are fissured so that drainage is not impeded. These micromorphological features of the A- and (B)-horizons are indicative of good aeration and water retention; the production of mull denotes high biological activity, and organic matter

---

* A (B)-horizon is one formed by deep-seated chemical weathering and simultaneous oxidation of the iron complexes; it is not an enriched horizon of illuviated substances (Kubiena, 1953).

is, therefore, quickly incorporated; furthermore, burrowing animals can penetrate the soil deeply and, by mixing it, some of the losses of nutrients by leaching are replaced. The soil is relatively stable and unlikely to suffer erosion or to develop an impermeable layer.

The (B)-horizons of some soils derived directly from limestone have a quite different micromorphology from those described above and are found in several soil types, but it will be sufficient to describe some features of the terra fusca (Kubiena, 1953) developed on limestone in a humid climate. The humus form of the A-horizon is mull (or mull like moder) formed of worm-casts with a low humus content so that they are easily destroyed, and, on drying, the peds are somewhat angular. When examined microscopically the (B)-horizon looks waxy, very dense and non-porous, though fissured by shrinkage cracks, small ripple-like markings can often be seen on the ped surfaces, mineral grains appear to be embedded and when removed are clean and neither stained nor coated; the peds are angular or sub-angular and have sharp edges. Ferruginous or manganiferous concretions of various sizes have smooth surfaces and round shapes.

The micromorphology of this type of soil suggests that the surface would not retain its structure and under certain conditions it would not be easy to prepare a good seed-bed. The properties of the sub-soil indicate that natural drainage may also be somewhat impeded although water may pass along cracks and fissures between the peds.

Soil fabrics. One of the most important subjects of micromorphological investigations is the study and interpretation of the soil fabric, i.e., the arrangement of the components of the soil in relation to each other (Kubiena, 1938). The components of the fabric are mineral grains and the matrix, which is largely unresolvable and consists of clay minerals, finely divided sesquioxides and organic matter. The way in which the components are arranged determines the shapes and sizes of the cavities. From observations on soils from many parts of the world it appears that there may only be a few kinds of fabrics and attempts are being made to identify fabrics with specific soil types by attaching the soil name to the fabric, but, although a certain fabric may be associated with a particular soil, Kubiena (1938) has given examples of the same fabric occurring in different soils.

When seen in thin section the fabric of most brown earths is spongy with sinuous, rough-walled channels and holes. The mineral grains are in a brown-coloured matrix that is uniformly distributed and there are no areas of accumulation or concentration; iron or manganese concretions are also lacking. The clay matrix is granular and flocculated, though the degree of flocculation may be variable, and, when viewed between crossed polars, shows no birefringence phenomena. The properties of the fabric, particularly of the matrix, suggest that the constituents are immobile and that clay particles released or formed by weathering remain *in situ* in a flocculated state.

A very different picture is presented by the (B)-horizon of the terra fusca. The fabric is dense with very few cavities; channels and cracks are fairly straight and result from shrinkage. The mineral grains are embedded in a uniformly yellow-coloured matrix in which dark brown ferruginous concretions of circular cross-section and smooth outline often occur; similar manganiferous deposits are black. Viewed between crossed polars, brilliantly illuminated streaks displaying complete or partial (shadowy) extinction are strikingly visible along channels completely or partly filling holes and surrounding mineral grains, the appearance being caused by the regular arrangement of oriented clay particles.

The difference between these two fabrics is very striking and has been used by Kubiena (1948) as a broad distinction between 'erde' with a loose, porous fabric without obvious flow structures or doubly refractive streaks and 'lehm' having a dense, uniformly coloured, anisotropic matrix. The distinction originated from field observations of tropical soils, for those that are friable, loose, crumb-structured, stable and of low plasticity have an erde fabric (roterde, braunerde), while those that are easily dispersed and hence subject to erosion, and are plastic and form hard angular clods when dry, have a lehm fabric (rotlehm, braunlehm). In braunlehm and rotlehm, developed on siliceous rocks under tropical and subtropical conditions, the birefringent parts of the matrix do not exhibit marked common orientation, and the matrix is traversed by numerous, small, doubly refractive streaks. This matrix has been observed in a rotlehm from Cyprus (Osmond and Stephen, 1957).

Fabrics have been recorded by Peterson (1937), Frei and Cline (1949), McCaleb (1954) and Brewer (1956a), in which along channels and in cavities there is an accumulation of strongly oriented clay, which is often arranged in layers as shown by alternating lighter and darker coloured yellow or brown bands. Such a layered, strongly oriented accumulation of clay suggests clay migration in the profile and, indeed, it is difficult to think of any other mechanism by which it could have been produced.

Few micropedological investigations have yet been reported of the fabrics of lateritic soils, but there is little doubt of their complexity, for Humbert (1948), describing sections of laterites, writes of hydrous iron oxides orienting themselves around nuclei, losing water, reducing their surface and forming compact concretions; also of the growth of concretions and conducting veins to entangle crystalline constituents. More recently Alexander *et al* (1956) discussing the hardening of laterite point out that "pisoliths, concretions or definitely defined separate bodies are found in most laterites... there are bodies of different sorts where a particular kind of material or mineral mixture has separated." In laterites with saprolitic structures "a lattice-like network of oriented films of kaolin upon which is overlaid the impregnating iron mineral, usually goethite," is commonly found. According to Kubiena (1956b) laterization manifests itself in all kinds of pore spaces in the subsurface layers of soils with braunlehm or rotlehm characters. In addition to translucent areas, the fabric includes dense, opaque parts with a great variety of form—filaments, flakes, coatings, etc., and silica appears as opal, chalcedony and secondary quartz. He is thus able to distinguish between the effects of laterization and "rubefication" which refers to the tendency to the production of red colorations in tropical soils of braunlehm character. The fabric of rubified soils has little variety of form, is uniformly translucent, is fairly uniformly red in colour, and silica in the forms given above is absent. It appears that there is a development sequence from one fabric to another. Brewer (1957) gives a brief description of the sequence of fabrics from braunlehm to roterde via rubefied braunlehm, rotlehm and earthy rotlehm.

It is evident that the elucidation of the genesis of fabrics is of great interest and will throw light on many problems of soil formation.

Soil genesis and classification. Soil microscopy is being increasingly applied to problems of soil genesis, one of the commonest being to find reasons for the formation of B- or (B)-horizons. Microchemical studies and examination of thin sections of B-horizons of humus or iron podzols show that the grains are coated with iron-humus compounds of the same character as the matrix forming intergranular braces or cement between the grains and that their disposition is what would be expected if they were deposited from solution (Kubiena, 1938). After percolating aqueous leaf extracts through columns of quartz grains coated with ferric oxide, Bloomfield (1956a) found the appearance of the columns strongly reminiscent of a podzol profile and in all cases the "B-horizons" were 1-2 cm below the level to which the sand had dried out.

Thin sections of brown earths show clean grains in an immobile clay matrix and no evidence of illuviation of ferruginous compounds or clay. It can be assumed that oxidative weathering is responsible for the production of ferruginous coatings on clay minerals released or formed *in situ* and that podzolization does not occur because of the activity of the numerous mull-forming fauna whose activity in the surface does not allow a raw-humus to form which could provide the necessary reducing and solvent agents.

The B-horizon of grey-brown podzolic soils is characterized by an increased clay content compared with that of the A-horizon and also by the presence of clay skins. Frei and Cline (1949) described the micromorphology of a sequence of soils from grey-brown podzolic soils through intergrades to brown podzolic soils. From the non-uniform distribution of clay and its optically continuous orientation along channel walls they concluded it unlikely that the clay was residual but that its accumulation was due to percolating clay suspensions. Studies of the intergrade soils showed that as the A-horizons of the grey-brown podzolic soils became depleted of bases, a progressive degradation of the upper part of the B-horizon occurred and micromorphological features of the brown podzolic soils developed in the A-horizons. Furthermore, the mineral grains became coated similarly to those thicker coatings found in the B-horizons of podzols. A similar conclusion regarding the origin of some of the clay in the B-horizon of certain loess soils had been reached by Peterson (1937) who also noted the partly layered structure of such illuviated clay; the same phenomenon has been reported by McCaleb (1954) and Brewer (1956a).

Not every case of increased clay content in the B-horizon need be explained by clay illuviation, for Mick (1949), after calculating "weathering indices" from the ratios of certain resistant minerals to show the conformity of soil and parent material, concluded, from physical and mineralogical determinations, that the increase in clay content in the B-horizon is not necessarily caused by this process and that it is probably not so dominant in the formation of the grey-brown podzolic profile as has been commonly supposed. This view has been substantiated by Brewer (1955) by a careful study by similar methods of a yellow podzolic soil. He found that the higher clay content of the B-horizon can be explained by weathering of minerals and synthesis of clay *in situ*. Thin sections were used to demonstrate this and to show that the micromorphology associated with clay movement was lacking.

Raeside (1956), examining the micromorphology of a New Zealand yellow-grey earth with a hardpan and characteristic columnar structure, found the pores "lined with laminated birefringent colloid" and "all mineral grains coated with yellow-brown colloid which acts as a cementing agent." The amount of colloid, however, was insufficient to explain the physical character which apparently is attributed to close packing of the material. He considers the soil polygenetic and that "the maritime situation contributes soluble salts that promote clay shift in a manner analogous to that in solodized solonetz soils . . . .". In some solodized solonetz soils in Australia, however, Brewer (1956b) found that thick clay coatings, assumed to be due to illuviation from their position in the profile, were only weakly oriented and suggested that the clay suspension had been flocculated and the clay deposited without much orientation in a zone of high electrolyte content.

In an examination of two profiles developed on essentially the same parent material under different drainage conditions and classified as a red-brown earth and a grey and brown soil of heavy texture respectively, Brewer (1956a) found that in the free-draining red-brown earth, clay migration, for which good permeability is necessary, had played an important part in its development. Although migration had occurred in the upper part of the more poorly drained soil its role in soil formation was much smaller and in the impermeable B_{21}-horizon it only occurred where pore spaces were unusually large. The investigation also led to the conclusion that the soils differ genetically—the more poorly drained soil being polygenetic—and Brewer observes that although they had been originally separated on a morphological basis, the sections show that much greater differences exist and he remarks that evidence derived from sections should be considered in classification.

Further point is given to this comment in Brewer's report (1957) that Laruelle, working on some Belgian soils correlated with the grey-brown podzolic soils of the U.S.A., fails to find much strongly oriented clay coating in the B-horizon. Macromorphologically the soils are similar, but the microscope reveals an important difference which emphasises Raeside's plea (1956) for more care in correlating soils in widely separated geographical regions "unless it can be established that all morphological characteristics are identical and the environmental histories of the soils are similar." It is evident that the micromorphology should be considered in correlating and classifying soils. Kubiena (1956a) has suggested that in many cases the grey-brown podzolic soils of the U.S.A. are synonymous with the Central European lessivé soils, the B-horizons of which have a fabric resembling that in a braunerde with numerous, often broad, channels and cavity walls lined by optically oriented clay and generally so disposed as to suggest downward clay migration.

Experimental. It is not easy to verify experimentally some of the hypotheses put forward to interpret micromorphological observations and in many instances only the accumulation of evidence, supplemented by observations from other lines of work, may enable definite conclusions to be drawn.

The investigation of the production of humus forms by biological activity is not easy in view of the many species of both flora and fauna that take part. Murphy (1953) refers to several investigations of the palatability of litter of various tree species, and also confirms that broadleaf species are more readily and quickly eaten by mites than are coniferous species. Kubiena (1948) demonstrated the formation in six months of an artificial protorendzina by allowing collembola (*Sinella coeca*) to feed on plant remains. By adding dolomitic marl to the protorendzina and allowing earthworms to work upon the mixture for a few months, it was converted to a mass resembling the surface soil of a mull rendzina.

Experiments demonstrating the orientation of clay have been carried out by Brewer (1956b) who showed that doubly refractive films can be formed by infiltrating dilute clay suspensions into columns of sand grains. The nature of the dominant exchangeable cation had no effect on the orientation of a clay consisting of 60 per cent illite, 40 per cent kaolinite and some quartz. Flocculated clay formed oriented clay films when mixed with sand and silt and allowed to dry with continual stirring, provided the sand/clay ratio was high, but less orientation occurred if the ratio was low or if a significant amount of silt was present.

While there seems little doubt that clay migrates through the soil profile the dynamics of the movement are not clear. In many publications Kubiena has pointed to the protective action of humus colloids in those situations where they can be found, and there are many cases where this explanation is feasible. Where humus colloids are absent or present in only small amount, he has postulated that colloidal silicic acid may act in the same way, as indicated by Reifenberg's experiments on the action of silica sols on solid ferric oxide. Bloomfield (1956b) has shown that aqueous leaf extracts are capable of deflocculating clay and hence that they may play a part in the mobilization of clay in the upper soil horizons.

Conclusion. In this short review of the kind of problems investigated by micropedological methods it is evident that the application of microscopic techniques has an important part to play in elucidating the dynamics of soil formation and of the processes that occur inside the soil body. After visiting many workers in soil microscopy, Brewer (1957) remarks that "foremost among the processes studied . . . are:—

(a) Biological activity, humification and development of humus forms.
(b) Illuviation of clay-size material, humus and sesquioxides.
(c) Reorganization of clay-size material by differential movement caused by wetting and drying, root pressure, etc.
(d) Separation and concentration of sesquioxides within soil horizons.
(e) Development of structure forms.
(f) Separation and crystallization of secondary silica.
(g) *In situ* weathering of some primary rock-forming minerals."

This list by no means exhausts the problems that await solution, for many more arise if the later development of the subject of palaeopedology is taken into account. Many papers of great interest have been written on this subject, some of which have been reviewed by Osmond (1956). Nothing has been said of the contribution of petrological studies to the weathering of rocks and minerals under different soil-forming conditions to give the constituents building the soil fabric, nor of the insight to be gained by X-ray analysis of clay minerals and the determination of their shapes and sizes by electron microscopy. Soil is such a complex body that its problems will only be resolved by the combined efforts of workers in many branches of science.

References

ALEXANDER, L. T., CADY, J. G., WHITTIG, L. D., ET AL. 1956. Mineralogical and chemical changes in the hardening of laterite. *Trans. 6th int. Congr. Soil Sci.* E, 67-72.

BLOOMFIELD, C. 1956a. Experimental production of podzolization. *Trans. 6th int. Congr. Soil Sci.* E, 21-23.

BLOOMFIELD, C. 1956b. The deflocculation of kaolinite by aqueous leaf extracts. *Trans. 6th int. Congr. Soil Sci.* B, 27-32.

BREWER, R. 1955. Mineralogical examination of a yellow podzolic soil formed on granodiorite. *C.S.I.R.O. (Australia) Soil Publ.* No. 5, 1-28.

BREWER, R. 1956a. A petrographic study of two soils in relation to their origin and classification. *J. Soil Sci.* 7, 268-279.

BREWER, R. 1956b. Optically oriented clays in thin-sections of soils. *Trans. 6th int. Congr. Soil Sci.* B, 21-25.

BREWER, R. 1957. Report of overseas visit. *C.S.I.R.O. (Australia)*.

DALRYMPLE, J. B. 1957. The preparation of thin sections of soils. *J. Soil Sci.* 8, 161-165.

FENTON, G. R. 1945-47. The Soil Fauna: with special reference to the ecosystem of forest soil. *J. Anim. Ecol.* 14-16, 76.

FREI, E., CLINE, M. G. 1949. Profile studies of normal soils of New York II. Micromorphological studies of the gray-brown podzolic-brown podzolic soil sequence. *Soil Sci.* 68, 333-344.

HUMBERT, R. P. 1948. The genesis of laterite. *Soil Sci.* 65, 281-290.

JACOT, A. P. 1940. The fauna of the soil. *Quart. Rev. Biol.* 15, 28.

JONGERIUS, A. 1956. Étude micromorphologique des sols sableux secs des bois et bruyères aux Pays-Bas. *Trans. 6th int. Congr. Soil Sci.* E, 353-357.

KUBIENA, W. L.
1938. Micropedology. Ames.
1948. Entwicklungslehre des Bodens. Springer, Wien.
1953. The Soils of Europe. Murby, London.
1956a. Zur Mikromorphologie. Systematik und Entwicklung der rezenten und fossilen Lössboden. *Eiszeitalter u. Gegenwart* 7, 102-112.
1956b. Rubefizierung und Laterisierung. *Trans. 6th int. Congr. Soil Sci.* E, 247-249.

KÜHNELT, W. 1950. Bodenbiologie mit besonderer Berücksichtigung der Tierwelt. Herold, Wien.

MCCALEB, S. B. 1954. Profile studies of normal soils of New York IV. Mineralogical properties of the gray-brown podzolic-brown podzolic soil sequence. *Soil Sci.* 77, 319-333.

MICK, A. H. 1949. The pedology of several profiles developed from the calcareous drift of Eastern Michigan. *Mich. agric. exp. Sta. tech. Bull.* 212, 1-58.

MURPHY, P. W. 1953. The biology of forest soils with special reference to the mesofauna or meiofauna. *J. Soil Sci.* 4, 155-193.

OSMOND, D. A. 1956. Palaeopedology. *Sci. Progr.* 44, 682-687.

OSMOND, D. A., STEPHEN, I. 1957. The micropedology of some red soils from Cyprus. *J. Soil Sci.* 8, 19-26.

PETERSON, J. B. 1937. Micropedology of some loessal soils of Iowa. *Proc. Soil Sci. Soc. Amer.* 2, 9-13.

RAESIDE, J. D. 1956. Yellow-grey earths of South Island, New Zealand. An example of polygenesis in soil development. *Trans. 6th int. Congr. Soil Sci.* E, 665-672.

SWANSON, C. L. W., PETERSON, J. B. 1940. Differences in the microstructure of the Marshall and Shelby silt loam. *Proc. Soil Sci. Soc. Amer.* 5, 297.

THE CHANGING FACE OF SOIL MICROMORPHOLOGY[1]

P. Bullock

Soil Survey of England and Wales, Rothamsted Experimental Station, Harpenden, England

ABSTRACT

Some of the main landmarks in the development of soil micromorphology are reviewed. The main application continues to be in studies of soil genesis; some advances have been made in areas such as podzolisation, weathering and gleying but, despite the number of investigations relating to clay translocation, many questions still remain to be answered. The main changes in soil micromorphology have come through the development of new apparatus and techniques which now enable rapid accurate measurements of soil components to be made. These techniques have mainly been used to quantify soil structure and chemical elements in soil components and to determine the mineralogy of particular soil features. Micromorphology is being applied in a number of disciplines, particularly agriculture, archaeology and Quaternary geology. Now that micromorphology can be quantitative as well as observational and descriptive, a wider range of potential users is likely. It is particularly important that more attention be paid to representative sampling to provide statistically reliable results.

INTRODUCTION

At the 11th International Society of Soil Science Congress in Edmonton, Canada, 1978, soil micromorphology was given Sub-Commission status by the ISSS. The International Working-Meeting held in London is the first meeting of the Sub-Commission. It is also the sixth International Working-Meeting on Soil Micromorphology. Micromorphology has in a sense 'come of age' and has entered an exciting and challenging period. This Address reviews some important recent developments and considers some of the important changes taking place.

---

[1] Presidential address delivered at the International Working-Meeting on Soil Micromorphology, London, 1981.

LANDMARKS IN THE DEVELOPMENT OF SOIL MICROMORPHOLOGY

The first landmark, and perhaps the most important, was the publication of the book 'Micropedology' by W. L. Kubiena in 1938. The beginning of micromorphology can be said to be synchronous with this event for, although there had been some applications of micromorphology earlier, these had been somewhat piecemeal. Since 1938 micromorphology has become an established branch of soil science. Progress was at first slow and the accelerator appears to have been the publication of 'Fabric and Mineral Analysis of Soils' by Brewer in 1964. In this he set out to systematise the description and classification of features observed in soils, in particular using the optical microscope. One objective was to facilitate precise descriptions of soils and soil materials to any desired level of detail, an approach much needed to produce basic data.

The increasing number of soil micromorphologists saw a need for the exchange of ideas and this led to the first Working-Meeting of Soil Micromorphology, held in 1958 in Braunschweig-Volkenrode, Germany. The success of this led to a regular succession of such meetings at 4-5 year intervals, in Arnhem, The Netherlands (1964), Wroclaw, Poland (1969), Kingston, Canada (1973) and Granada, Spain (1977). The Proceedings published from each of these Meetings have served as a summary of progress in soil micromorphology and have provided valuable sources of information in a field with few textbooks.

At the Working-Meeting in Poland an International Working-Group was set up under the auspices of the International Society of Soil Science. Its initial brief was twofold. Firstly, because of the proliferation of new micromorphological terms and the confusion that often existed in the use of existing terms, it was thought necessary to produce a Glossary. The Glossary comprising 661 terms, and containing the original definition of each term and a subject index for each of five languages, was published in 1979 under the editorship of A. Jongerius and G. K. Rutherford. Secondly, the Working-Group was to review existing systems of description and classification in micromorphology and to develop internationally acceptable systems for both. This is nearly complete and is expected to be published soon under the title of 'A Handbook for Soil Thin Section Description'. The principal authors are P. Bullock, N. Fedoroff, A. Jongerius, G. Stoops and

T. Tursina. The International Working-Group ceased to exist when Soil Micromorphology became a Sub-Commission of ISSS in 1978 but the spirit and aims have been carried forward.

Looking back over this period of development it is clear that micromorphology traditionally has been an observational, descriptive and interpretative branch of science. The main studies have been directed towards soil genesis and this area is a suitable one with which to begin a survey of recent developments and the changes that have been taking place.

MICROMORPHOLOGY IN SOIL GENESIS STUDIES

Kubiena's book 'Micropedology' established the direction of micromorphology for many years. Its main theme was the use of the microscope and microchemical methods to study the morphology of soils as a basis for interpreting the processes involved in their formation. This approach was further developed in his 'Entwicklungslehre des Bodens', published in 1948, in which he used micromorphological evidence to support a genetic approach to soil development In his book 'The Soils of Europe', published in 1953, micromorphological criteria were used to differentiate the various soil types and humus forms. This represented the first systematic use of micromorphology in soil classification.

Until recently, micromorphological interpretations of soil genesis have been based on many assumptions and hypotheses about soil behaviour and soil-forming processes. As Brewer (1973) indicated 'recognition of microscopic features begins with accumulating data on the optical properties of the constituents that form the features. Interpretation of the processes involved in the formation of features depends on hypotheses of soil-forming processes and some knowledge of the chemical and physical properties'.

Currently, there are two main aims in applying micromorphology to study soil genesis:
1) The simplest is to supplement field observations and description in order to more fully characterise the morphological properties of soils and to understand the interaction of different soil processes. When used for this purpose, it is important that the link between field and laboratory observations are maintained. On the one hand,

this allows the micromorphologist to relate features observed in thin sections to the field environment, undoubtedly facilitating surer interpretations of genesis than would be the case if the micromorphologist worked 'in vacuo'. For the field investigator, the detailed information produced by the micromorphologist on fabric, feature composition, etc., leads to a better understanding of the relationships between morphological features and soil-forming processes and experience gained from such data can be related to other soils and other areas of land.
2) The second main aim is to contribute to a fundamental understanding of soil-forming processes and the morphological features to which they give rise. This approach involves in-depth studies using a number of methods of which micromorphology is only one, albeit a very important one. The combination of micromorphology with biological, physical and chemical methods is essential if real progress is to be made in fully comprehending soil processes.

Advances in knowledge of soil genesis attributable to micromorphology over recent years have been gradual rather than spectacular. They have however, been significant in understanding the processes of podzolisation, weathering and gley phenomena. In contrast, relatively little progress has been made in improving knowledge about clay translocation.

Podzolisation and the Spodic Horizon

The types of microstructure associated with spodic horizons, pellety and coated grains, have been recognised for some time (Soil Survey Staff, 1960). Both types have been used as criteria for the spodic (podzol) B horizon in some classification systems. Micromorphologists in western Europe, particularly De Coninck, Righi and co-workers, have attempted to establish the origin of these contrasting microstructures. The coated grain structure consists of plasma made up of organic matter and Al, or Al and Fe, which not only coats grains but also bridges them to form a more or less cemented mass. The coatings are characterised by a marked crack pattern and by a large number of pores $<0.2\mu m$ diam. which De Coninck and co-workers have interpreted as evidence for their deposition from a hydrated gel with subsequent desiccation and cracking. Relatively large amounts of organic matter, Al and Fe are extractable from the coatings with Na- or K- pyrophosphate. The organic matter extracted is rich in fulvic acids with a high content of COOH and phenolic OH groups, leading De Coninck and co-workers to suggest that the Fe and Al move as a complex with the fulvic acids. It is generally agreed that the morphology of the coatings is consistent with a direct process of eluviation followed by illuviation.

There is less agreement about the nature and origin of pellety aggregates. These, in contrast to the relatively undisturbed grain coatings, may occur in horizons of more intense biological activity and De Coninck and co-workers have linked their presence and form to faunal activity. Using the ^{14}C method, De Coninck (1980) measured the mean residence time (MRT) of organic matter in the two kinds of spodic horizon. The results show that the MRT of organic matter in the pellets is always less than that in the coatings, indicating a faster turnover in the case of pellets, consistent with greater biological activity. Although there is a strong school supporting a biological origin for the pellety horizon, some workers prefer a physico-chemical origin. Bruckert and Selino (1978), for example, have noted that some pellety aggregates, with the appearance of faunal excrements, seem to form by adsorption of soluble organic complexes on to the surfaces of clays and amorphous hydroxides, the whole developing a pellety structure.

The last few years have seen a much more convincing characterisation of the properties of the two kinds of spodic horizon, using micromorphology as well as a number of other techniques. Although considerable advances have been made, a completely satisfactory explanation for the pellety structure is still lacking. It is important that its origin should be established; if the pellety structure is associated with biological activity, then it is unlikely to be confined to podzolic soils. There is much stimulus and interest in this field of research and more convincing evidence is likely to be produced in the near future.

Weathering Studies
There have been some interesting investigations into the weathering of minerals in recent years. Some stimulus has come from the Advisory Group on Weathering and Neoformations of the ISSS Sub-Commission of Soil Micromorphology. The Group has published a system for the description of

mineral alterations (Stoops et al., 1979) and a detailed study of olivines, their pseudomorphs and secondary products (Delvigne et al., 1979). Its future programme includes a systematic study of the weathering of some other important soil minerals including mica and chlorite.

Recent work by Meunier (1976) and with Velde (1978, 1979) has highlighted the value of an integrated approach to the study of weathering at the fresh rock-saprolite interface, in which micromorphology is one of several methods used. Meunier used optical microscopy to divide two granites in western France into a number of weathering zones. Within each, a number of micro-environments were selected and the weathering phases associated with them examined in detail. Not only did Meunier establish the secondary phases produced for each initial granite mineral but proceeded a stage further to relate the particular petrographic zones to geochemical environments. The particular type of pore space which affects the flow of solutions, the concentrations of the solution and the nature of the dominant ion, was defined for each environment.

The study was carried out at a number of levels from field description through microscopic studies to X-ray and electron microprobe analyses. It highlights the importance of recognising that the soil is made up of a number of microsystems each with its own weathering and geochemical environment. Such a careful approach can provide a wealth of information on weathering, which is unlikely to be produced from analyses of large bulk samples in which several micro-environments may be compounded. There is much scope for further investigations of this kind.

Some advances have also been made in studying the processes that transform saprolite, in many cases the C horizon of the overlying soil, into subsoil B horizons. Flach et al.(1968) reported that the transition from saprolite to oxic horizon, for example, consisted of two types of morphological rearrangements. The contrasting components of the saprolite first break down and combine into a uniform, near unoriented matrix in the lower part of the B horizon. In the upper part, the matrix assumes a new orientation pattern in response to stresses generated in the soil. The transition from saprolite to oxic horizon is associated with a marked increase in clay. The authors proposed the term 'pedoplasmation', the formation of plasma, for the transformation process and considered the main agents to be shrink-swell, wetting and drying, root action and soil fauna. Pedoplasmation is thought to be the dominant process forming soil horizons from weathered rock in Oxisols, Inceptisols and Vertisols. Further investigations into pedoplasmation in tropical soils have been made by Beaudou and Chatelin (1979), but there is a need for the process to be examined in a wider range of soils.

Gley Morphology in Relation to Water Regimes

Direct monitoring of soil moisture regimes, although important, is costly. Instead, morphological features are being used in soil mapping to indicate particular moisture regimes, a knowledge of which is important for determining the potential use of soil for a number of purposes. There have, however, been very few detailed studies relating moisture regimes to morphological features on which to base this extrapolation. Bouma and co-workers (Veneman et al., 1976) have attempted to remedy this with a detailed investigation of the potential of macro- and micromorphological features as indicators of moisture regime by (i) monitoring soil moisture regimes in a representative toposequence in Wisconsin, (ii) investigating and describing the morphological features in the soils and (iii) by determining the relationships between the features and the soil moisture regimes. Three broad categories of moisture regimes and associated macro- and micromorphological features were distinguished: (a) Saturation for only short periods: Periods of saturation not exceeding one day were sufficient to cause reduction of manganese but not iron. The main morphological features associated with this regime are ped mangans and Mn nodules. Mottles of chroma of two or less were absent. (b) Saturation for periods of several days: Where the soil was saturated for periods of several days and also had a high matric potential for several months, reduction of both iron and manganese occurred. The soil was characterised by low chroma ped interiors due to migration of iron to ped edges and large air-filled planar voids, and ped and channel neo-ferrans where the mobilised iron had oxidised adjacent to these voids. There were few Mn segregations and it was assumed that Mn had been leached from the profile. (c) Saturation for periods of several months: Iron was reduced adjacent to the large pores and migrated towards more aerated ped interiors where it became oxidised. Two morphological features resulted: grey iron-depleted zones (neoalbans) adjacent to voids and on ped edges and quasi-ferrans or nodules at some distance into the ped. Many reduced iron and manganese compounds were removed from the profile by receding groundwater and low chromas, apart from ferrans and nodules, resulted.

Although in many cases it is impossible to ascribe gley morphology solely to current moisture regimes, the approach of Bouma and co-workers is useful and could be extended to an experimental situation in which ungleyed soil material is subjected to a range of known water regimes. Such an approach in which the conditions producing particular morphological features can be established with more certainty is important in improving criteria for soil classification and in the interpretation of soil maps for particular purposes, such as field drainage.

Clay Translocation and the Argillic Horizon

Much emphasis has been given to clay translocation in recent years and thin sections have been used widely to identify clay coatings and to measure amounts of illuvial clay. Micromorphology remains the most sensitive tool with which to attempt to identify translocated clay and accurate recognition of it can be made in most cases. Fedoroff (1972, 1974) has identified several types of illuvial clay concentrations and has related some of these to particular soil environments, for example, distinguishing concentrations formed under hydromorphic conditions from those formed under well-drained conditions.

Despite the large input of micromorphology into this area, several problems still remain. It is difficult, if not impossible, in some soils, to distinguish between illuvial, stress-oriented and neoformed clay. Translocated clay does not appear to possess unique optical properties by which it can be recognised. In particular, little is known about the importance and optical properties of neoformed clay. The fact that it is sometimes difficult to identify illuvial clay suggests, on the one hand, that not all translocated clay may be identifiable as such, and, on the other, that over estimates can sometimes occur. It is hardly surprising, therefore, that there is little consistency between operators in recognising and quantifying illuvial clay. McKeague and co-workers have shown this clearly in recent papers (1980, 1981).

This problem might be remedied to some extent by sets of standard slides but, until problems concerning identification are solved, operator inconsistency will continue. Considering the amount of attention that has been given to clay translocation, it is disappointing that there are few or incomplete answers to the questions: How, Why, When and Where does the clay move?

Unlike the products of some of the other soil processes, those that arise from translocation of clay are often not mineralogically or chemically distinguishable from clays in the adjacent matrix. If clay translocation is to continue to be an important diagnostic criteria in soil classification systems of the world, then some rigourously defined research projects need to be undertaken to solve some of these outstanding problems.

In summary, there has been considerable progress in some areas of micromorphology applied to soil genesis. It is significant that most of the advances have been made using micromorphology as one of a number of techniques. This bears out Cady's (1974) observation that 'the potential for interpreting and understanding features seen in thin section is increased manyfold according to the amount of other information on the sample, on the profile and on the whole geomorphic and biological setting'.

One particularly neglected area of research in soil genesis has been in the use of experimental pedology. Controlled conditions and known inputs need to be set up so that features formed under such conditions can be related directly to them. This can be a very useful aid in identifying the environmental conditions under which particular soils have formed.

There have been a few experimental approaches. Brewer and Haldane (1957) have studied translocation of clay into and through columns of sand, but the work has not been extended to finer textured materials in which the identification of translocated clay is more difficult. Greene-Kelly and Mackney (1970) experimented with the effects of wetting and drying on plasmic fabrics and Dalrymple (1972) attempted to artificially create lehm and erde fabrics using a series of clay-iron complexes. None of these experiments has been taken further.

Potentially, the experimental approach is a powerful one and it is disappointing that it has been so little used.

There has been a considerable input of micromorphology into the recognition of some soil types and diagnostic horizons. Temperate soils have received much attention whereas cryogenic, tropical and arid soils have received much less. This balance is likely to be redressed in the future judging by the interest shown in micromorphology, particularly in many Third World countries.

DEVELOPMENTS IN TECHNIQUES

If advances in the application of micromorphology in soil genesis have been gradual rather than spectacular, the opposite is true in the case of the developments of techniques. These developments are responsible for the main changes taking place in micromorphology. The most important influence of the new techniques is that rapid and accurate measurements can now be made on features in thin sections, polished blocks or on pieces of unimpregnated soil or rock. In the past, micromorphology has been considered almost solely a descriptive and interpretative branch of science. There have been few attempts to obtain quantitative data, but these represented an extravagant expenditure of time and manpower and perhaps viewed in terms of the natural heterogeneity of soils were of limited significance.

The development of several new techniques applicable to micromorphology provides the potential for accurate detailed measurements of soils in their undisturbed structure and fabric setting. Most of the techniques fall into the category of Submicroscopy. Much basic work has been done through the International Working-Group on Submicroscopy of Undisturbed Materials and the recent publication 'Submicroscopy of Soils and Weathered Rocks', edited by Bisdom (1981), is essential reading for soil scientists interested in this field. It provides an excellent account of developments in submicroscopic techniques and their applications. The most important uses of the new techniques are, at present, in studies of structure, fabric, mineralogy and chemical composition, but potentially they are applicable to most areas.

Structure and Fabric

Two new techniques in particular facilitate structure and fabric measurement:

1. Image Analysis: The introduction of image analysing computers has made possible the accurate measurement of pore space and structural units in soil thin sections or polished blocks (Jongerius et al., 1972; Murphy et al., 1977). A number of measurements can be made on features selected for measurement including area, size, shape, perimeter, number, orientation and irregularity. The method has been most successfully applied to soil structure because techniques have been developed for the satisfactory discrimination of pores. Its application to other soil features depends on obtaining sufficient contrast, in terms of grey level, between features to be measured and all other features in the specimen. The time needed to make the variety of measurements listed can be about 2 min. compared with at least 2 hours to make a more limited range of measurements by point counting or other manual techniques. Until recently the size of units that could be measured related to the optical resolution of the input, normally about $20 \mu m$. Now it is possible to measure pores $< 1 \mu m$ diam. using back scattered electron images derived from the scanning electron microscope (Jongerius and Bisdom, 1981).

2. Ultra-Thin Sections by Ion Thinning: This technique involves the production of sections that are sufficiently thin to be used directly on a transmission electron microscope. For this purpose, the sections should not exceed a few hundred to a few thousand Å thickness. In most soils this thickness cannot be obtained by microtomes because of the heterogeneous nature of soil materials, many of which are hard. Thinning is achieved down to electron transparency by ion bombardment. The following plasma characteristics can be obtained by TEM on ultra-thin sections: microporosity, mineralogy, morphology and related distribution of constituents, including constitution and association of oriented domains, and the relationship between iron compounds and the clay minerals. The method, although not quantitative at present,

allows a much more detailed examination of the fabric in situ than has hitherto been possible in thin sections (Bresson, 1981a and b).

Chemical and Mineralogical Composition

A number of instruments and techniques have been developed which can be used to micro-analyse the chemical and, in some cases, the mineralogical composition of grains and other morphological units in soil thin sections or polished blocks. These techniques fall into the following three categories:

1. Electron Probe Micro-analysis (EMA) and SEM coupled with Energy Dispersive X-ray Analyser (EDXRA) (Bisdom 1981): The electron probe micro-analyser has been used by micromorphologists for some time to determine elemental composition of soil features. Using thin sections and polished blocks it is possible to accurately determine elements of Atomic No. > 5 in features in situ. Using SEM coupled with EDXRA it is possible to analyse features in loose soil as well as in thin sections. It has the advantage over EMA in that elements can be analysed at magnification of x10,000 compared with x500 in the case of EMA, and several elements can be measured simultaneously. Its disadvantage compared with EMA is that only elements with atomic number $\geqslant 11$ can be measured. Linear traverses as well as point analyses can be made.

2. Auger Electron Spectroscopy (AES) and Electron Spectroscopy for Chemical Analysis (ESCA) (Bisdom 1981): Peaks in Auger spectra obtained by bombarding a sample in a thin section or impregnated block with high speed electrons can be used for chemical analysis by comparison with peak intensities obtained from known standards. Auger electron spectroscopy is essentially a surface analytical technique by which all elements other than hydrogen and helium can be detected.

Electron spectroscopy in which a sample is irradiated with X-rays which generate electron emission by the photo-electric effect, is also a surface analytical technique. The technique serves not only to identify elements but also their chemical bonding. All elements other than hydrogen can be detected.

3. Ion Microscopy (SIM), Ion Microprobe Analysis (IMMA) and Laser Microprobe Mass Analysis (LAMMA) (Bisdom,1981): The principle of these instruments involves the erosion of the surface of the material, either by ion bombardment or by irradiation,with a laser beam. The sensitivity of the instruments is some 1000 to 10,000 times better than with X-ray analytical techniques. All elements can be detected and local concentrations of as little as 10 mg/kg can be measured. These methods are particularly useful because trace elements and isotopes can be detected as well as major elements.

Some work has still to be carried out in testing the suitability of some of these techniques for accurate quantitative analysis of components in thin sections and polished blocks, but with many the suitability is already established. Most of the instruments are, however, expensive and are thus likely to be available in only a few laboratories. The limited availability of many of these techniques brings to the forefront the important question of satisfactory sampling. Soils are very heterogeneous, both vertically and laterally, and far too little attention is paid to representative sampling. The significance of much previous micromorphological data based on one, or only a few, small samples is often of limited or no value. There have been some attempts to determine the number of samples needed to adequately characterise a particular feature or horizon. One important paper, that of Milfred et al. (1967), investigated the number of samples needed for pedographic modal analysis of an argillic horizon by point counting. Thin sections were made of 124 samples taken from an area of 61 x 61 cm between depths of 69 and 130cm. Six constituents were counted in each thin section: matrix, glaebules, pores, papules, argillans and skeleton grains. A number of statistical analyses were applied to the point count data for each constituent. The results indicated that a sample adequate to measure the six constituents in an argillic horizon, with a standard error of 10% of the mean, would involve taking 21 cores (100cc), making 2 thin sections from each core, and counting 1000 points per thin section.

A study conducted by Murphy and Banfield (1978) to examine pore space variability in two clayey soils led to similar conclusions. They concluded that six tin samples should be taken from each horizon, with one section (7.5 x 6cm) from each tin, for reasonably accurate measurement by image analysis of pore space in clayey soils.

The findings of Milfred et al. have been largely ignored. Although satisfactory sampling entails more labour and cost, it is nevertheless essential that more attention be paid to obtaining sufficient samples for the eventual production of statistically acceptable results from any project.

APPLICATIONS OF MICROMORPHOLOGY

Micromorphology has become important to scientists in a number of disciplines, particularly now that, in addition to observation and description, features can also be quantified accurately and rapidly. Although micromorphology is still involved in soil genesis studies to a great extent, there is a fast growing interest in applying it to more practically-oriented problems. The most striking example is its use in research into agricultural problems. In his review in this volume, Jongerius lists thirteen different types of application of micromorphology in agriculture. These include comparisons of the effects of different land use, effects of machinery, influence of crop rotations, effects of fertilisers on microfabrics, effects of compaction and of drainage. Micromorphology is being used in a number of laboratories for the measurement of soil structure and its component aggregates and/or pores. There is growing confidence that this technique can be used to classify and measure different structural patterns as a basis for a more functional treatment of differences in soil structure.

The application of micromorphology in archaeology is at a more embryonic stage, but there is a growing interest in its use to differentiate soils from archaeological fills and for determining the petrography of pots. Inevitably, progress has been relatively slow because many features are unfamiliar and time to build up experience is required.

There has been much progress in applying soil micromorphology in Quaternary geology, in particular to identify and characterise paleosols. In some cases this has been extended to investigate the possibility of using paleosols as soil stratigraphic marker beds. Attention has been given to distinguishing between features associated with stable periods of soil development and unstable periods characterised by erosion and formation of sediments and slope deposits.

The application of micromorphology in ecological studies has been disappointing apart form a few notable exceptions (e.g. by Babel, Bal, Barrett and Jeanson). After much attention in the early days of micromorphology to humus forms, there is now little research in this field. There have been some investigations into the effects of soil fauna on humus forms and soil structure but much more needs to be done.

In the discipline of soil mechanics, there have been some interesting investigations into fabric and structure of soils and soil materials and their possible relationships with soil strength.

CONCLUSIONS

There have been some gradual and some very rapid changes in soil micromorphology. The most spectacular changes have resulted from the new apparatus and techniques that are now available. Submicroscopic techniques in combination with micromorphological studies with the optical microscope bring an exciting new era to soil micromorphology. Rapid, accurate physical and chemical analyses can now be carried out on undisturbed soil material. It is becoming increasingly possible to determine the structure, fabric and chemical composition of microenvironments in the soil and to examine these in relation to movement of water, air, roots, etc. This often enables more insight to be obtained than by the analysis of bulk samples.

Advances in micromorphological research into soil genesis, the main applications, have been gradual but considerable information has been obtained for a number of soil-forming processes and diagnostic horizons. Advances in this particular field are likely to be greater in the next few years as the new techniques become more widely applied.

Although micromorphology is often used as the only approach in some investigations, particularly those relating to soil genesis, there is no doubt that more significant advances are made when it is used with a number of other techniques.

Micromorphology is being more widely applied in a number of disciplines particularly agriculture, archaeology and Quaternary geology. There is also some interest in its use in engineering geology but the extent of its use in ecology remains disappointing.

As soil micromorphology enters a more quantitative era, it is important that attention is given to sampling to ensure that the results obtained are statistically sound. There is also a need to restrict the amount of new complex terminology if soil micromorphology is to attract an even larger number of users.

REFERENCES

Beaudou, A. G. and Chatelin, Y. 1979. La pédoplasmation dans certains sols ferrallitiques rouges de Savane en Afrique Centrale. Cah. ORSTOM, ser. Pedol., 17, 3-8.

Bisdom, E. B. A. (Ed) 1981. Submicroscopy of Soils and Weathered Rocks. Pudoc, Wageningen, The Netherlands, 320 pp.

Bresson, L. M. 1981a. Etude ultramicroscopique d'assemblages plasmiques sur lames ultraminces de sols réalisées par bombardement ionique. In: E. B. A. Bisdom (Ed), Submicroscopy of Soils and Weathered Rocks. Pudoc, Wageningen, The Netherlands, 173-189.

Bresson, L. M. 1981b. Ion micromilling applied to the ultramicroscopic study of soils. Soil Sci. Soc. Amer. J., 45, 568-573.

Brewer, R. 1964. Fabric and Mineral Analysis of Soils. J. Wiley and Sons, New York, 470 pp.

Brewer, R. 1973. Micromorphology. A description at the chemistry-mineralogy interface. Soil Sci. 115, 261-267.

Brewer, R. and Haldane, A. D. 1957. Preliminary experiments in the development of clay orientation in soils. Soil Sci., 84, 301-309.

Bruckert, S. and Selino, D. 1978. Mise en évidence de l'origine biologique ou chimique des structures micro-agregées foissonnantes des sols bruns ocreux. Pédologie, 28, 46-59.

Bullock, P., Fedoroff, N., Jongerius, A., Stoops, G. and Tursina, T. (in press). Handbook for Soil Thin Section Description. Waine Research Publ., Wolverhampton, England.

Cady, J. G. 1974. Applications of micromorphology in soil genesis research. In: G. K. Rutherford (Ed), Soil Microscopy. The Limestone Press, Kingston, Ontario, 20-27.

Dalrymple, J. B. 1972. Experimental micropedological investigations of iron oxide - clay complexes and their interpretation with respect to the soil fabrics of paleosols. In: St. Kowalinski (Ed), Soil Micromorphology. Warsaw, Poland, 583-594.

De Coninck, F. 1980. Major mechanisms in the formation of spodic horizons. Geoderma, 24, 101-128.

Delvigne, J., Bisdom, E. B. A., Sleeman, J. and Stoops, G. 1979. Olivines, their pseudomorphs and secondary products. Pédologie, 29, 247-309.

Fedoroff, N. 1972. The clay illuviation. In: St. Kowalinski (Ed), Soil Micromorphology. Warsaw, Poland, 195-208.

Fedoroff, N. 1974. Classification of accumulations of translocated particles. In: G. K. Rutherford (Ed), Soil Microscopy. The Limestone Press, Kingston, Ontario, 695-714.

Flach, K. W., Cady, J. G. and Nettleton, W. D. 1968. Pedogenic alteration of highly weathered parent materials. Trans. 9th Int. Congr. Soil Sci., Australia, Vol. IV, 343-351.

Greene-Kelly, R. and Mackney, D. 1970. Preferred orientation of clay in soils: the effect of drying and wetting. In: D. A. Osmond and P. Bullock (Ed), Micromorphological Techniques and Applications. Soil Surv. Tech. Monogr., 2, Harpenden, England, 43-52.

Jongerius, A., Schoonderbeek, D., Jager, A. and Kowalinski, St. 1972. Electro-optical soil porosity investigation by means of Quantimet B equipment. Geoderma, 7, 177-198.

Jongerius, A. and Rutherford, G. K. 1979. Glossary of Soil Micromorphology. Pudoc, Wageningen, The Netherlands, 138pp.

Jongerius, A. and Bisdom, E. B. A. 1981. Porosity measurements using the Quantimet 720 on backscattered electron scanning images of thin sections of soils. In: E. B. A. Bisdom (Ed), Submicroscopy of Soils and Weathered Rocks. Pudoc, Wageningen, The Netherlands, 207-216.

Kubiena, W. L. 1938. Micropedology. Collegiate Press, Ames, Iowa, 243 pp.

Kubiena, W. L. 1948. Entwicklungslehre des Bodens. Springer-Verlag, Wien, 215 pp.

Kubiena, W. L. 1953. The Soils of Europe. Thomas Murby & Co., London, 317 pp.

McKeague, J. A., Guertin, R. K., Valentine, K. W. G., Belisle, J., Bourbeau, G. A., Michalyna, W., Hopkins, L., Howell, L., Pagé, F. and Bresson, L. M. 1980. Estimating illuvial clay in soils by micromorphology. Soil Sci. 129, 386-388.

McKeague, J. A., Wang, C., Ross, G. J., Acton, C. J., Smith, R. E., Anderson, D. W., Pettapiece, W. W. and Lord, T. M. 1981. Evaluation of criteria for argillic horizons (Bt) of soils in Canada. Geoderma, 25, 63-74.

Meunier, A. 1977. Les mecanismes de l'alteration des granites et le rôle des microsystèmes. Thèse, Univ. de Poitiers.

Meunier, A. and Velde, B. 1976. Mineral reactions at grain contacts in early stages of granite weathering. Clay Miner., 11, 234-240.

Meunier, A. and Velde, B. 1979. Weathering mineral facies in altered granites: the importance of local small-scale equilibria. Min. Mag. 43, 261-268.

Milfred, C. J., Hole, F. D. and Torrie, J. H. 1967. Sampling for pedographic modal analysis of an argillic horizon. Soil Sci. Soc. Amer. Proc., 31, 244-247.

Murphy, C. P., Bullock, P. and Turner, R. H. 1977. The measurement and characterisation of voids in soil thin sections by image analysis. Part I. Principles and Techniques. J. Soil Sci., 28, 498-508.

Murphy, C. P. and Banfield, C. F. 1978. Pore space variability in a sub-surface horizon of two soils. J. Soil Sci., 29, 156-166.

Soil Survey Staff, 1960. Soil Classification, 7th Approximation. USDA, Washington, D.C., 265 pp.

Stoops, G., Altemüller, H-J., Bisdom, E. B. A., Delvigne, J., Dobrovolsky, V. V., FitzPatrick, E. A., Paneque, G. and Sleeman, J. 1979. Guidelines for the description of mineral alterations in soil micromorphology. Pédologie, 29, 121-135.

Veneman, P. L. M., Vepraskas, M. J. and Bouma, J. 1976. The physical significance of soil mottling in Wisconsin toposequence. Geoderma, 15, 108-118.

Part II

GENERAL CONCEPTS OF FABRIC

Editors' Comments
on Papers 4, 5, and 6

4A   KUBIËNA
Introduction

4B   KUBIËNA
Elementary Fabric

5   BREWER and SLEEMAN
Soil Structure and Fabric. Their Definition and Description

6   ALTEMÜLLER
Excerpt from *Gedanken zum Aufbau des Bodens und seiner begrifflichen Erfassung*

The concept of soil fabric, as proposed first by Kubiëna in *Micropedology* (Paper 4), was most probably inspired by the petrographic concept of fabric introduced by B. Sander (1930). As defined, it is a very open and broad concept: "the arrangement of the constituents of a soil in relation to each other." As a matter of fact, Kubiëna was concerned essentially with the "elementary fabric:" the arrangement of the soil components of the lowest order, that is, the least complex constituents of the soil, such as sand grains and clay. He distinguished two such types of components, which are considered to be opposites—namely, the skeleton grains, which are thought to be relative stable, coarse components (fragments of minerals, rocks, or organisms), and the plasma, which is the more mobile, colloidal fraction of the soil. The plasma can exist in either a peptized or a pectized (more or less flocculated) state. Ten major types of elementary fabrics are recognized: bleached sand, intertextic, chlamydomorphic, agglomeratic, plectoamictic, porphyropeptic, porphyropectic, rendzina, magmoidic, and mortar. The elementary fabric is the most essential feature described in thin sections in all micromorphological studies made until the 1940s, when Kubiëna's morphogenetic system became generally accepted; and even in this system, the elementary fabric was one of the most important diagnostic criteria for differentiating among several types of soil materials. In the actual descriptive systems the concept still retains a very important place (see Papers 7, 8, and 9). The concept of plasma and skeleton survived until the mid-1970s, and played an essential role in the development of micropedology.

In the early 1960s, R. Brewer gave a new stimulus to micropedology by elaboràting a strongly analytical system for describing the microscopic features of a soil (Paper 5; Brewer, 1960 and 1964). Although Brewer retained most of Kubiëna's concepts (Paper 4) in broad terms, he placed more emphasis on the systematic study of new formations in the soil; this increased emphasis played a large role in spurring on the evolution of soil micromorphology. Brewer's concepts and terms show a strong petrographic influence, and organic materials are practically excluded from the system. Unlike Kubiëna (Paper 4), Brewer clearly defined the concepts. One of his most important innovations was the distinction between the soil matrix and the so-called pedological features (e.g., glaebules and cutans). Paper 5, like Brewer's later handbook (1964), shows that he attached a great deal of importance to the analysis of the fabric elements (distribution and orientation patterns) and to the levels of fabric and structure.

Both in soil science and in the earth sciences, the terms *fabric, structure,* and *texture* have different meanings for different authors. Altemüller (Paper 6) gave a clear, profound analysis of these concepts and thus had a major influence on later definitions. His definition of soil fabric was based on the Sander's (1948-1950) general concept of fabric. The particular soil is limited in time (a given moment), in space (both vertically and horizontally) and in components (solid, liquid and gaseous). (§3.2). All soil components can be considered fabric elements, either as individuals or as more complex units (§3.3). The fabric concept is independent of size, but sizes have to be indicated (§3.4). According to the complexity of the material, different levels of fabric can be distinguished (§3.5), for example, the internal fabric of an aggregate and the fabric formed by the arrangement of aggregates. Sander distinguishes between morphological (or geometric) and functional fabrics; in micropedology, usually only the morphological fabrics can be studied (§3.6). Although fabric analysis is an effective way to discover the evolution of the soil material, the objective description of the fabric and its genetic interpretation should be separated (§3.7). Sander proposed the following definition: The soil fabric is to be understood as the disposition of the soil constituents from a morphological, functional, and genetic point of view, and independent of size and composition.

REFERENCES

Brewer, R., 1960, The Petrographic Approach to the Study of Soils, *Internat. Cong. Soil Sci. 7th. Trans. Madison* **1**:1-13.

Editors' Comments on Papers 4, 5, and 6

Brewer, R., 1964, *Fabric and Mineral Analysis of Soils,* J. Wiley & Sons, London, New York, and Sydney, 470p.

Sander, B., 1930, *Gefügekunde der Gesteine,* Springer, Vienna, 352p.

Sander, B., 1948-1950, *Einführung in die Gefügekunde der geologischen Körper,* 2 vols. Springer, Vienna.

4A

Copyright © 1938 by Collegiate Press, Inc.
Reprinted from pages 125-128 of *Micropedology,* Collegiate Press, Inc., Ames, Iowa, 1938, 242p.

Introduction

W. L. Kubiëna

1. PRINCIPLE AND GOAL OF FABRIC ANALYSIS

By "soil fabric" we mean the arrangement of the constituents of a soil in relation to each other. Some soil horizons may show no changes in arrangement when examined in different directions with the microscope, while some others may exhibit very marked changes. The differences in arrangement are caused by past differential movements of all or some constituents. From the study of the arrangement we can draw conclusions concerning the kind and the direction of the movements.

The totality of movements of components, which occur in a given soil, is called the dynamics of the soil. Since the soil can change its dynamics remarkably during the course of the year and during the different stages of its development, the modern pedologist does not classify soils according to their properties, but according to their dynamics. The term, "soil type," is an expression for the totality of its dynamics, that is, of all the component movements possible in a given soil. "Soil type," as an expression of its dynamics, refers not only to the regional and local climates, but also to the position, surface relief, plant growth, water conditions, and internal construction of the soil, mainly the grain sizes, density, permeability and specific type of the fabric, as well as the chemical nature of the original constituents. The final goal of microscopic fabric analysis is the recognition of the dynamics of a given soil, the creation of refined research methods by which it is possible to disentangle the complexity of microscopic events, in order to recognize the detailed movements of the different components represented in chemical and morphological transformations or in displacements. The detailed knowledge of the different processes involved will contribute to our understanding of the genesis of soils as well as to their classification. In coherent soils almost every microscopic happening, as far as it was not succeeded and covered by another happening, has left its picture in the fabric. In the microscopic fabric we are able to read the history of a soil as we may a book. We need only to learn to interpret that which we find in it.

However, not only can conclusions as to the past processes of the soil be drawn from the fabric, but also the details of its behavior in the future, primarily in the sense of its agricultural efficiency, can be predicted. It will require some time and detailed studies to pass only theoretically through the domain of fabric analysis. It has been shown, however, on several occasions, that a successful application to practical soil science is possible at the very beginning. With almost every kind of practical problem we can receive new and valuable data by supplementary microscopic investigations. In the earlier stages of medicine the diseases of the interior of the body were diagnosed only by outer symptoms. Only in the later medical science were the organs and tissues opened for independent study with regard to their formation and their function in the living body. In like manner we begin now to open the natural soil as an organism-like body, and to study the details of its formation and the function of its different parts in relation to the whole.

2. MEMBERS OF SOIL FABRIC

The components which constitute a fabric may be of simple or of compound construction. Both represent fabric units. We distinguish fabric units of lower and of higher order. A building is constructed of numerous single stones which represent some of the fabric units of the building. Every single stone, however, also shows a construction of its own. Suppose the building were constructed by brick-shaped artificial stones consisting of round gravel, angular grains of sand, and a dense binding substance of lime filling the spaces between the sand particles and the gravel. Gravel, sand grains and cementing substance represent the fabric units in the single building stone. The way they are arranged would represent a counterpart of what we call the elementary fabric in the soil. A fabric unit, similar to the building stones, is found in the aggregates in the soil. Besides the building stones, units of higher order are distinguishable. If the building were a dome, we would distinguish different parts which are arranged together to make up the whole. Such units of higher order would be the arched buttresses and the arch-masonry. By analogy we are able to state fabric relationships in the soil which are based on units of higher order. The different grades of fabric relationships in the soil—the elementary fabric, the fabric of aggregates and wall complexes, the fabric of cleavage blocks or of other fabric units of higher order—are called members of the soil fabric. The different fabric members are contained in each other; they are present at the same time and fill the same space.

3. NOMENCLATURE

The term, "soil fabric," is a counterpart of the term, "rock fabric," created by B. Sander for the arrangement of the mineral grains in rocks determined by their optical orientation or their shape orientation (36, 37). Although in soil fabrics other relationships and other materials are more prominent, the denotation and sense of the idea are the same in both fields. Also the science of soil fabric, with its observations and results will be able to contribute to a science of general fabric to be created in the future.

The author, therefore, found much justification for using the term "fabric" for the arrangement of the constituents of the soil, and their role in relation to each other, instead of "structure" or "texture" (the latter in the European meaning). Another reason for this decision was the fact that the terms "structure" and "texture" in soil science in the different countries have different meanings.

In the old petography, and in the old soil science in Europe, the one comprehensive term "structure" was used, which referred to all the ideas later developed in petrography and expressed by the terms "Gefüge," "Textur" and "Absonderung."

In the new German petrographic literature the terms have come to mean for the most parts the following: By "structure," the form, relative size and reciprocal boundary lines of the components (minerals) are understood; by "texture," their arrangement and distribution in the space; and by the term "Absonderung," complexes formed by the natural cleavages, as in the formation of columns and plates.

In pedology, as in German petrography, it will become advantageous to use several detailed expressions for the different conceptions enumerated above instead of using only the term "structure" for all. A beginning in this respect was made by the adoption of the term "texture" in soil science in Europe and in America. Unfortunately, the same expression was used in the opposite sense; in America it meant the grain sizes, and in Germany it had several meanings, the chief of which was the form and relative size of the soil spaces (the German term for grain sizes is "Körnung").

Of all the terms, "texture," applied in the sense of German petrography, comes closest to the author's conception of soil fabric. It was avoided, however, because "texture" in the United States and other countries has been in general use for many years in the sense of grain sizes. The term, "structure," in the agricultural sense, means the general appearance of the soil, as consisting of crumbled or flocculated aggregation complexes or, in the opposite case, as lacking in such complexes. The microscopic investigation shows that the term, "structure," used in this sense, describes only the macroscopic appearance of different fabric relationships which show in microscopic dimensions pictures other than the conjectured "crumb" or "single grain" arrangements. By the introduction of the new expression, "soil fabric," the meaning of the term, "soil structure," used in agricultural soil science over almost all the world, will not be influenced. The author also will use "structure" occasionally in microscopic investigations where he does not mean the arrangement of components and their relation to other constituents, but the mere appearance of a component in regard to its outlines, the designs observable in its interior, etc. Thus, the old terminology remains untouched, and no old term is used in a sense contrary to its application in any country at present.

Elementary Fabric

W. L. Kubiëna

1. GENERAL

By elementary fabric (German—Elementargefüge, French—assemblage élémentaire) is meant the arrangement of the constituents of lowest order in the soil and their relation to each other.

These constituents represent more or less independent units, and, in most cases, more or less homogeneous bodies. Not all soil constituents of these kinds can be considered as fabric units in the sense of elementary fabric. We have to distinguish between the principal building elements and accessory constituents. Fabric units may be grains or fragments of rock minerals, particles of raw humus (which may also serve as skeleton), films, coatings, grain deposits, intergranular braces and space deposits of salts, colloids and suspended substances. The accessory constituents may be of particular importance for the character of the soil. They are, however, of minor importance in the construction of the soil, and the regular arrangement of its building elements, which represent genetically, in most cases, the primary constituents. Accessory elements, in this sense, would be locally limited crystal formations and accumulations of substances, space-fillings of fungus mycelium or organism remnants, cysts of protozoa, etc.

According to the role of the different fabric units of lowest order, we are able to distinguish two different groups. One forms the *skeleton* of the fabric, which consists mainly of the residues of rock minerals and organisms not decomposable or which are only slowly decomposing. The units of the other group are much more easily moved, changed in composition and shape, and redeposited. The latter are generally identical with the finely dispersed and highly active, newly-formed compounds in the soil. They are pedogenetically the most important ingredients of the soil, and represent the real carriers of the typical properties of the soil. In order to characterize their nature in comparison to the skeletal units we call them, as a whole, the *fabric plasma*.

Factors influencing the formation of fabric types. The regular formation of definite elementary fabric types is caused, in the first place, by the manifold influence of flocculation and peptization on the morphology and mobility of the fabric plasma. Flocculation and peptization may act upon the different substances of the fabric plasma in different ways, that is, the substances react differently to treatment by one and the same natural reagent. The kind and strength of electric charge, and its influence on the different soil colloids, the ability of alkalies and acids to dissolve components of the plasma, the protective effect of hydrophilic colloids, and finally the action of the eluvial and illuvial processes in general, bring about remarkable divergences in the fabric of soil types. The repeated drying out of the soil, especially if accompanied by irreversible processes, effects to a great extent the microscopic fabric. Finally, the morphological properties of the components, as sizes and shapes of particles and the quantitative relationship between the constituents, are important in determining the type of arrangement.

Stability of elementary fabric. Like the formation of horizons, the formation of elementary fabric is also an expression of the soil type produced by the continuous influence of certain soil forming processes over prolonged periods of time. As long as these processes are not changed the elementary fabric will show no changes, except those which are characteristic for the particular soil type in the different stages of its development.

The elementary fabric is not obliterated by mechanical destruction of the soil produced by breaking, crushing or tilling. It is preserved in the smallest aggregations or fractions; in many cases it may be recognized even on single soil grains and in the way the plasma residues are arranged on its surface. By knowledge of the elementary fabric we may draw conclusions as to a number of other properties of the soil correlated with it and with the soil type.

As it is possible to influence a given dynamic of a soil artificially by treatment with chemical substances, or by changing its water and air conditions, it will be possible also to influence its elementary fabric.

Advantages to expect from fabric analysis. The fabric investigations concern not only the mere arrangement of the components, but also the study of the components themselves, their morphology and chemical nature. The variety shown by the elementary fabric in general is so great that its knowledge will contribute not only to the genesis and classification of soils, but also to the recognition of soils present only in smallest quantities in regard to the respective soil type or subtype and horizon. It will be possible to conclude, even from the minute soil particles adhering on the roots of plants after years in herbariums, as to the climate and general nutrient conditions of the location where they had been growing.

2. EXPERIMENTS IMITATING THE FORMATION OF SOME FABRIC TYPES

Experiments imitating chlamydomorphic fabric. About 7 grams of pure glass sand (quartz sand) of an average grain size of 0.5 mm. are mixed with 0.2 gm. iron hydroxide, and 2.5 cc. hydrochloric acid (1:10) added to the mixture. The suspension is shaken until the iron hydroxide is entirely dissolved. The suspension is then poured into a small glass dish and left to dry out at room temperature. The microscopic picture of the dried sand shows that every glass or quartz grain is covered by a uniform lemon yellow coating of $FeCl_3 + 6\ H_2O$ (Ferrum sesquichloratum. On slight heating the color is changed to brown, caused by loss of water ($FeCl_3 + 3\ H_2O$). On ignition the coatings turned into dark gray; on cooling, however, they are changed into an intense bright red (Fe_2O_3). The grains are combined with each other at the junction points and form partly coherent but very friable complexes. In the angles of the intergranular spaces the coatings are thicker and have deeper color. In the center of the junctions a round bare spot can be observed. Instead of intergranular braces, we find only a ring-shaped thickening of the coatings, a round accumulation of iron oxide of a shape resembling the margin of a crater. Accumulation of iron oxide can be seen also in the cracks and dents of the grain surface. The arrangement thus produced represents a construction similar to that which will be discussed later as chlamydomorphic fabric in soils.

A similar result can be obtained by application of distilled water in place of the hydrochloric acid above, particularly when it shows a more acid reaction or when some carbon dioxide is introduced into it. After drying, the sand grains are surrounded by iron coatings which readily become visible after ignition. Besides the coatings some granular deposits of undissolved iron oxide particles may be generally observed on the surface of the grains.

Experiments imitating agglomeratic fabric. Seventy parts of pure quartz sand are mixed with two parts of iron hydroxide and suspended with twenty-five parts of water. Then one part of calcium oxide is added, the suspension is shaken and poured into a small dish where it is dried by slight heating. The microscopic investigation shows the absence of coating formations. Instead deposits of flocculated iron hydroxide and some deposits of calcium hydroxide may be found in the intergranular spaces. The cracks and dents of the grain surfaces are free from iron accumulations. The sand does not form complexes, and the iron hydroxide does not act as a binding substance in this case. The arrangement corresponds to an arrangement in soils which we designate as agglomeratic fabric.

A similar effect may be produced by using tap water without addition of calcium oxide.

Experiment imitating chernozem fabric. John B. Bartlett, of Ames,[*] mixed 25 g. pure quartz sand with 70 mg. fine particles of lignin and placed the mixture in a small petri dish of a diameter of 1.5 inches. To the dry mass 6 cc. of 5 per cent NaOH were added. A part of the lignin went into solution at once, producing a deep brown coloration on the bottom of the petri dish. The next day the deep brown coloration had moved to the evaporating surface of the sand, while the deeper layers showed a lighter degree of brown. The material was left to dry out at room temperature.

The microscopic investigation showed that the undissolved grains of lignin were deposited in the angles of the intergranular spaces and cemented to the quartz grains by dissolved lignin, part of which was seen in the form of slightly brownish films on the surface of the quartz grain.

A very constructive picture of the fabric relationships could be obtained by the thin sections, which were made somewhat thicker for this purpose (about 0.08 mm.). The quartz grains were ground until only their caps on both sides were taken off. Almost all grain-like lignin was deposited in the angles of the intergranular spaces. The film nature of the dissolved lignin was clearly shown. In the center of each grain, where the caps had been ground off, appeared a white spot which was round and regular in the case in which the contours of the caps on both sides of the section were congruent.

This arrangement corresponds to the intertextic fabric in chernozems and other steppe soils, designated as chernozem fabric.

[*] Unpublished experiments on the solubility and decomposition of lignins in soils.

The complexes thus formed are rather coherent in spite of the fact that only a very small quantity of lignin was applied (about three parts per thousand). They retain their shape for some time when placed in water. Their stability would be much greater if comparatively high amounts of NaOH were not deposited in the intergranular braces and films, since this causes solution of the binding lignin to a large extent.

The formation of films on the surface of the quartz grains could be shown more clearly by another experiment in which the 70 mg. of lignin particles were shaken with dilute sodium hydroxide, and the solution brought into the petri dish filled with white quartz sand. Again on the next day the deepest coloration of the sand was observed on the surface. Microscopic investigation showed that only a small amount of lignin was present in the form of grain-like particles. Most of the lignin was found in the form of uniform films. The nature of the films could be made most visible by staining with malachite green; the dye was absorbed to such a degree that the films lost their transparency and took on the appearance of heavy coatings.

Experiment imitating magmoidic fabric. Sodium hydroxide added to a solution of an iron salt causes the normal production of a flocculated precipitation of iron hydroxide. If some sodium silicate (sodium waterglass) were added to the iron solution, the precipitation of iron hydroxide would not occur. J. Bastisse, in Versailles, placed fine quartz sand in a small glass cylinder standing in a somewhat larger glass dish. Some iron solution, protected with sodium waterglass from precipitation by added alkalies, was poured into the glass dish. This resulted in a rise due to surface tension of the liquid in the glass cylinder filled with sand. After some time the liquid had moved from the dish and migrated into the sand cylinder, forming a deeper brown illuvial zone in its surface layer.

A microscopic investigation of the dried sand mass revealed that the whole fabric plasma was entirely peptized. Still, the quartz grains showed no coatings and were united by the peptized intergranular braces which filled a part of the spaces.

The absence of coatings was shown particularly by the debris preparation. On the surface of the quartz grains could be seen deposits of peptized fabric plasma. The surface between the deposits was entirely bare. After washing, the preparation was treated with malachite green by which all deposits were brought out as intense blue-green.

Particularly remarkable was the arrangement on the surface of the undestroyed sand cylinder. The upper parts of the grains sticking out of the complex body were quite bare and striking, due to their white color. The fabric plasma in the intergranular spaces did not quite reach the surface. It came up as far as the narrowest point of the spaces where the curvature of the surfaces of the upper grains came closest together. From above it gave the impression of a tough lava stuck in a rock flue. The menisci, however, were somewhat deeper than normal, obviously formed by partial creeping of the suspension at the margin.

The arrangement corresponds to a soil fabric which will be discussed later as magmoidic fabric. The colloid chemist understands a magmoid to be a solid-liquid (festflüssig) body, i. e., one which behaves in slow movement as a liquid, in fast movement as a solid body. It is characterized by a peptized but comparatively thick-flowing fabric plasma, particularly by increasing concentration in the stages before the drying out of the soil. The genesis of the particular formation has not yet been studied.

3. GENERAL GENESIS OF DIFFERENT FORMATIONS ON THE GRAIN SURFACES

A general explanation of the genesis of the formation of films, coatings, grain deposits and intergranular braces may be given as follows. Due to evaporation taking place on the soil surface the interior soil gradually loses water. The soil solution, which has filled almost all soil spaces after excessive moistening, retires to the angles of the intergranular spaces and to the surface of the grains, forming only a thin coating on the latter. Constituents of the fabric plasma suspended in the soil solution are moved with it. If these constituents were present in the form of larger flocculated complexes there would be room for them only in the water accumulations in the angles of the spaces. They cannot spread out over the soil surface. They are pulled either into the intergranular angles or remain attached somewhere at some point on the grain surface. They form intergranular braces on these places if they are cemented to the soil grains or to each other. They become space deposits if no kind of cementation takes place.

In the case in which the constituents of the fabric plasma are peptized in the soil solution, they are distributed uniformly in the angles and in the liquid coatings of the grains. After drying out, they will form a uniform layer around grains except in the angles where a larger amount of former soil solution is collected.

If almost no colloids or binding substances are present, and the fabric plasma consists only of rock powder, the latter may be suspended uniformly in the soil solution also. In the dried soil they will be found distributed and adhering all around the grain surface (Bröselhüllengefüge). Here, also, large numbers of particles are collected in the angles of the spaces.

One of the experiments of J. B. Bartlett, cited on page 132, showed that coatings can be formed also by capillary creeping of salt solutions or colloid suspensions. In this experiment the lignin solution poured into the petri dish previously filled with quartz sand, formed films on all quartz grains except those on the surface. These grains remained, for the most part, bare for the first two days of the drying out process. On the third day, however, most of them were covered with coatings. At different stages of these coating formations it could be seen that they were formed by creeping of the lignin on the grain surfaces. The lignin coating formed a system of fine capillaries. Due to the higher evaporation on the upper border of the coating the strongest flow of the lignin solution was in this direction. New layers of deposited lignin were formed constantly by efflorescence and the system of capillaries was thus lengthened for the movement of the lignin solution.

4. SOME TYPES OF ELEMENTARY FABRIC

The following formulation of fabric types is far from being completed. It is just a beginning and, in some cases, probably only a tentative establishment. More detailed investigations have shown that there is an even greater variety in the finer differences of elementary fabrics than can now be described, and this is likely to become more apparent as additional soils from the different parts of the earth's crust are thoroughly examined. The establishment and naming of new fabric types will be justified only if a particular elementary arrangement is found to be characteristic for a particular soil type or subtype.

In view of the difficulties in language, and possibilities of translation, the author followed the example of petrography by using the Greek and Latin language in some cases for the naming of types of elementary fabric. Where English or German expressions are easy to translate, however, they will be used in preference.

PORPHYROPECTIC FABRIC

Description. The strongly reflecting skeleton minerals appear at broken planes, entirely clear and free from coatings, and are imbedded in a dense ground mass showing almost no spaces. Between the ground mass and mineral grains almost no affinity is to be found; the grains easily fall out of the complex. The dried ground mass is often interspersed with cracks. It is of special significance that the cracks go around the soil grains frequently so that the ground mass breaks loose easily from the grains.

In the debris preparations the mineral grains are found almost entirely bare. Surface deposits on the grains adhere only loosely. They are not cemented on it and are easily removed. Beside the grains, fragments of the ground mass are found in great preponderance.

The picture of this arrangement resembles somewhat the picture of the porphyric fabric or the poikilitic fabric of igneous rocks.

Varieties of porphyropectical fabric. (A) The spaces are entirely filled with a mass consisting of very fine, more or less flocculated substances which act as filling but not as binding substances.

The arrangement is found frequently in alluvial soils, particularly those which contain a certain amount of calcium carbonate to flocculate the colloids, but not so much that the lime can act as a cement. The filling substances are washed into the spaces by flowing water.

This is one of the cases in which coagulation is producing, not crumb structure, but the opposite, an entirely dense structure. The floccules are not big enough to be seen as such with the naked eye.

(B) The ground mass appears more or less peptized; it acts as a binding substance in itself but has no affinity for the mineral grains. The cracks in the dried mass often end in a marginal cleft which is formed between the grain surface and the adjacent ground mass. The dry soil is coherent, not because the soil grains are combined by a binding substance, but because the ground mass forms a compact connection in which the mineral grains are loosely imbedded. The ground mass is not pectized in this case; the arrangement has merely a similar appearance to porphyropectic fabric.

Occurrence. Fabric relationships of both variation (A) and (B), as described above, have been found in a number of lateritic red loams. The arrangement of variation (A) can be found in certain alluvial soils or in finely dispersed muck soils rich in calcium carbonate.

Genesis. The term, "porphyropectic fabric," was derived from soils where the flocculated state of the ground mass was more or less evident. It was believed that the lack of affinity between ground mass and the mineral grains always was due to the effect of the pectization (flocculation) of the fabric plasma. This is evident in alluvial soils with flocculated colloids. However, as well as the fabric relationship described as variation (A), relationships of variation (B) also were found in some lateritic soils where the effect of pectization could not be seen by microscopic investigation. It is probable that soils with a fabric showing variation (B) represent younger lateritic soils. The colloidal silica content was always much higher than commonly found in laterites and lateritic soils. The electronegative character of the colloid could be demonstrated by staining with malachite green. It is possible that the silicic acid present in large amounts may act here as a protective colloid for the aluminum and iron hydrates. The matter requires more thorough investigation, especially as to the cause of the lack of affinity between ground mass and mineral grains. A greater number of lateritic soils than has been possible up to the present should be examined.

The soil in fig. 65, a lateritic red loam on lipartic tuff from Kwala Bingei, Sumatra, showed a brownish-red ground mass which, with a microspatula, could be cut easily like a waxy substance. The mineral grains had no coatings and showed a strong glass-like reflection. In the debris preparation they were colorless; treated with malachite green they showed smears of an intense blue-green on their surface. The staining of the smears could not be produced with congo red. The fragments of the filling mass were also intensely stained by malachite green.

Short heating of a small specimen with hydrochloric acid (1:3) dissolved a great part of the iron compounds. After filtration, the almost white precipitate showed colorless, jelly-like gel packets. After removing the acid, a part of every gel packet could be stained with malachite green, while a part remained unstained. A much greater part could be stained with congo red. By igniting with cobalt nitrate, well outlined sections of Thenard's blue could be produced. In the gel packets a great number of mineral fragments were present which appeared clearly as points of light between crossed Nicols.

PORPHYROPEPTIC FABRIC

Description. The mineral grains are cemented into a dense ground mass. When isolated, they almost always show adhering colloids on the surface as if they were coated (fig. 66).

Occurrence. This fabric type is found in desert crusts, in some Mediterranean red earths, and in illuvial horizons of some podsols.

Genesis. The ground mass consists mainly of colloids in peptized state which have a marked affinity for the surface of the soil grains.

A desert crust on the surface of rocks or cemented soils shows, under the microscope, dense dark-colored accumulations of colloids on the highest points of the microscopic relief. The crust on these points is shining and wax-like. The debris preparation of the material of these points, of a desert crust on sandstone near Gillette, Wyoming, showed all mineral grains surrounded by toughly adhering orange-colored colloid masses, which could be stained an intense blue-green with malachite green. The reaction of suspensions of the colloid mass was alkaline.

The action of colloidal silica as a protective colloid may be observed also in the case of some very dense red earths showing porphyropeptic fabric. The plasma substances are intense red in color, generally forming very tough, hard masses which have a waxy appearance under the microscope, and which can be cut like wax with microtools. The mineral grains imbedded in them are almost entirely invisible. The whole ground mass is stained intense blue-green color by malachite green.

Porphyropeptic fabric was reported in the illuvial horizons of a ground water podsol in Rumania by Christache V. Oprea (15). The soil profile, situated near Pantelimon Ilfov, showed, at a depth of about 35 cm., a whitish-gray horizon of extreme toughness. On microscopic examinations it could be seen that the soil grains were cemented in a wax-like, shiny, dirty white-colored ground mass locally tinted with ochracious shades. Since most of the substances were light colored or almost without color in microscopic preparations, the nature of the fabric was not so striking as shown in the ordinary debris preparation. The marked affinity of the ground mass to the mineral grains became very evident, however, in the ignition preparation. The ignited debris appeared as an intense orange-red color, indicating that iron colloids were present in the soil mass in the form of ferrous compounds. The mineral grains were covered with parts of the ground mass cemented to their sufaces. The ground mass was peptized, but on examination under high power in the dry state exhibited laminated structures and a very fine granulation in its interior. In thin layers the orange-red marks were ochracious in color.

INTERTEXTIC FABRIC

Description. The mineral grains appear bare and free of coating. They are united with each other by intergranular braces or are imbeded in a porous ground mass of flocculated or crumbled colloids (fig. 67). In soils particularly poor in skeletal material, the uncoated grains are cemented into a ground mass showing cavities of many shapes.

In the debris preparation, uncoated mineral grains showing surface deposits and fragments of intergranular braces are seen.

Variations. (A) The fabric plasma is dark brown to blackish in color and consists of blackish particles of humus, coagulated inorganic colloids, fine rock fragments, and a peptized brown humic substance which is almost insoluble in water but soluble in dilute alkalies. The same humic substance is found in the form of a brownish film around the mineral grains. This variation is designated as chernozem fabric.

(B) The fabric plasma consists mainly of brown to yellow-colored, mostly coagulated inorganic colloids. Compared with those in variation (A) the intergranular braces are considerably less stable in water.

(C) The intergranular braces are formed of flocculated colloids, mineral fragments and insoluble splinters of humic substances, interspersed and united by lime accumulations. This variation will be described in detail in a special paragraph on the fabric of some Austrian muck soils.

Occurrence. Intertextic fabric of the variation (A) is to be found in chernozems, chestnut brown soils, and brown and gray earths of the semi-deserts. Agriculturally, it is the best type of component arrangement. The slow solubility of the peptized fraction of humic substances explains the great stability of the aggregations and wall bodies of chernozems.

Intertextic fabric of the variation (B) is found in most of the brown and gray-brown earths of the humid regions, and also in many lateritic soils.

Chernozem fabric sub-variations. The intertextic fabric of the chernozems, chestnut brown soils, and brown earths of the desert steppe varies somewhat according to the soil type, as indicated by Christache V. Oprea in his studies of soils formed on loess in Rumania. Oprea describes the differences as follows:

The debris preparation of the desert brown earth showed humus films as well as grain deposits. The grain deposits, however, were much denser, so that they appeared, in many cases, almost like coatings. The intergranular braces and grain deposits were strongly interspersed with salt accumulations with a high content of calcium carbonate. By treatment with dilute hydrochloric acid, all complexes of fabric plasma fell apart rapidly and entirely. Humic substances were present more in dissolved form than in the form of leaves and scales. In spite of this the fabric was less stable in water.

The chestnut brown soil showed fewer but thicker grain deposits in which more leaf-shaped humus occurred than in the brown earth. The intergranular braces were less interspersed with salts; therefore, the union was stronger, and the complexes fell apart less as a result of treatment with hydrochloric acid. Their stability in water was also greater.

The debris preparation of the so-called chocolate-colored soil of Rumania showed grain deposits which consisted of crumbled, blackish, somewhat granulated humic substances which were imbedded in yellowish aureoles of highly peptized humic substances. The grain surfaces were covered by brownish films. Most of the complexes were remarkably stable in water.

Similar grain deposits of blackish and granulated elements were also shown in chernozem. The aureoles and films were brownish. Almost all complexes were stable in water.

Genesis of chernozem fabric. As indicated on page 132, chernozem fabric is produced because most of the inorganic and organic colloids are flocculated or in an undissolved state, although some of the humic substances are dissolved or peptized in the soil solution. The latter act as binding material in the dry soil. Since these humic substances cannot be dissolved in water experimentally, but can be dissolved in diluted alkalies, the mineral grains and flocculated deposits are combined with each other in such a way that the alkaline soil solution, concentrated in the last stages of drying of the soil, is able to dissolve the substances which act later as binding material. That the soil solution in some stages is able to dissolve a part of the humic substances may readily be seen from the fabric of the chernozem aggregations which will be discussed later.

The more alkaline a humus soil may be in reaction, the more humic substances we can find dissolved. This may be seen by comparison of thin sections of soils different in alkalinity. That the substances soluble in alkali are not lignins can be shown by treatment with acetyl bromide which dissolves lignins quite rapidly, but attack only very slowly the brown-colored binding substances in the chernozems.

Intertextic fabric in brown earths (Braunerden). The central European brown earths and the gray-brown earths of the humid regions of America show mostly intertextic fabric. Most of the fabric plasma appears in typical flocculated state. The grain surfaces between the deposits are almost bare. There is no film formation around the mineral grains. The appearance of characteristic aureoles, however, in which the grain deposits are imbedded, indicates that a part of the colloids must have been dissolved or peptized temporarily. The binding substances are much less stable in water, and the complexes lose their shapes much more readily than do the complexes of the chernozems.

Fabric of calcareous muck soils in summer dry climate. It was indicated by Georg Hardt (Vienna) that many muck soils in summer dry climate show intertextic fabric. Such muck soils developed in the southern Viennese Basin showed, in the debris preparation, mineral grains with crumbly deposits of blackish humus particles. The deposits were interspersed with lime acumulations. Among the inorganic colloids, silica was much in preponderance and could be found in great abundance throughout the soil. Staining of aggregations, fragments and debris with malachite green (after removal of the organic substances by ignition) showed that the accumulations, efflorescences and concretions of lime particularly were covered with finely flocculated silica layers (1).

The accumulation of silica could be shown also in ignited debris preparations which were lightly washed with dilute hydrochloric acid and treated with ammonia. After removal of the ammonia by slight heating, nearly all deposits took the dye and appeared as a very intense dark green. In only some cases, in which a larger amount of iron hydroxide was present, did the color turn to a more yellowish green. The potassium ferrocyanide reaction showed that the iron colloids were flocculated in a manner similar to the silica. By treatment with congo red only local staining could be observed. Here, also, the stained substances appeared in a flocculated state. The shape of the humus particles could be seen best in the charred preparation. They were mostly present in form of splinters and scales. The grain deposits varied much in shape, thickness and color. In some places calcium carbonate accumulations were preponderant. The shape of the crystal formation could partly be recognized, showing outlines of very small rhombohedrons. By treatment with dilute hydrochloric acid the deposits were partly removed from the grain surfaces and floated in the liquid in the form of more or less coherent flakes.

Elementary fabric and resistance to wind erosion. The muck soils investigated by G. Hardt were characterized by their strong inclination to lose their coherence on the soil surface in the dry summer period, and to change into a very loose material which could easily be carried away by the wind. Every year, this property causes very severe damage to cultivation and is a constant danger to the farmer growing sugar beets. The microscopic investigation of the windblown material shows a picture similar to that obtained by a debris preparation; the intergranular braces, mostly undestroyed as such, are broken off and separated from the mineral grains.

G. Hardt showed by comparative chemical and physical investigations on soils subject to wind erosion on the one hand, and resistant to wind erosion on the other, that neither the grain sizes nor the amount of humus, nor even a change in the nature of the humus substances by heating and drying, had any appreciable influence on erodibility by wind. A very marked influence, however, was due to the lime content, especially in the clay fraction (particles smaller than 2 μ obtained by mechanical analysis). The lime content of the clay fraction of the windblown soils was 20-50 per cent higher than in soils resistant to wind erosion. The statements could be confirmed by rigidity tests with a pressure apparatus.

Microscopic investigations showed that the main binding substance was calcium carbonate which proved to be a minor binding material in soils showing intertextic fabric. The muck soils investigated were characterized by excessive water content in the cold and wet season, while in the dry summer period they dry out almost entirely. The fabric analysis showed a strongly marked movement of the calcium carbonate from the subsoil to the surface layer where it was deposited in the form of efflorescences on the soil surface, in the soil spaces in the interior of the surface layer, and in the form of interspersions in the fabric plasma. In these illuvial zones the intergranular braces and grain deposits became lighter in color; the precipitation of calcium carbonate took place in the form of flowery elements or minute crystals, which had a loosening influence on the complexes.

In addition to the above explanation of the loosening effect of the lime precipitation, it may be mentioned also that salts crystallizing out of the solutions are able to burst open solid bodies similar to the effect of freezing water. It is a matter of fact that the growth of crystals can overcome considerable resistance as shown by the splitting effect of salts on rocks infiltrated with salt solutions in desert regions.

CHLAMYDOMORPHIC FABRIC*

Description. Every mineral grain is surrounded by a uniform colloidal coating. The grains are generally united into coherent, but very fragile, complexes. The complexes are produced by the growing together of the coatings at the points where the grains touch each other, forming a ring-shaped thickening caused by the larger acumulation of fabric plasma in the space angles. The intergranular spaces are generally entirely empty (fig. 68).

The debris preparations show only coated mineral grains and no fragments of intergranular braces or space deposits.

The presence of coatings around the mineral grains in some soils was observed by many investigators long ago, for instance, by P. E. Müller (38). J. Dumont in Paris, (39) was of the opinion that coatings occur in every soil. He explained that the formation of aggregations in soils is possible only if the mineral grains have coatings so they can adhere to each other. Aggregates which consist of uncoated mineral grains cannot form sufficiently coherent and stable complexes.

---

* From Greek: χλαμυδος = mantle, cloak, sheath.

A. Demolon and S. Hénin (Paris) stated, in objection to this, that the most stable aggregations, as may be found in chernozems, have no coatings but flocculated accumulations of colloids between the soil grains. (40).

Occurrence. Chlamydomorphic fabric can be found in illuvial layers of sandy humus or iron podsols particularly. It occurs, furthermore, in some sandy ground water soils of more or less acid reaction, the soil solutions of which contain large amounts of dissolved iron or manganese compounds.

Genesis. Coatings are produced where all colloid or plasma substances in a soil have been present in the peptized state or true solution. The formation of coatings is only possible if the soil has dried out from time to time, as a result of which the soil solution is forced to retire in a more concentrated state to the surfaces of the mineral grains in the form of adhesion water. Genuine chlamydomorphic fabric will be found only in sandy soils showing a large preponderance of skeletal material and a comparatively dilute soil solution. The growth of the coatings is due on the one hand to the easy and uniform deposition, and on the other to the much more difficult re-solution of the coating substances. The procedure of the formation of the coatings in general was discussed on page 134.

The picture in fig. 68 shows mineral grains of the illuvial horizon of a humus podsol near Dorum, Bezirk, Bremen, Germany. The grain surfaces are covered by a dark brown coating which shows numerous cracks caused by drying. On slight heating and charring, the sections of the coating between the cracks shrink considerably. After ignition the sections are shrunk to about one-half of their original size. Their color is whitish, by stronger magnification slightly yellowish. On treatment with malachite green the ignited sections remain almost entirely unstained. On treatment with congo red, however, an intense coloration results. Sometimes the whitish sections appear deposited on a yellow-red film of iron oxide. Accumulations of iron oxide also can be found deposited in cracks and depressions of the grain surface.

Many dried ground water soils show coatings of iron hydroxide. The original coatings of a groundwater soil near Neuhof of the Marchfield, in Lower Austria, had an intense orange-red color. Before ignition, they could be stained a dark green with malachite green. After ignition, however, they remained practically unstained by the same treatment. The ignited preparation showed charred spots, and sometimes a charring of the whole coating could be noticed. The staining with malachite green before ignition was caused by the presence of organic matter in the coatings, evidently substances which had been dissolved in the soil solution and organisms, both of which were deposited on the grain surfaces after evaporation of the soil water. The coatings consisted of almost pure iron hyroxide, produced by oxidation of ferrous compounds present in the dissolved state in the former soil solution.

The same soil showed, in other zones of the soil profile, almost opaque blackish-brown coatings of manganese hydroxide. In other parts, coatings of manganese hydroxide, of iron hydroxide (generally in the minority), or of a mixture of both could be noticed.

PLECTOAMICTIC FABRIC*

Description. The mineral grains are entirely imbedded in peptized fabric plasma which forms not only coatings, but also intergranular braces. In soils particularly poor in skeletal material the coated grains are cemented in a ground mass showing many-shaped cavities. The cavity walls exhibit, in many cases, a shiny surface layer (fig. 69).

Occurrence. Plectoamictic fabric is found mostly in the illuvial horizons of iron podsols and podsolic brown earths. Furthermore, in many soils which show porphyropeptic fabric there also can be found plectoamictic fabric in some parts where the fabric is less dense. In this case, the plectoamictic fabric may be looked upon as a preliminary stage of the porphyropeptic fabric. A similar arrangement may be found also in some ground water soils rich in iron compounds and clay substances.

Some red earths and brown earths in limestone show a fabric which represents an intermediate position between the intertextic and the porphyropeptic fabric.

Genesis. Plectoamictic fabric may be formed if most of the colloids were in a peptized state, particularly in those cases where the fabric skeleton is slightly in excess of the fabric plasma or, sometimes, even if the reverse is the case. The evaporating soil solution, rich in peptized colloids, forms

---
*From Latin: plecto=interwoven (used as a prefix); amictus=coated.

coatings after retiring to the grain surfaces; the colloid substances will be deposited next in the angles of the spaces, forming intergranular braces instead of a mere thickening of the coatings at these places. Coatings and intergranular braces grow by continuous influx of plasma substances and continuous drying and deposition. Finally, cavities are formed which tend to develop a more rounded shape. The walls of the cavities grow towards the center of the empty space.

In order to give an example of this arangement, the elementary fabric of the illuvial horizon of an iron podsol near Pilsen in Bohemia may be described. Its mineral grains were united by means of ochraceous coatings and intergranular braces in such a way that a great number of many-shaped spaces were formed which showed dense wall formations with smooth and shining surfaces. In some larger spaces the smooth shiny surfaces were much darker in color, evidently showing a composition different from the interior of the coatings and intergranular braces. On cross sections against the margin a deep brown coloration of the plasma substances is found, indicating a more densely constructed marginal zone. Mineral grains projecting from the wall complex into the space were less coated, and therefore lighter in color, sometimes almost whitish. Round-shaped zones of deep brown color, which indicated the accumulation of iron colloids, could be seen also in the interior of the wall complexes.

The debris preparation showed that the grains were mostly covered with thick coatings. The more or less homogeneous plasma substance had the appearance and deep ochraceous color of rosin. When investigated with high magnification it showed a fine granulation in the interior. With malachite green it was generally stained a uniform blue-green. In the presence of larger amounts of iron hydroxide a yellow-green color was sometimes obtained. In cases in which iron hydroxide was in preponderance the substance remained unstained. With congo red only a slight change in the original ochre of the iron hydroxide color could be obtained, giving it a more orange tint. Very small parts of a pronounced red color could be observed only sporadically.

AGGLOMERATIC FABRIC

Description. The mineral grains are entirely bare and show at the most only sporadically crumbly, loose and easily removable deposits. All substances of the fabric plasma are present in a flocculated state or represent insoluble bodies. They are generally to be found only in the form of loose space deposits (fig. 70). A formation of somewhat coherent complexes or aggregates takes place only sporadically or cannot be found at all. Therefore, material of this type often gives rise to windblown soils.

Occurrence. Typical agglomeratic fabric was found in a number of sandy prairie soils in northwestern Minnesota. One of these prairie soils, near Richdale, had a surface layer which was very easily eroded by the wind. In this layer the strongly reflective mineral grains were almost entirely bare and glass-like. In the intergranular spaces somewhat coarse, brownish-gray to blackish-colored, loose deposits were found which consisted of flocculated colloids interspersed with silt particles. In the deeper layers of the humus horizon a slight tendency toward aggregate formation could be observed, though the arrangement was similar to that in the surface layer. The surfaces of the mineral grains were frequently spotted with silt particles which were easily washed out of the space deposits where they were not held by binding substance.

Agglomeratic fabric may be found in many windblown humus soils, also in the surface layer of the calcareous muck soils mentioned in the paragraph concerning intertextic fabric.

Genesis. The general genesis has already been discussed with reference to the description of the experiments imitating agglomeratic fabric (p. 131). Agglomeratic fabric generally occurs only in sandy soils under special circumstances. These sandy soils would not be so open to wind erosion if they were built up another way such as is demonstrated by the coherent and stable complexes produced by J. Bartlett's experiments with lignin solutions and quartz sand.

The formation of agglomeratic fabric in the surface layer of calcareous muck soils well dried out in the hot season is due to the loosening effect of the precipitation, and crystallization of accumulated calcium carbonate as described on page 142.

BLEACHED SAND FABRIC

Description. The bleached horizons of podsols show by microscopic examination, in many cases, some raw humus between the mineral grains. If the raw humus particles are present to some extent, the picture of the fabric may resemble the agglomeratic fabric. The mineral grains are entirely naked. The space deposits of humus, however, contain residues of plants with structures preserved, particularly roots and root hairs. In the splits and dents of the grains, accumulations of peptized colloids may be found by examination of debris preparations, particularly when higher magnifications are used. They represent residues of former coatings which are preserved in places and more protected against washing. Sporadically, parts of the coatings or almost undestroyed coatings covering the whole grain surface may be found. By ignition of the humus particles, residues of inorganic colloids which were accumulated in the pores of the plant residues and thus protected from being carried away by the soil water, may be noticed.

Occurrence. Bleached sand fabric similar to the above description generally occurs only in humus podsols. In iron podsols, in most cases, only residues of coatings or some remaining accumulations of inorganic colloids, mostly iron hydroxide, are found in the more protected cracks and dents of the grain surfaces.

Figure 71 shows a debris preparation of the whitish-gray-colored layer (4-10 cm.) of the humus podsol near Dorum, Bezirk, Bremen, mentioned on page 144. The mineral grains consisting of glassy quartz throughout the soil were almost entirely bare. Only some parts showed almost red-colored accumulations of colloids in the irregularities of their surfaces. Dark brown coatings, or parts of them, could be seen in only very rare cases. The humus deposits in the spaces consisted, in general, of leaf-shaped, sepia-brown tissue fragments which were much shrunk together. Soaked in water and prepared in the wet state they showed, in most cases, the original cell structures. Furthermore, somewhat reddish-colored, wooly remnants of plant roots were to be found. After ignition, the humus particles showed, in all cases, brownish-red, and by higher magnification yellowish-colored residues of inorganic colloids. They could be stained as well with congo red as with malachite green.

Genesis. The difference between the bleached sand fabric and the agglomeratic fabric lies chiefly in their genesis. In the case of the first type, soil constituents of originally coated mineral grains were washed out. The colloid coatings were so far removed in most cases that only very minute residues remained in the cracks and dents. In the case of the latter type, the lack of coatings is due to the lack of peptized colloids.

RENDZINA FABRIC

Description. The mineral grains are bare and, in debris preparations, show no films and almost no surface deposits. Many of the mineral grains are calcite. They are embedded in a loose ground mass of humus particles. There is almost no binding substance to unite grains and humus particles. The grains fall out of the complex easily. Fine sand and silt may be easily carried away by the soil water and frequently may be found deposited in large numbers on the wall surfaces of the larger soil spaces in the dried-out soil. Also, the humus particles of the ground mass show no binding between one another. They are loosely combined, the whole fabric being held together only by the interlocking of the splinter-like particles. The fabric of the humic ground mass is more stable in the wet stage, while the smaller mineral grains are easily washed away. In the dry state the humus particles become very friable, the complexes fall apart very easily and may be dispersed by slight friction into a blackish powder.

The humus particles show, in almost all cases, their original structure as plant residues. They generally consist of reddish-brown to ochraceous-brown (in thin sections translucent) interior showing original plant cell structures, and an outer layer of blackish to dark brown, crusty, opaque, sometimes coal-like substances. In many cases, efflorescences of these substances may be found in the form of protuberances on the surface of humified stem or root fragments or even on the surfaces of space walls. Similar bodies shrunk together and somewhat broken up may be found removed from their positions and distributed throughout the soil.

Occurrence. The arrangement occurs with some variations in rendzina soils. Humus-carbonate soils of the same fabric and the same fabric constituents were found by the author not only on limestone, but also were observed on dolomite in Austria, on dolomite and silicious limestone in eastern Iowa, and, furthermore, even on crystalline schists rich in lime (Kalkglimmerschiefer) in Austria (Glocknergebiet). They do not cover large areas on these rocks but occur only locally.

Genesis. The rendzina is characterized by the fact that the peptized, chernozem humus fraction stable in water and soluble in alkali, and the highly hydrophilic acid humus fractions of the podsols soluble or highly dispersible in water, are both practically absent in the process of soil formation. They cannot be discovered as fabric units by fabric analysis. This property is due to the fact that rendzinas are formed on limestone or calcareous rocks in pronounced humid or podsol climates. The humid climate suppresses the formation of the alkali soluble humus fraction of the chernozem. The presence of large amounts of lime prevents the formation of acid humus. When lime is entirely removed from the soil, acid humus is formed. Its action as a protective colloid is made possible and podsolization of the rendzina begins.

Of particular interest is the blackish-colored, opaque humic substance produced in the form of crusts and many-shaped efflorescences on the surface of humus particles and plant residues. The fact that they are efflorescing indicates that they occur temporarily in a dissolved or peptized state inside the plant residues or humus particles. The absence of films on the mineral grains, however, indicates that they are not present in a dissolved state in the soil solution. A real knowledge of the relations and microdynamic processes of this extremely interesting soil type will be obtained only by further detailed investigations.

MAGMOIDIC FABRIC

Description. The naked mineral grains are combined by intergranular braces, generally of ochraceous to orange-red, peptized inorganic colloids. Formations of films, coatings or flocculated deposits are almost entirely absent.

The most important features of the debris preparation are the sharp-edged fragments of intergranular braces and the bare mineral grains with peptized inorganic deposits. Of significance, also, is the formation of many-shaped tongue-like irregularities on the marginal contours of the deposits or residues of intergranular braces on the grain surfaces (fig. 72).

The fabric generally is very unstable in water. By washing a suspension in water the grain deposits are in many cases easily removed. The material obtained consists of entirely denuded grains and residues of intergranular braces which are changed to more or less round-shaped complexes as a result of the washing away of their edges (fig. 73).

The name "magmoidic fabric," explained on page 134, was given to this arrangement because of the magmoidic state of the fabric plasma in those stages of the soil in which the fabric originally acquired its characteristic features. Details of the fabric which indicate the temporarily magmoidic nature of plasma substances are to be found in fabric members of higher order more than in the elementary fabric as will be shown later.

Occurrence. Magmoidic fabric was found in some Mediterranean red earths. It was observed first, however, in the B-horizon of some intense dark ochraceous-colored soils near Versailles, in France. The soils were named and mapped as brown earths. Microscopically, however, they differ so greatly in their dynamics, constituents, and fabric arrangement that they form a type of their own, resembling the degraded yellow earths of the southeastern United States. A detailed study of one of these soils in the "Sablière" near Versailles was begun by the author in the summer of 1935. The work will be continued with American soils.

Genesis. The fragments of the illuvial horizons of the sandy soil of the "Sablière," in their extreme hardness when in a dry state, resemble rocks. On slight moistening, however, they lose their coherence very rapidly and fall apart. After ignition the fragments are not only stable in water, but also resistant when treated with dilute acids and alkali.

The main binding substance in the soil fragments is a form of colloidal silica, which is so easily soluble in water that it may be obtained by evaporation of a filtrate produced by filtering a soil suspension which has been shaken for a few minutes in a shaking apparatus. It may definitely be seen by microscopic investigations that the silica is acting as a protective colloid for the iron hydroxide and probably, also, in a similar way for the aluminum hydroxide, as was indicated by Reifenberg in Mediterranean red earths. The colloid fraction, although entirely free of acid humus substances, is dispersed so easily in water that the soil develops a whitish-gray, bleached horizon at a depth of 35-45 cm., consisting, in the main, of more or less naked mineral grains. The eluvial process could be observed even by microscopic investigation on soil clods exposed to the rain. These clods developed eluvial zones on the upper surface, and well marked illuvial zones showing accumulation of the colloid fraction on the lower surface.

To what particular processes the arrangement of the elementary fabric is due cannot be decided because of the lack of detailed data. It may be derived from the magmoidic nature of the fabric plasma produced by the higher contents of active and water soluble silica, or it may have originated because the coatings or smaller deposits are easily removed by washing as could be seen by the investigation of washed debris preparations mentioned above. Both possibilities may prove to be right.

MORTAR FABRIC

Description. This fabric type occurs mostly in connection with plectoamictic or intertextic fabric. It differs from both in that the intergranular spaces contain a crusty or flowery white mass. It consists, in the main, of very small (generally only 1-2μ) round crystals of calcite. In larger spaces the microcrystals cover the surface of the space walls (fig. 74).

In the debris preparations, coatings (or grain deposits) and intergranular braces are found covered with microcrystals, and also fragments of the lime accumulations which had been growing into the interior of the air-filled soil spaces. The microcrystals, seen with higher magnifications, are grain-like, colorless, and transparent. In some cases they show in definite but visible crystallographic contours.

The name "mortar fabric" was chosen because it resembles very much the fabric of a dry and cured mason mortar. In both cases the storage of complexes of microcrystals of calcium carbonate are observed. If a yellow sand consisting of coated quartz grains should be used in the manufacture of mortar, the similarity of the fabric picture would be so close that it almost could not be distinguished from the microscopic picture of the lime horizon of a Central European brown earth.

Occurrence. Mortar fabric may be found in the lime horizons of brown earths, podsolic brown earths, chernozems and calcareous ground water soils.

Genesis. The microcrystals are formed largely by an efflorescence process and only in a few cases, mostly in ground water soils, are they formed by direct precipitation and sedimentation from the evaporating soil solution.

The calcium carbonate is transported by the soil solution in a dissolved state in the form of calcium bicarbonate, into the deeper horizons of the soil profile. The evaporating soil solution retires to the pores of the fabric plasma. By continuous drying of the plasma substances the dissolved lime begins to effloresce in the form of calcite on the surface of the intergranular braces and coatings. If the concentration of the soil solution is lower, the efflorescence will take place in the form of very thin but long needles as will be shown later. If the soil solution reaches a high concentration of dissolved salts, then the efflorescences will be formed as very small round crystals. Precipitations of these microcrystals will be found not only on the surface, but even inside of the fabric plasma, i. e., in the intergranular braces and coatings. These interior precipitations make the fabric plasma appear lighter in color. By treatment with dilute hydrochloric acid all intergranular braces or coatings effervesce and fall apart.

The formation of the lime accumulations could be studied in all its stages on a brown earth on loess near Altenhof, Lower Austria. Soil samples in an undisturbed condition were taken in the moist state from the lime horizon which was found at a depth of 110-130 cm. The samples were taken to the laboratory and the changes due to drying observed microscopically. Under the microscope the fragile, whitish-ochraceous soil mass showed, on freshly torn planes, whitish areas which were filled with large quantities of lime efflorescences, as well as dark ochraceous-colored parts which showed almost no efflorescences. The surfaces of the dark ochraceous-colored space walls were covered in a short time with microcrystals which could be observed microscopically growing out of the colloid pores. The growth began on the highest points of the micro-relief of the wall surfaces. The efflorescences covered the whole surface in a few minutes, but the growth did not stop. The white layer, consisting of a large number of crumbly complexes of microcrystals, gained in thickness and developed into the empty space. From the appearance of the growing crystal complexes in the dry state, and from the observations of the efflorescing process in its earlier stages and the formations of the first crystal complexes, it may be concluded that the efflorescences are developing on the original wall surface and not on the new surface of the lime accumulations. The old crystals are pushed away from the wall surface by the newly developing efflorescences. The small crystals adhere to each other in the first stage when they are not entirely dried. The crumbly complexes still may be seen in all debris preparations of the horizon.

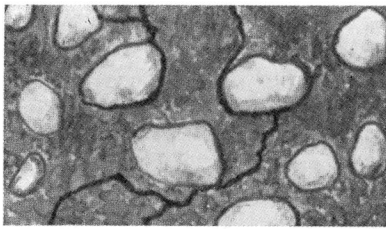

Fig. 65. Porphyropectic fabric. Non-coated mineral grains embedded in a dense ground mass from which they may be removed easily.

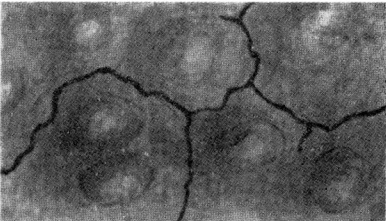

Fig. 66. Porphyropeptic fabric. Mineral grains cemented with a dense ground mass. Isolated grains always more or less coated.

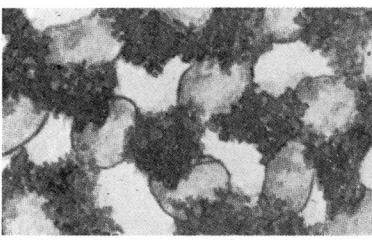

Fig. 67. Intertextic fabric. Bare mineral grains united by intergranular braces. In the case of chernozem fabric the grains show transparent humus films.

Fig. 68. Chlamydomorphic fabric. Mineral grains surrounded by a uniform colloidal coating. The intergranular spaces are empty.

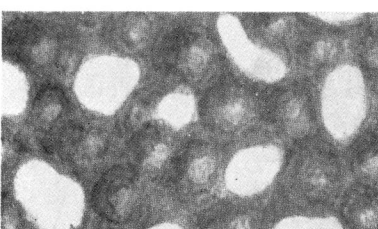

Fig. 69. Plectoamictic fabric. Mineral grains coated and united by intergranular braces.

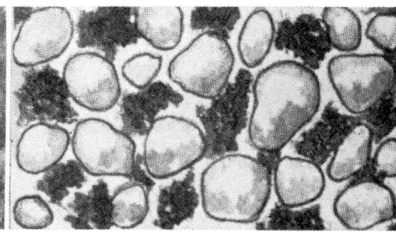

Fig. 70. Agglomeratic fabric. Mineral grains bare. In the intergranular spaces are loose deposits of flocculated or insoluble plasma substances.

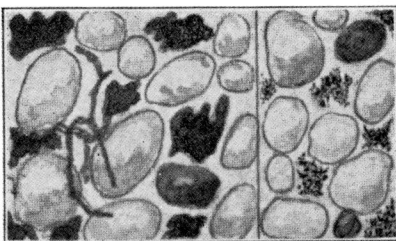

FIG. 71. Bleached sand fabric. Mineral grains bare but occasionally showing residues of coatings or deposits of peptized colloids in the dents and cracks of the grain surface. Right part of the preparation ignited, showing residues of inorganic colloids which were protected from leaching in the pores of the raw humus particles.

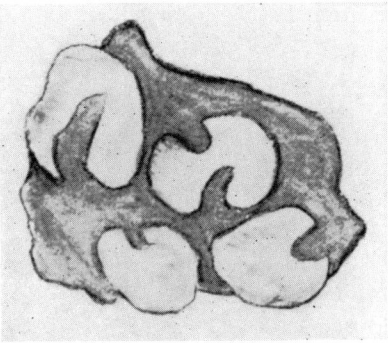

FIG. 72. Fragment of a magmoidic soil showing tongue-like marginal contours of the fabric plasma (soil of the "Sablière" near Versailles, France).

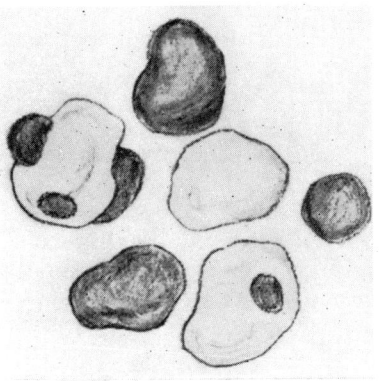

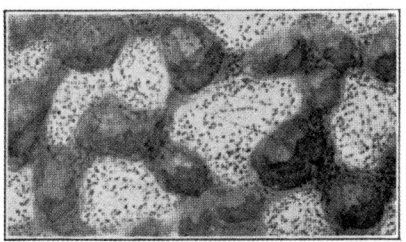

FIG. 74. Mortar fabric. The intergranular spaces are filled with microcrystals of calcite.

FIG. 73. Debris of the Sablière soil after washing in water.

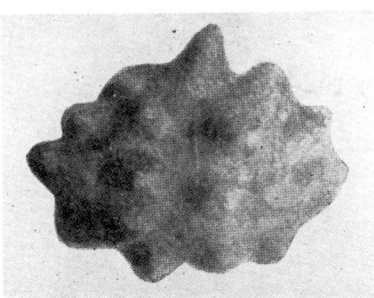

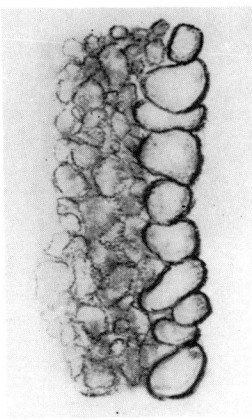

FIG. 75. Mammilated aggregate, from black earth near Tower City, North Dakota.

FIG. 76. Wall formation on the surface of an aggregate with a granular shell.

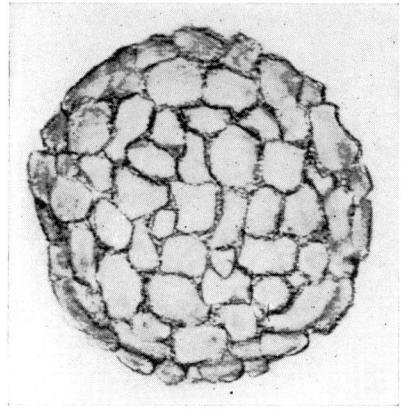

Fig. 77. Artificial granular shell made from quartz grains and chloroform drops.

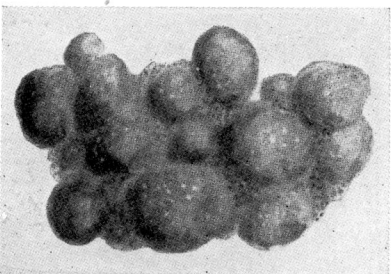

Fig. 78. Aggregate of a brown earth on limestone, Anninger, Lower Austria.

[*Editors' Note:* Only the references cited in the preceding excerpts are reprinted here.]

Demolon, A. & Henin, S. 1932. Recherches sur la structure des limons, etc. Soil Research Vol. III.

Dumont, J. 1913. Etude sur le sol I. Agrochimie, Paris.

Muller, P. E. 1887. Studien uber naturliche Humusformen, etc. Berlin.

Oprea, Chr. V. 1936. Mikroskopische Untersuchungen an verschiedenen Bodentypen auf Loess. Tipografia Bucovina, Bukarest.

Sander, B. 1930. Gefugekunde der Gesteine. Wien.

Copyright © 1960 by Blackwell Scientific Publications Ltd.
Reprinted from *Jour. Soil Sci.* **11**(1):172-185 (1960)

SOIL STRUCTURE AND FABRIC

THEIR DEFINITION AND DESCRIPTION

R. BREWER AND J. R. SLEEMAN

(*C.S.I.R.O. Division of Soils, Canberra*)

Summary

Definitions of structure, fabric, and texture, as used in pedology and geology, are proposed in which structure conforms to the broader concept of the soil physicist, fabric conforms to the concept as used in pedology by Kubiena (1938) and in sedimentary petrology (Pettijohn, 1949), and texture conforms to the broad concept as used in petrology (Rice, 1954). Fabric is a part of structure, and structure a part of texture. The term pedality is proposed for the study of what has been called structure in the field, and field grading for field texture. Kubiena's (1938) concepts of skeleton and plasma are defined more specifically, and the concepts of pedological features, and basic, elementary, primary, secondary, and tertiary structure introduced. Descriptive terms for structure are suggested and a descriptive system for soil materials proposed. Nomenclature of pedological features and the various levels of structure and fabric are discussed.

1. *Introduction*

THE study of soils, whether aimed at soil classification or the interpretation of soil genesis, involves the description of a great number of characteristics. Of late, the realization that soils are extremely complex materials has indicated the need for more detailed and comprehensive description. To make this mass of descriptive data understandable and manageable it is necessary to organize it into logical units which deal with the characteristics of some particular aspect of the soil material. One such aspect is its physical constitution. Although the importance of this aspect has been recognized in field pedology, soil physics, and soil mechanics, the terms used are ambiguous and the present systems of soil description are inadequate for the more recent detailed studies.

It is the object of this paper to define concepts and terms to cope with the description of the physical constitution of soil materials having regard to current concepts as used in pedology and the related science of geology.

In petrology the essential characteristics of the physical constitution of rocks have been enumerated and grouped in various ways. Although these groupings do vary they all deal with all or some of the following features of the solid particles: their crystallinity, shape, size, and arrangement. Since this approach is logical and has been proved in petrology and to a lesser extent in pedology it is proposed to apply it for developing concepts and terms for the description of the physical constitution of soil materials having due regard to their special nature. It should be emphasized at the outset that these concepts concern only the qualitative description of the physical constitution of soil materials as distinct from quantitative estimations and descriptions of complete soil profiles.

2. Terminology and Definitions

It is proposed to consider the current terminology as applied to the physical constitution of rocks and soil materials and so attempt to arrive at a systematic, universally applicable terminology for soil materials.

(a) Fabric

The simplest concept for describing the physical characteristics of soil materials and rock is that of fabric. In igneous petrology, fabric is defined as 'that part of texture* which depends on the shapes and arrangement of the constituents of a rock' (Rice, 1954). In sedimentary petrology, fabric is 'the orientation in space of the elements of which a rock is composed' (Pettijohn, 1949). These two petrological definitions differ only in that no mention of shape is made in the latter. Kubiena (1938) was the first to speak of soil fabric as a counterpart of rock fabric. He later (1953) used the term as 'the arrangement of the constituents, not only in the narrow sense of aggregate formation but in a general sense, referring also to the variable minor fabric of dense soil masses etc.' There is general agreement then that fabric deals primarily with *arrangement* of the constituents. However, in its application to soil materials it is considered that the definition should deal specifically with the arrangement of voids which has been considered only as a complementary result of the arrangement and packing of the solid particles. Thus:

Soil fabric is the physical constitution of a soil material as expressed by the spatial arrangement of the solid particles and associated voids.

Since soil fabric is concerned with spatial arrangement it is described in terms of the orientation and distribution patterns of the primary particles, compound particles, and voids.

(b) Structure and pedality

The petrological concepts of structure (Krumbein and Sloss, 1955; Pettijohn, 1949) which deal with regional features of sedimentary rocks or specific arrangements (the sedimentary features of Shrock (1948)) are generally unsuitable for soil materials. In pedology there are three variations of the concept of soil structure which overlap in their treatment of the physical characteristics of soil materials.

1. Baver's (1948) concept—'the arrangement of the soil particles'—considers 'the arrangement of sand, silt and clay, and of secondary particles (aggregates or structural elements) into a particular structural pattern'; size and shape of the particles are not considered. This concept is almost identical with that of fabric defined earlier.

2. Zakharov's (1927) concept—'the very fragments or clods into which the soil breaks up'—considers only the size and shape of Baver's secondary particles and ignores arrangement.

3. The concept of the U.S.D.A. Soil Survey Manual (1951)—'the aggregation of primary soil particles into compound particles, or clusters of primary particles, which are separated from adjoining aggregates by

---

* Texture is used here in its specific petrological sense.

surface weakness'—considers the size, shape, and arrangement of Baver's secondary particles but ignores primary particles.

These three variations encompass most of the shades of meaning implied by definitions put forward by various pedologists from time to time. Butler (1955) states that Zakharov's (1927) concept is the one applicable to the study of soils in the field. Kubiena (1953) uses structure in a similar way. Jongerius (1957) uses it in a similar way to Baver (1948) except that he considers also the arrangement of the voids. It is apparent that these variations are due to the different interests of the various workers and the sensitivity of their techniques but they may be combined logically in a single definition.

Soil structure is the physical constitution of a soil material as expressed by the size, shape, and arrangement of the solid particles and associated voids, including both the primary particles to form compound particles and the compound particles themselves. Thus fabric is the element of structure which deals with arrangement.

Soil-structure studies in the field, because of the techniques available, are concerned chiefly with the size, shape, and arrangement of compound particles to which the term pedality has been applied.

Pedality is the physical constitution of a soil material as expressed by the size, shape, and arrangement of the compound particles or peds.

(c) *Texture and field grading*

In igneous petrology texture is an extension of structure to include the degree of crystallization (crystallinity) while in sedimentary petrology it is synonymous with the definition of structure given earlier (Pettijohn, 1949). In pedology, texture has been used with various shades of meaning as an estimation of the percentage of clay-size particles, the complete particle-size distribution, the effects of structure, and consistence on the feel of a soil material of particular particle-size distribution, and other allied properties (Robinson, 1949; U.S.D.A. Soil Survey Manual, 1951; Leeper, 1952, &c.). Such concepts are quite incompatible with those used in petrology which come closer to normal English usage of the term.

It is realized that texture as used in pedology, although ill defined, is strongly established, but a very strong case exists for redefining the term in line with the petrological meaning and substituting another term for the present pedological concept. The following terms and definitions are suggested:

Soil texture is the physical constitution of a soil material as expressed by its structure and by the degree of crystallization (crystallinity) of the solid particles. Thus fabric is a part of structure which is a part of texture in this context.

Field grading is the assessment of the particle-size distribution by the method of moistening and manipulation of the soil material in the hand.

The relationship between the proposed pedological terms and those currently used in pedology and geology is summarized in Table 1.

TABLE I

Relationships of Proposed Concepts to Current Concepts

Characteristics	Soil components*	Proposed Pedological Usage					Current Pedological Usage							Geological Usage				
		Field grading	Pedality	Fabric	Structure	Texture	Field texture (U.S.D.A. manual)	Fabric (Kubiena)	Structure (Zakharov)	Structure (Bauer)	Structure (U.S.D.A. manual)	Grade	Packing	Fabric (sedimentary)	Fabric (igneous)	Structure† (sedimentary)	Texture (sedimentary)	Texture (igneous)
Crystallinity	Skeleton grains					×												×
	Peds					×												×
	Pedological features					×												×
	Plasma					×												×
Size	Skeleton grains	×				×	×										×	×
	Peds	×	×		×	×	×		×		×	×				×	×	×
	Pedological features				×	×											×	×
	Plasma grains				×	×											×	×
	Voids				×	×											×	×
Shape	Skeleton grains					×											×	×
	Peds		×		×	×			×		×	×				×	×	×
	Pedological features				×	×											×	×
	Plasma grains				×	×											×	×
	Voids				×	×											×	×
Arrangement	Skeleton grains			×		×							×	×	×	×	×	×
	Peds		×		×	×				×	×							
	Pedological features			×	×	×		×		×					×			
	Plasma			×	×	×		×							×			
	Voids			×	×	×		×		×			×	×	×	×		

* For geological usage substitute rock components, viz. solid elements and voids.
† Restricted usage e.g. oolitic structure, equivalent to sedimentary features.

3. Application to Soil Materials

(a) The nature of soil constituents

To apply the concepts of structure and fabric to soil materials it is necessary to consider the nature of the materials and the kinds of phenomena which may occur. In considering soil materials it must be recognized that they are composed of two broad groups of constituents with very different properties; one which is relatively stable and one which is capable of remarkable movement, concentration, and reorganization during soil formation. Each of these groups is composed of diverse constituents but it is the activity of the latter which causes many of the significant features of soil materials. Kubiena (1938) has referred to the relatively stable soil constituents as 'fabric skeleton' and to the mobile, active constituents as 'fabric plasma'. The adjective 'fabric' seems unnecessary and rather misleading since the terms really refer to the components of the soil materials. Since Kubiena has not defined these terms concisely they are defined here retaining his original concepts:

1. *Skeleton grains* of a soil material are individual grains which are relatively stable and not readily translocated, concentrated, or reorganized; it includes mineral grains and resistant siliceous and organic bodies larger than colloidal size. Complex grains are not considered as skeleton grains but as pedological features (see later). The skeleton, as such, is relatively immobile except for processes such as washing down cracks, but it is capable of weathering to form plasma.

2. *Plasma* of a soil material is that part which is capable of being, or has been moved, reorganized, and/or concentrated by the processes of soil formation. It is the mobile active part of the soil material. The plasma includes all the material, mineral or organic, of colloidal size and relatively soluble material which is not bound up in the skeleton.

It is proposed that living roots, soil fauna, and flora should not be considered as part of the soil materials as regards description of their physical constitution in terms of structure and fabric; the soil material is regarded simply as the medium for their activities although its structure may be affected by their activities. Similarly, recognizable undecomposed organic remains are not considered. These are materials with characteristics so different from the mineral constituents that they require a different method of treatment. Humified organic matter, however, can be considered in structure and fabric descriptions. Humification releases mobile organic constituents, which are part of the plasma, and resistant siliceous and organic bodies which can be included in the skeleton.

(b) The organization of soil constituents

The reorganization of constituents within soil materials due to the nature of the constituents and of the soil-forming processes, including the activity of soil fauna and plant roots, leads to the development of a series of features with specific fabrics and structures. Other features may be inherited from the parent rock or from the processes of deposition of transported parent materials. All these features are the pedological equivalents of what Shrock (1948) refers to as sedimentary features such

SOIL STRUCTURE AND FABRIC

as cone-in-cone structure, oolitic structure, animal tracks and trails, &c. Following this nomenclature it is proposed to refer to these features, as a group, as *pedological features* which can be subdivided into various kinds.

Pedological features are recognizable units within a soil material which are distinguishable from the enclosing material for any reason such as origin, differences in concentration of some fraction of the plasma, or differences in arrangement of the constituents (fabric).

These pedological features can be divided into two broad groups on the basis of origin:

1. *In situ* pedological features which are formed in the soil profile. Of these the three chief features which have been recognized so far are:

Plasma concentrations which are concentrations of any of the fractions of plasma in various parts of the soil material, due to soil formation. Examples of these are carbonate nodules, iron-oxide nodules, clay-mineral coatings.

Plasma separations which are features characterized by a significant change in the arrangement (fabric) of the constituents, rather than a change in concentration of some fraction of the plasma. An example of this is the change in orientation of the clay-mineral fraction near the surface of slickensides.

Fossil formations which are preserved features resulting from biological activity such as burrows and root channels. These may or may not be filled with various materials.

2. Inherited pedological features which are, in fact, relicts of the soil parent rock or parent material. The three inherited pedological features recognized so far are:

Litho-relicts which are derived from the parent rock and usually recognizable by their rock structure and fabric, for example, so-called 'floaters' of parent rock.

Pedo-relicts which were formed as *in situ* pedological features in a previously existing soil material and later transported and deposited with the present parent material, for example, transported abraded nodules and concretions.

Sedimentary relicts which are formed during deposition of a transported soil parent material, for example, clay galls and clay flakes.

Besides the formation of pedological features, soil materials may become more highly organized by the development of peds. A highly aggregated soil material may have primary peds packed or bound together to form secondary peds, which in turn may form tertiary peds, and so on; in this context primary peds are the smallest recognizable natural aggregates in the soil material. The structure of such a soil material can be described systematically at several levels of organization:

(i) The size, shape, and arrangement of the constituent grains and voids within the primary peds.

(ii) The size, shape, and arrangement of primary peds and the interpedal voids to form secondary peds, and the relationships of the primary peds with materials which may occur between the peds.

(iii) Similarly, higher levels can be described in terms of the characteristics of the peds of the appropriate level, the interpedal voids, and their relationships with each other and with materials which may occur between the peds.

The structure of interpedal material can be described separately; usually it belongs to one or other of the pedological features.

In a similar way, pedological features can exhibit levels of organization. For example, it has been observed that plasma concentrations, such as carbonate nodules, may be made up of a group of dense bodies of strongly concentrated carbonates set in a larger body of less strongly concentrated carbonates. Thus, the structure of the smaller dense concentrations can be described as well as that of the larger composite concentrations. Similar complexities have been observed in other kinds of pedological features.

The description and classification of these complex soil structures and fabrics can be approached in a logical sequence. The following definitions split the observations into usable units, any one of which can be studied separately or in relation to any other one or more.

1. *Basic structure* is the size, shape, and arrangement of simple grains (skeleton grains and plasma) and voids in primary peds, or in an unaggregated soil material within which there are no observable pedological features other than plasma separations. Complex pedological features have a basic structure for each part which shows a distinctive structure or fabric.

2. *Elementary structure* is a characteristic size, shape, and arrangement of *specific* pedological features with regard to the basic structure of the material associated with them and to specific reference features.*

Elementary structure does not necessarily deal with all the pedological features present but only those kinds which contribute to a characteristic structure. For example, strongly oriented coatings of clay minerals in root channels may occur frequently associated with a particular basic structure. This characteristic association constitutes an important elementary structure which can be described and named. Thus elementary structure is a concept of convenience. It is useful also for complex pedological features.

3. *Primary structure* is the size, shape, and arrangement of all of the pedological features present within the primary peds, or in an unaggregated soil material, with regard to the basic structure and to specific reference features. It is the complete structure of the primary peds or the unaggregated soil material.

4. *Secondary structure* is the size, shape, and arrangement of the primary peds, interpedal voids, and associated interpedal pedological features. Unaggregated soil materials do not have secondary structure or structures of higher order.

5. *Structures of higher order* deal with the size, shape, and arrangement of the peds of one order lower, and the associated interpedal voids and pedological features. Thus tertiary structure deals with the size, shape, and arrangement of secondary peds, interpedal voids and associated pedological features, quaternary structure with tertiary peds, and so on.

* Reference features are features which can be specified such as the surfaces or shapes of peds and pedological features, the walls of voids, inter-horizontal contacts, &c.

6. *Complete structure* of a soil material is the size, shape, and arrangement of all the pedological features in relation to the structure of the soil material, at all levels of organization.

The schedule of description of soil structure is set out in Table 2. This is not the only possible sequence of description and individual workers may find it convenient to rearrange the order of treatment for specific purposes.

4. *Nomenclature*

From the foregoing concepts of structure and fabric it is apparent that full descriptions will be unavoidably complex and lengthy (see Appendix). These descriptions can be abbreviated and simplified by the judicious application of names to specific, important structures and fabrics. This emphasizes the urgent necessity for a system of nomenclature. It is not intended to set out here a series of names and corresponding descriptions but rather to propose a systematic method of naming structures and fabrics at various levels of organization. Since the same principles apply to naming both structure and fabric, structure only is considered here for conciseness.

At the lowest levels of organization basic structures and pedological features are relatively simple and limited in number. Once these have been characterized and named the structures at higher levels, which will be more numerous and complex, can be described in terms of specific combinations of these features. Thus, elementary structure is described in terms of basic structure and one or more specific pedological features and primary structure in terms of basic structure and all included pedological features. Since structures at still higher levels are classified on the characteristics and arrangement of the peds, rather than in terms of basic structures and pedological features, they can be named independently. It is essential, therefore, that standard common basic structures, pedological features and structures at the secondary and higher levels be classified and named.

Pedological features can be classified and named on characteristics of shape, sharpness of separation, kind of material (which is not really a part of the structure description), orientation and distribution patterns, and internal structure. A few of these features, such as 'amygdali' (McMillan and Mitchell, 1953) have been described and subdivided according to their structure and fabric. Others such as 'crystal tubes' have been described by Kubiena (1938).

Basic structures are named solely on the characters of the elements of structure within the material. Kubiena's (1938) elementary fabrics apparently are akin to basic fabrics.

Elementary and primary structures can be described in terms of basic structure and associated pedological features. Characteristic ones should be named.

Secondary structures and structures at higher levels can be classified and named on the characters of the elements of structure considering the peds as skeleton units equivalent to skeleton grains in structures at lower levels. Kubiena (1938) has applied names to 'fabric members of higher

TABLE 2
Description of Structure

Structure level	Material	Components	Characteristics	
Pedological features	Individual pedological features*	Skeleton grains Plasma grains Voids	Size	Absolute and relative
Basic structure (including plasma separations)	Material within primary peds but excluding pedological features other than plasma separations	Skeleton grains Plasma grains Voids	Shape	Shape and roundness classes; absolute and relative
Elementary structure	Material within primary peds including that with basic structure and specific characteristic pedological features	Skeleton grains Plasma grains Voids Specific pedological features as units	Arrangement	Orientation pattern
Primary structure	All material within primary peds (material with basic structure and all pedological features)	Skeleton grains Plasma grains Voids Pedological features as units		1. Orientation of individuals with each other. 2. Referred orientation—with regard to a reference. 3. Orientation between groups of individuals.
Secondary structure	Primary peds as units and interpedal pedological features	Primary peds Interpedal voids Interpedal pedological features		Distribution pattern
Tertiary structure	Secondary peds as units and pedological features between secondary peds	Secondary peds Interpedal voids Interpedal pedological features		1. Distribution of individuals with regard to each other. 2. Referred distribution—with regard to a reference. 3. Distribution between groups of individuals. 4. Sharpness of boundary.

* Pedological features may have levels of organization requiring description in terms of several basic structures and their relations to each other.

83

order' but these do not seem to fulfil consistently the requirements for application to structures or fabrics of higher order proposed here.

Finally, *complete structures* can be classified and named but these will be very numerous and very complex. Kubiena's (1956) fabric names for whole soil materials do not seem to be suitable for this concept of complete structures or fabrics; they appear to be too generalized to succeed for complete soil materials.

In the actual naming of these structures at the various levels of organization there are two important points to consider:

1. Since structure and fabric analyses are not considered in any of the classification systems in use at the present time, the names of any of the soil groups at any level in any of these classification systems should not be used. An additional reason for this is that structure and fabric can be described only for a soil material and not a profile, since they may vary down a soil profile. Names should refer specifically to the characteristics observed and described; they should be coined, or selected from geographical names, and given specific meaning.

2. It is possible that fabrics at different levels may be identical. For example, a well-aggregated soil material may have its primary peds organized to form secondary peds in the same manner as the secondary peds are organized to form tertiary peds and so on. Thus, the secondary, tertiary, and higher level fabrics may go under the same name.

5. Discussion

It is considered that the proposed definitions of structure and fabric given above are justified by the fact that they include all the concepts which have been used in pedology and fulfil the needs of workers in all branches of pedology. In addition, the concepts have a systematic relationship to each other: structure embraces all the characteristics of the physical constitution of the soil material except crystallinity, while fabric and pedality are specific parts of structure. Field grading is considered a more appropriate term than 'field texture' for particle-size distribution assessments in the field; 'texture' should be reserved for the complete physical constitution, i.e. structure plus crystallinity.

One of the major difficulties in dealing with the description of soil structure and fabric, is the scale at which the observations are made. It cannot be emphasized too strongly that the definitions of fabric and structure place no limitations on the scale of development of phenomena so that observations at all possible magnifications must be considered. In this paper, the descriptive terms have been defined to cover as wide a range of scales as possible, but it is obvious that some of the characteristics can be observed only under the petrological microscope, not only because of the scale of development but also because this is the only possible technique for observing them. Again, no characteristics beyond the resolving power of the petrological microscope have been used. This is a limitation which may be partly overcome by electron microscopic studies, e.g. it is known that up to 60 per cent. of the voids in a specific soil material may be less than $1\,\mu$ in diameter (Katchinski, 1956); such voids are not observable under the microscope. As the magnification at which observations are made is increased it becomes possible to describe the detail of arrangement (fabric) so that descriptions of the same material may vary considerably according to the magnification used. This difficulty has been met in the schedule for description of structure and fabric by allowing for descriptions at all levels of organization; the proposed system of naming also allows for names to be attached to fabrics and structures at all levels.

Fabric and structure, as such, do not take account of the frequency of occurrence of the constituents of a soil material; they are not measurable quantities. However, they do include descriptions of size and shape of the constituents together with an estimate of the variability of these characteristics. Such descriptions indicate the probability of sorting, when applied to skeleton grains, but not absolute frequency of occurrence. Thus, the descriptions are chiefly qualitative. Quantitative studies of particle-size distribution, void-size distribution, &c., must be made separately by the accepted techniques if they are to be accurate and if the results are to be analysed in terms of populations and distributions. As Low (1954) points out, such measurements are not measurements of structure.

Again, fabric and structure are not actually concerned with the mineralogical or chemical identification of the constituents of the soil material, or the colour of the constituents. However, in making observations, especially at the microscopic stage, variations in colour may reflect variations in mineralogy which in turn reflect variations in concentration and separation of the plasma and perhaps in orientation and distribution of the plasma grains (fabric). Thus, these characteristics are used in classifying and naming pedological features, as a matter of convenience, although such features can be described without reference to such characteristics.

It should also be noted that the descriptive system and schedule are directly applicable to texture as defined earlier. If crystallinity is included, the word 'texture' can be read for 'structure'. Thus, all of the terms basic texture, elementary texture, primary texture, secondary texture, &c., are applicable.

Acknowledgements

The authors are indebted particularly to Mr. B. E. Butler, Officer-in-Charge of the South Eastern Region of the Division of Soils, and Mr. J. K. Taylor, Chief of the Division, for helpful discussions and criticisms of the manuscript.

REFERENCES

BAVER, L. D. 1948. Soil Physics. Chapman and Hall, London.

BUTLER, B. E. 1955. A system for the description of soil structure and consistence in the field. J. Aust. Inst. Agric. Sci. **21**, 239–49.

JONGERIUS, A. 1957. Morphologische Onderzoekingen over de Bodemstructuur. Stichting voor Bodemkartering, Wageningen.

KATCHINSKI, N. A. 1956. Soil structure and differential porosity. Rept. 6th Int. Congr. Soil Sci. **B**, 127–34.

KRUMBEIN, W. C., and SLOSS, L. L. 1955. Stratigraphy and sedimentation. Freeman, California.

Kubiena, W. L. 1938. Micropedology. Collegiate Press, Ames, Iowa.
—— 1956. Rubefication and laterization (their micromorphological distinction). Rept. 6th Int. Congr. Soil Sci. B, 241–9.
—— 1953. The Soils of Europe. Murby, London.
Leeper, G. W. 1952. Introduction to Soil Science. Melbourne Univ. Press.
Low, A. J. 1954. Study of soil structure in the field and the laboratory. J. Soil Sci. 5, 57–74.
McMillan, N. J., and Mitchell, J. 1953. Microscopic study of platy and concretionary structures in certain Saskatchewan soils. Sci. Agric. 33, 178–83.
Pettijohn, F. J. 1949. Sedimentary Rocks. Harper, New York.
Rice, C. M. 1954. Dictionary of Geological Terms. Edwards Bros., Ann Arbor, Michigan.
Robinson, G. W. 1949. Soils. Their Origin, Constitution and Classification. Murby, London.
Shrock, R. R. 1948. Sequence in Layered Rocks. McGraw-Hill Book Co., New York and London.
U.S.D.A. 1951. Soil survey manual, U.S. Dept. Agric. Handbook No. 18.
Zakharov, S. A. 1927. Achievements of Russian science in the morphology of soils. Acad. Sci. U.S.S.R. Russ. Ped. Invest. No. 2, 47.
Zingg, Th. 1935. Beitrag zur Schotteranalyse. Schweiz. Min. u. Pet. Mitt., 15, 39–140.

(*Received 7 July 1959*)

APPENDIX

DESCRIPTIVE SYSTEM

Many of the terms used in the description of rock texture are directly applicable to soil structure but some additions and modifications are necessary. The characteristics requiring description are summarized in Table 1. The proposed terms are based mainly on megascopic and microscopic examination.

(a) Size and Shape

1. *Absolute size*

Extremely fine	< 0.005 mm.
Very fine	0.02–0.005 mm.
Fine	0.1–0.02 mm.
Medium	0.5–0.1 mm.
Coarse	2.0–0.5 mm.
Very coarse	10–2 mm.
Extremely coarse	> 10 mm.—specify actual size

2. *Absolute shape*

    (i) *Shape classes*: after Zingg (1935) with the following additions and modifications.

Prolate (rods)	$b/a < 2/3$ but $> 1/10$; $c/b > 2/3$
Acicular (needles or channels)	$b/a < 1/10$; $c/b > 1/10$
Planar (planes)	$b/a > 1/10$; $c/b < 1/10$
Other shapes.	

    (ii) *Roundness classes*: after Pettijohn (1949).

3. *Relative size and shape*

Very even:	a strong peak in the frequency distribution curve.
Even:	a distinct peak in the frequency distribution curve.
Uneven:	a slight peak in the frequency distribution curve.
Very uneven:	irregular distribution.

(b) Arrangement

1. Orientation pattern

(i) *Orientation of individuals with each other*

Skeleton grains, peds, pedological features, and voids

Strongly oriented: more than 60 per cent. of the individuals have their principal axes within 30 degrees of each other.
Moderately oriented: 40–60 per cent. have their principal axes within 30 degrees of each other.
Weakly oriented: 20–40 per cent. have their principal axes within 30 degrees of each other.
Unoriented: There is no preferred orientation.

Plasma. If individual plasma grains are large enough the terms used for skeleton grains are applicable. However, they are usually ultramicroscopic, and only aggregates (plasma aggregates) of individual grains are large enough to be recognized. Thus the orientation of clay-size particles can be stated only for anisotropic materials where it is reflected in the extinction phenomena in thin section at various magnifications. The most useful magnifications have been found to be approximately $30\times$, $80\times$, $200\times$, $800\times$, $1200\times$. The descriptions are made at each of these magnifications and the increasing observable degree of orientation with increasing magnification is noted as a comparative measure of orientation.

Strongly oriented: at the magnification used the plasma aggregates are continuously birefringent in thin section, that is, under crossed nicols dark extinction lines run across the aggregates or they extinguish as a unit.

Moderately oriented: at the magnification used the plasma aggregates have rather indistinct boundaries and have an incomplete mottled or fuzzy extinction reminiscent of the extinction of some micas.

Weakly oriented: at the magnification used the plasma aggregates are anistropic but too small to observe in detail so that the mass has a flecked appearance.

Unoriented: at the magnification used the plasma appears isotropic because of the low degree of orientation.

Indeterminate: the plasma is isotropic because of its crystallographic nature or opacity.

Crystallized: the plasma occurs as crystals large enough to be described in the terms used for skeleton grains.

Besides the orientation of individual plasma grains with each other, the plasma aggregates, especially if they are not equidimensional, may have various patterns of orientation observable at various magnifications. The degree of development of these arrangements is recorded at each magnification as a further measure of orientation.

Continuous orientation. This applies especially to pedological features which consist entirely of a single unit of strongly oriented plasma.

Striated orientation. The plasma aggregates have a lineal arrangement:

Strongly striated: more than 60 per cent. of the material has a lineal arrangement.
Moderately striated: 20–60 per cent. of the material has a lineal arrangement.
Weakly striated: less than 20 per cent. of the material has a lineal arrangement.

Flecked orientation. The plasma aggregates are arranged randomly.

(ii) *Referred orientation.* If a group of individuals have a preferred orientation then this can be referred to prominent directions or surfaces within the soil material, such as ped faces, systems of planes, &c., or to the ground surface:

Parallel orientation: the principal axes of the individuals are parallel to the reference.
Normal orientation: the principal axes are normal to the reference.

Transverse orientation: the principal axes are at an angle to the reference; this angle should be stated.

Unrelated orientation: there is no constant relationship with the reference.

(iii) *Orientation between groups of individuals.* The orientation of one group of oriented individuals can be stated with regard to other groups. The terms parallel, normal, transverse, and unrelated are applicable.

2. *Distribution pattern.* Individual plasma grains are usually too small, unless crystallized, to describe their distribution patterns. However, where pedological features have been formed these can be treated as entities whose size and shape can be described according to the terms already listed, and whose distribution pattern can be described. Pedological features also require a description of the sharpness of their boundaries which is an expression of the distribution of the plasma fractions within the pedological features.

(i) *Sharpness of boundary of pedological features.*

Very diffuse: no boundary is observable.
Diffuse: gradual transition over a distance greater than one-quarter of the shortest dimension of the feature.
Rather diffuse: gradual transition over a distance between one-quarter and one-tenth of the shortest dimension of the feature.
Rather sharp: gradual transition over a distance of less than one-tenth of the shortest dimension of the feature.
Sharp: the boundary is quite sharp.

(ii) *Relationship of plasma grains and skeleton grains in the basic fabric.*

The terms used follow those used by Kubiena (1938) as far as possible.

Porphyritic: the plasma occurs as a dense matrix in which the skeleton grains are set after the manner of phenocrysts in a porphyritic rock.
Agglomeratic: the plasma occurs as loose or incomplete fillings in the intergranular spaces between the skeleton grains.
Intertextic: the plasma occurs as intergranular braces linking the skeleton grains.
Granular: there is no plasma, or all the plasma occurs as pedological features.

Undoubtedly additional terms will be necessary to describe other specific arrangements as more data are collected.

(iii) *Distribution of individuals with regard to each other* (skeleton grains, voids, peds, pedological features).

Regular: distributed evenly throughout the soil material.
Clustered: concentrated in clusters or groups.
Banded: concentrated in bands or layers.
Other patterns: other possible patterns are radial, concentric, &c.

(iv) *Referred distribution.* The distribution of individuals or groups can be referred to prominent directions or surfaces within the soil material or to the ground surfaces. For elongated patterns the terms parallel, normal, transverse, and unrelated, as used for the orientation pattern can be used to indicate elongation in regard to a specific reference. In addition, the following terms are useful applied especially to pedological features.

Cutanic: associated with natural surfaces.
Sub-cutanic: the pattern is elongated parallel to adjoining natural surfaces.
Non-cutanic: unrelated to natural surfaces.

(v) *Distribution between groups of individuals.* Within the one soil material, certain groups of individuals (shape classes, oriented groups, &c., of voids, skeleton grains, peds or pedological features) may have a particular distribution pattern with regard to other groups. Such patterns, if characteristic, should be described and given a descriptive name.

GEDANKEN ZUM AUFBAU DES BODENS UND SEINER BEGRIFFLICHEN ERFASSUNG

H.-J. Altemüller

[*Editors' Note:* In the original, material precedes and follows this excerpt.]

3. Vorschlag einer Gliederung des Bodengefüges

In der Definition des Bodengefüges nach KUBIENA (25) ist der Weg aufgezeigt, auf dem eine Erweiterung und Zusammenfassung der verschiedenen Vorstellungen vom Bodenaufbau möglich ist. Eine außerordentliche Hilfe im einzelnen bietet darüber hinaus die am Beispiel der Gesteine von SANDER (39, 40) ausgearbeitete allgemeine Gefügekunde. Hier wurde ein Begriffsgebäude von außerordentlicher gedanklicher Klarheit aufgebaut, das in seinen Grundlagen weit über den Bereich der Geologie und Petrographie hinaus Bedeutung hat. Durch eine sinnvolle, dem jeweiligen Stoffgebiet angepaßte und dessen Eigengesetzlichkeit respektierende Übertragung in andere Forschungsbereiche ist hier die Möglichkeit zum Ausbau einer breit angelegten G e f ü g e k u n d e gegeben, die auch dazu beitragen wird, die zwischen den Fachrichtungen entstandenen Grenzen überwinden zu helfen. Die Bodenkunde, als junge Wissenschaft im Grenzgebiet der klassischen Disziplinen, sollte auf eine solche Hilfe nicht verzichten

Der hier vorgelegte Versuch, den wichtigsten gefügekundlichen Grundbegriffen eine bodenkundliche Auslegung zu geben und sie mit anderen Vorschlägen und bewährten Begriffen zu verbinden, kann zunächst nur in kurzer Fassung unternommen werden. Eine Erweiterung und Vertiefung muß in weiteren Arbeiten folgen.

3.1 *Allgemeine Gefügedefinition nach* SANDER

Die Definition des Gefüges nach SANDER ist umfangreich. Ein grundlegender Satz, der die allgemeine Bedeutung deutlich macht, ist folgender:

„Die Raumdaten im Innern eines betrachteten Bereiches beschreiben dessen Gefüge; ihre Änderung beschreibt die Änderung des Gefüges; die (statistische) Symmetrie der

An English summary appears at the end of this excerpt.

Raumdaten in ihrer Gesamtheit beschreibt die Symmetrie des Gesamtgefüges. Die Wahl der Raumdaten erfolgt in Auslese für ein bestimmtes Interesse, auf welches die Beschreibung des Gefüges bezogen wird."

3.2 Abgrenzung des Bereiches „Boden"

3.21 zeitlich

Das Bodengefüge ist einer mehr oder weniger raschen Veränderung unterworfen. Die Untersuchung des Bodengefüges gilt daher immer nur für einen bestimmten Zeitpunkt — im allgemeinen für den Zeitpunkt der Probenentnahme. Die zeitliche Erfassung von Gefügeveränderungen ist von grundlegender Bedeutung für die Bewertung von Gefügemerkmalen.

3.22 räumlich

Die äußere Umgrenzung des Bereiches Boden ist, wie es unlängst LAATSCH und SCHLICHTING (29) dargelegt haben, schwierig zu beschreiben. Sie ergibt sich aus der Unterscheidung gegen die Pflanzendecke und den Gesteinsuntergrund in der Vertikalen und ist in der Horizontalen meist durch Übergänge in andere Bodenformen gegeben, wenn nicht äußere geographische Gegebenheiten eine deutliche Abgrenzung erlauben. Es versteht sich von selbst, daß Teile, die zum Bodenaufbau gerechnet werden, gleichzeitig auch Teile anderer Naturkörper sein können. Daher wäre es wenig fruchtbar, eine scharfe Grenzziehung anzustreben. Für die Untersuchung des Gefüges im hier vorgetragenen Sinne ist es auch von zweitrangiger Bedeutung, wo z. B. die Grenze zwischen Sediment oder Boden angesetzt wird. Das Abgrenzen des betrachteten Bereiches muß im wesentlichen eine Frage des Forschungszieles sein.

Die innere Abgrenzung des Bereiches Boden erfolgt in Anlehnung an die herrschenden Vorstellungen vom Bodentyp (MÜCKENHAUSEN, EHWALD, KUBIENA, LAATSCH und SCHLICHTING) und dessen Untergliederung. An der Ausgestaltung dieser Vorstellungen hat die Gefügekunde aktiven Anteil.

3.23 stofflich

Abgesehen von der gegenseitigen Durchdringung von belebter und unbelebter Materie und der Zusammensetzung aus Einzelteilen verschiedener Herkunft und Beschaffenheit ist der Boden vor allem aus festen, flüssigen und gasförmigen Stoffen aufgebaut, die wir im ganzen wie im einzelnen als Gemengteile bezeichnen. Diese greifen aufs engste ineinander und bilden in ihrer räumlichen Anordnung das Gefüge des Bodens. Bodenwasser und Bodenluft können also als Teile des Bodengefüges verstanden und grundsätzlich in die Gefügebetrachtung einbezogen werden, wie es von BOLT, JANSE und SCHUFFELEN (2) unter „Struktur" zu einem Teil versucht wurde.

Bedingt durch ihre besondere physikalische Natur, werden Wasser und Luft im Boden jedoch meist nicht selbst als Gefüge, sondern in Abhängigkeit von den festen Gemengteilen untersucht. Im einzelnen sind aber die Vorstellungen über die gegenseitigen Beziehungen nur wenig über schematisch angenommene Modelle hinausgekommen. Eine gefügekundliche Betrachtung wird hierzu noch wertvolle Beiträge liefern können.

3.3 Bodengemengteile und Gefügeelemente

Nach dem Vorangehenden können die Bodengemengteile allgemein verstanden werden als die am Bodenaufbau beteiligten, unterscheidbaren Elemente, unabhängig von ihrer physikalischen Zustandsform und stofflichen Beschaffenheit. Für die Gefügekunde ist es wesentlich, sie nicht nur als Einzelteile, sondern, wie KUBIENA (27) betont, auch in ihrem Anteil am Bodenaufbau als „Bauelemente" zu betrachten. Von dieser Vorstellung ausgehend werden als Gefügeelemente auch zusammengesetzte Teile betrachtet, die entweder als begrenzte Körper (z. B. Aggregate) oder mehr nach ihrer Funktion (z. B. Plasma und Skelett) zu unterscheiden sind.

3.4 Größenordnung

Der Gefügebegriff gilt für alle Größenordnungen. Elemente, die nach cm und dm gemessen werden, bilden ebenso wie Elemente von mikroskopischem oder submikroskopischem Bau eine räumliche Ordnung und werden als Gefüge untersucht. Einen gefügelosen Boden gibt es nicht. Allerdings wirken mit abnehmender Größe zunehmend andere Kräfte. Die Schwerkraft tritt gegenüber den Oberflächenkräften zurück. Auch mechanische Eingriffe wirken sich in größeren Dimensionen der Schollen und Bröckel anders aus als im Bereich der Tonteilchen. Infolgedessen treten in verschiedenen Größenbereichen teilweise auch verschiedene Formprinzipien in den Vordergrund. Es ist daher wichtig, die Größenbereiche zu beachten und in Untersuchungen anzugeben.

Eine einfache und klare Aussage ist mit der Angabe der Maßverhältnisse möglich. Ein Bodenverband aus ton- und schluffreichem Material kann in säulige Absonderungen von dm-Größe gegliedert sein, während er im mm- und cm-Bereich kompakt oder feinporig erscheint und im my-Bereich ein geschichtetes Anlagerungsgefüge aus vorherrschend blättchenförmigen Elementen aufweist. Erst alle diese Merkmale zusammen machen das Bodengefüge im ganzen aus. Auch die Geräte zur Untersuchung des Bodengefüges (z. B. Lupe, Mikroskop) oder die Präparationsverfahren (Handstück, Trümmerpräparate, Schliffe usw.) richten sich teilweise nach den Größenbereichen.

Man unterteilt zweckmäßigerweise einen makro-, mikro- und submikroskopischen Bereich und kann dementsprechend vereinfacht ein Makro- oder Mikrogefüge usw. unterscheiden. In diesem Sinne benutzt auch JONGERIUS (18) die Ausdrücke Makro- und Mikrostruktur. Bei anderen Autoren ist aber unter demselben Wort nicht nur der Größenbereich angesprochen, sondern wiederum die Verknüpfung „Struktur" und „Aggregatbildung" vorhanden, was wohl auf GEDROIZ (13) zurückgeht. JONGERIUS gibt eine Übersicht über verschiedene Auffassungen. Es ist also erforderlich, auf die Bedeutung im Sinne einer Angabe zur Größenordnung stets klar hinzuweisen.

Ergänzend sei hingewiesen auf die Arbeiten von TERZAGHI (46, 47), VON NITZSCH (35) und BRINCH-HANSEN und LUNDGREN (5). Ein Teil dieser Vorstellungen jedoch bringt nicht Größenordnungen, sondern Rangordnungen zum Ausdruck.

3.5 Rangordnungen

Im Gegensatz zur Angabe der Größenbereiche ist hier mehr die „soziologische" Gliederung entscheidend. Bezeichnungen wie „Aggregate 1. oder 2. Ordnung" (SEKERA [43]) sind typische Beispiele. Auch BREWER und SLEEMAN (4) bringen ähnliche Ordnungen zum Ausdruck. Die Festlegung einer „1. Ordnung" ist jedoch häufig angreifbar und vor allem auch mißverständlich, wenn nicht sauber von zeitlichen Vorstellungen getrennt wird.

Deshalb müssen Gliederungen, die mehr auf das Objekt gerichtet sind, günstiger erscheinen, wie z. B. der Begriff K o r n g e f ü g e von SANDER, der die Beziehungen zwischen den Gefügekörnern ausdrückt. Ergänzend hierzu wären z. B. A g g r e g a t g e f ü g e und P o r e n g e f ü g e zu nennen, die als Bezeichnungen im Hinblick auf den Zusammenbau aus Aggregaten oder die Anordnung, Form und Gliederung der Poren zu verstehen wären. Schwierigkeiten ergeben sich aber noch aus dem andersartigen Gebrauch des Ausdrucks Korngefüge in der Bodenkunde (vgl. MÜCKENHAUSEN in diesen Beiträgen), so daß weitere Überlegungen erforderlich sind.

Gut eingeführt ist der Begriff E l e m e n t a r g e f ü g e von KUBIENA (25), der ähnlich wie der Begriff Korngefüge nach SANDER zu verstehen ist, aber stärker die eigene Zusammensetzung des Bodenmaterials berücksichtigt. Er bringt die Beziehungen zwischen den Gefügeelementen „Skelett" und „Plasma" zum Ausdruck und ist in vielen Fällen gut anwendbar.

Zur Klärung von Mißverständnissen sei gesagt, daß Elementargefüge, Einzelkorngefüge, Primitivgefüge (FREI [11]) und Mikrogefüge keine identischen Begriffe sind! Als weitere Ordnung neben dem Elementargefüge kann man allgemein von G e f ü g e n   h ö h e r e r   O r d n u n g sprechen.

3.6 *Gestaltliches und funktionales Gefüge nach* SANDER

Mit der Betrachtung der Formen der Gefügeelemente, ihrer gegenseitigen Bindung und Anordnung im Raum, einschließlich ihrer Darstellung nach Maß und Zahl, erfassen wir das g e s t a l t l i c h e   G e f ü g e. Umgekehrt läßt sich der Boden auch vorstellen als ein Feld physikalischer Kräfte und Wirkungen. Die Ordnung dieser Kräfte und Wirkungen, die sich theoretisch formelmäßig ausdrücken läßt, bildet das f u n k t i o n a l e   G e f ü g e. Beide Gefüge stehen in direkter Abhängigkeit voneinander, und eine Änderung des gestaltlichen Gefüges hat stets auch eine Änderung des funktionalen Gefüges zur Folge.

Wie SANDER (40) selbst betont, bleibt die Untersuchung funktionaler Gefüge noch auf lange Zeit Aufgabe der Physik. Die Bodenkunde muß sich vor allem am gestaltlichen Gefüge orientieren. Aber auch hier gilt die begriffliche Trennung in eine gestaltliche und funktionelle Seite. Die Aufgabe der Bodenkunde liegt also in der Untersuchung gestaltlicher Gefüge und ihrer Funktionen. Beide Seiten werden bis heute noch vorherrschend getrennt voneinander untersucht. Permeabilitätsmessungen, Festigkeitswerte, Bestimmungen des Porenvolumens u. dgl. gewinnen jedoch ganz wesentlich an Wert, wenn sie nicht nur gedanklich, sondern auch methodisch einem bestimmten gestaltlichen Gefüge zugeordnet werden. Hier kommt der Weiterentwicklung von Methoden, die vom gestaltlichen Gefüge ausgehen, eine besondere Bedeutung zu (vgl. KUBIENA und Mitarbeiter [26]). In der Petrographie sind solche Methoden hoch entwickelt. Die morphologische Untersuchung erscheint dadurch nicht als spezielles Sachgebiet am Rande, sondern bekommt eine zentrale Stellung. Eine Untersuchung gestaltlicher Gefüge in Beziehung zu ihren physikalischen Funktionen wird sich für beide Arten der Betrachtung, die morphologische wie die funktionelle, als ebenso anregend, wie regulierend erweisen.

3.7 Genetische Gefügebetrachtung

Gefügeuntersuchungen dienen nicht nur zur Feststellung von momentanen Zuständen und Eigenschaften eines untersuchten Objektes, sondern sind in besonderem Maße geeignet, Bildungs- und Entwicklungsprozesse im Boden erkennen zu lassen.

Im Interesse einer sauberen Begriffsbildung ist jedoch eine deutliche Trennung zwischen beschreibender (oder messender) und genetischer Untersuchung zu beachten. Die Sammlung von Tatsachen und ihre vergleichende Ordnung und Deutung sind zwei Schritte, die stets m i t einander, aber im gedanklichen doch g e t r e n n t voneinander zu unternehmen sind. Auf diese Weise können auch morphologische Untersuchungen als objektiv anerkannt werden. Eine wesentliche Hilfe bietet zudem auch das Experiment, auf welches jedoch hier nicht eingegangen werden soll.

Im gewissen Sinne kann man aus dem Gesagten auch drei Stufen der Untersuchung ableiten. Angewandt auf ein praktisches Beispiel könnte man den Begriff „Krümel" in folgender Weise erklären:

(3. 81) beschreibend:
Ein Haufwerk von Bodengemengteilen von rundlichen und gelappten Formen mit rauhen Oberflächen und zahlreichen verschiedenartigen Hohlräumen; oft reich an organischen Stoffen; —

(3. 82) beschreibend und messend:
Ein Haufwerk usw. ... Größe bis etwa 10 mm $\emptyset$, von bestimmter Stabilität gegen Wasser, bestimmtem Wasserhaltevermögen, Porenvolumen, usw.

(3. 83) beschreibend, messend und deutend:
Ein Haufwerk usw. ... bestimmter Größe und physikalischer Eigenschaften usw. Als Ergebnis aktiver Bodentätigkeit aus dem Zusammenwirken von menschlicher Bearbeitung, Tätigkeit der Bodentiere, Mikroorganismen und Pflanzenwurzeln gebildet und stabilisiert.

Aus der dargestellten Folge ergibt sich neben der zunehmenden Erkenntnis aber auch eine Einengung des Krümelbegriffs. Aus dem morphologischen Bild lassen sich über Stabilität und Luftdurchlässigkeit nur qualitative Angaben machen. Wird also der Krümelbegriff mit der Angabe der Wasserbeständigkeit verknüpft, wie es KULLMANN (28) wieder vorschlug, dann können Krümel morphologisch nicht mehr allein bestimmt werden. Es wäre dann die Unterscheidung von verschiedenen Krümelarten notwendig, wie etwa:

„Krümel"	Bestimmung nach der Form
„wasserbeständige Krümel"	Bestimmung nach der Form und der Stabilität gegen Wasser
„echte Krümel"	Angabe als Ergebnis erwiesener genetischer Zusammenhänge.

So, wie hier vom Gestaltlichen ausgehend zur Messung und Deutung vorangegangen wird, muß auch dort, wo man vom Messen ausgeht, das gestaltliche Gefüge berücksichtigt werden. Handelt es sich z. B. im Stabilitätsversuch um beliebig aus einem kohärenten Verband gewonnene Fragmente, so können diese auch bei bester Wasserbeständigkeit nicht als wasserbeständige „Krümel" bezeichnet werden. Ebenso kann ein beliebiges Gefüge mit hohem Porenvolumen nicht als „Schwammgefüge" bezeichnet werden, wenn es nicht morphologisch bestimmt wurde.

Die völlig ungenetischen Gefügenamen, die KUBIENA (22) für das Elementargefüge prägte, sind deshalb auch mit Vorteil zu verwenden, wo oder solange die genetischen Zusammenhänge noch nicht geklärt sind.

4. Definition des Bodengefüges

Aus den dargestellten Aspekten kann man versuchen, eine Definition des Bodengefüges zu geben, die etwa folgendermaßen lauten soll:

Das Bodengefüge ist zu verstehen als
die Ordnung der Bodengemengteile, betrachtet als Bauelemente des Bodens
  in gestaltlicher,
  in funktioneller und
  in genetischer Hinsicht,
unabhängig von den Größenordnungen und
unabhängig von der stofflichen Beschaffenheit.

5. Schlußbetrachtung

Nach der vorliegenden Darstellung erscheint das Wort Gefüge als Ausdruck für den Aufbau des Bodens im weiten, umfassenden Sinne ebenso wie im Detail als besonders zweckmäßig. Teilbereiche oder Teilaspekte können ohne Schwierigkeit untergeordnet und mit verständlichen Ausdrücken versehen werden, z. B. Elementargefüge, Aggregatgefüge oder Porengefüge usw. Funktionelle, gestaltliche und genetische Angaben werden begrifflich voneinander getrennt.

Ein Nebeneinander von Ausdrücken ähnlichen Inhalts oder unklarer Definition wirkt verwirrend. Die weitere Anwendung des Wortes Struktur muß daher im rein wissenschaftlichen Gebrauch problematisch erscheinen. Im Hinblick auf verschiedene Auffassungen wären zwar folgende Zuordnungen möglich, die aber kaum miteinander zu vereinen sind:

(5. 1) mit Rücksicht auf den landwirtschaftlichen Gebrauch:
  für den makroskopischen Aufbau des Bodens —
(5. 2) mit Rücksicht auf die Mehrzahl der Strukturdefinitionen:
  auf das Vorhandensein von Aggregaten (unabhängig von der Größenordnung) —
(5. 3) nach BREWER und SLEEMANN (4):
  auf Anordnung, Größe und Form (unabhängig von der Aggregatbildung)
(5. 4) oder synonym mit Gefüge im hier dargelegten Sinne nach einer völligen Veränderung der bisherigen Strukturbegriffe.

KUBIENA hat 1938 den Gefügebegriff so definiert, daß die Begriffe Struktur und Textur davon unberührt bleiben konnten. In der weiteren Entwicklung sind aber durch neue Beiträge und engere Kontakte verschiedener Forschungsrichtungen — wie schon eingangs erwähnt — neue Verhältnisse entstanden, die eine Zusammenfassung angezeigt sein lassen. Es muß deshalb als das Beste erscheinen, vor allem in wissenschaftlichen Darstellungen, den Gefügebegriff auf der Grundlage der allgemeinen Gefügekunde SANDERS zu gebrauchen und weiter im bodenkundlichen Sinne zu entwickeln. Das Wort Struktur wird man danach — will man es nicht ganz fallen lassen — zweckmäßig nur soweit verwenden, wie es ohne Definition rein sprachlich für „Bau" oder „Aufbau" verstanden werden kann.

Im Rahmen dieser Arbeit ist es nicht möglich, weiter ins einzelne gehende Ausführungen mit praktischen Beispielen, vor allem in den mikroskopischen Bereich hinein, vorzulegen. Für die Beschreibung des Bodengefüges im Felde geben die in späteren Ausgaben der Zeitschrift für Kulturtechnik folgenden Arbeiten von KOEPF und MÜCKENHAUSEN weitere Unterlagen.

[Editors' Note: Only the references cited in the preceding excerpt are reproduced here.]

Literaturverzeichnis

BOLT, G. H., A. R. P. JANSE und A. C. SCHUFFELEN, 1959: Definition and Determination of "Soil structure". — Med. Landbouwhogeschool, Gent, 24, Nr. 1, S. 251—256.

BREWER, R., und J. R. SLEEMAN, 1960: Soil structure and fabric. — J. soil science 11, 172—185.

BRINCH-HANSEN, J., und H. LUNDGREN, 1960: Hauptprobleme der Bodenmechanik. — Springer-Verlag, Berlin-Göttingen-Heidelberg.

Der große Brockhaus, 1954: Bd. 4, Wiesbaden.

FREI, E., 1948: Gefügeuntersuchungen an landwirtschaftlichen Kulturböden. — Landw. Jb. d. Schweiz, S. 1—17.

GEDROIZ, K. K., 1926: Zur Frage der Bodenstruktur und ihrer Bedeutung für die Landwirtschaft. — Leningrad 1926. Zit. in: KRAUSE, M.: Russische Forschungen auf dem Gebiet der Bodenstruktur. — Landwirtsch. Jb. 73 (1931), 603—690.

JONGERIUS, A., 1957: Morfologische onderzoekingen over de Bodemstructuur. — Wageningen.

KUBIENA, W. L., 1935: Über das Elementargefüge des Bodens. — Bodenkundl. Forsch. 4, 380—412.

KUBIENA, W. L., 1953: Bestimmungsbuch und Systematik der Böden Europas. — Ferd. Enke Verlag, Stuttgart.

KUBIENA, W. L., W. BECKMANN und E. GEYGER, 1961: Zur Methodik der photogrammetrischen Strukturanalyse des Bodens. — Z. Pflanzenern., Düng., Bodenkunde 92, 116—126.

KUBIENA, W. L., 1962: Wesen, Ziele und Anwendungsmöglichkeiten der Mikromorphologie. — Z. Pflanzenern., Düng., Bodenkunde 97, 193—205.

KULLMANN, A., 1958: Zur Problematik der Krümelstabilitätsmessungen und zur Methodik des Durchflußverfahrens. — Tagungsber. Nr. 13 d. Dt. Akad. d. Landwirtsch. Wiss., Berlin, 7—34.

LAATSCH, W., und E. SCHLICHTING, 1959: Bodentypus und Bodensystematik. — Z. Pflanzenern., Düng., Bodenkunde 87, 97—108.

NITZSCH, W. v., 1939: Porengrößen im Boden, ihre Beziehungen zur Bodenbearbeitung und zum Wasserhaushalt. — Habil.-Schrift, Halle (Saale).

SANDER, B., 1930: Gefügekunde der Gesteine. — Berlin.

SANDER, B., 1948/1950: Einführung in die Gefügekunde der geologischen Körper. 2 Bde. — Wien u. Innsbruck.

SEKERA, F., 1951: Der allgemeine Bauplan der Bodenstruktur und die Dynamik der Bodengare. — Z. Pflanzenern., Düng., Bodenkunde 52 (97), 57—60.

TERZAGHI, K. v., 1931: Festigkeitseigenschaften der Schüttungen, Sedimente und Gele. — In: Handb. d. physikal. u. techn. Mechanik. Herausg.: AUERBACH, F., u. W. HORT, Bd. 4/2, 518—578, Leipzig.

SUMMARY*

The definition of soil fabric proposed here is based on Sander's general fabric concept: The spatial data of the interior of the domain in question describe its fabric; their change describes the change of the fabric. The selection of spatial data is done according to a specific interest, to which the description of the fabric is related.

The extent of a "soil" is limited in time (because of fast changes of the fabric, a fabric analysis is only valuable for a given moment), in space (both vertically and horizontally, e.g. with reference to other soil types) and by its composition (solid material, soil water, and air). The fabric concept is valuable for all sizes, but as with decreasing sizes other forces will be active (e.g. gravity versus mechanical forces), it is important to consider the size range of the material studied and to report it. In addition different levels of organization may be distinguished based on the object, for example a grain fabric (relation between fabric grains) an aggregate fabric, a pore fabric, and so forth.

Considering the shape of the fabric elements, their mutual bond and arrangement in space, including their size and frequence distribution, one defines the *formal fabric*. On the other hand, a soil can be represented as a field of physical forces and actions, whose arrangement forms the *functional fabric*. Both fabrics are directly interdependent. Study of the functional fabric will remain for long time the task of physicists. The task of soil science is to study the formal fabric and its function, for example measurements of permeability, consistence, porevolume, all of which gain in value when related to a given formal fabric.

Fabric studies are especially suitable to investigate the processes of soil formation and development. In order to keep the concepts clear, it is necessary to distinguish between purely descriptive (or measuring) and genetic research.

Based on the above discussed aspects the following definition of soil fabric is proposed: Soil fabric has to be understood as the arrangement of soil components, considered as building elements of the soil, from a formal, functional and genetic point of view, independent of size considerations and composition.

*This summary was prepared by the volume editors.

Part III
SPATIAL RELATIONSHIPS OF BASIC SOIL COMPONENTS

Editors' Comments
on Papers 7, 8, and 9

7 STOOPS and JONGERIUS
Proposal for a Micromorphological Classification of Soil Materials. I. A Classification of the Related Distributions of Fine and Coarse Particles

8 BREWER and PAWLUK
Excerpt from *Investigations of Some Soils Developed in Hummocks of the Canadian Sub-Arctic and Southern-Arctic regions. 1. Morphology and Micromorphology*

9 ESWARAN and BAÑOS
Related Distribution Patterns in Soils and Their Significance

Since the early days of micropedology, the distribution of fine material (e.g., clay) with respect to coarser material (e.g., sand) has been one of the most characteristic microscopic features of a soil material.

As mentioned in Part II (Paper 4), Kubiëna's system of micropedology was essentially based on the recognition of elementary fabrics, it is the arrangements of plasma and skeleton grains. Ten major fabric types are recognized; their distinction, however, is based not only on differences in distribution patterns, but also partly on composition of the components (e.g., calcite in rendiza and mortar fabric) and the character of the plasma (peptized or pectized). Considering only the arrangement of plasma and skeleton grains, Brewer (1960, 1964) distinguished four types of related distribution patterns (see Part II, Paper 5).

As more different and peculiar types of soils were studied micromorphologically at the end of the 1960s, disadvantages and imperfections of the existing concepts became obvious—for example, with respect to the related distribution patterns (e.g., Brewer 1973). In response to these criticisms, three papers dealing with the related distribution patterns of soil components were published in 1975 and 1976 (Papers 7, 8, and 9). All three are based on Brewer's concepts of related distributions, all try to overcome the problems inherent to the

Editors' Comments on Papers 7, 8, and 9

plasma–skeleton grain concept, and all extend the related distribution types to the arrangement of more complex units such as microaggregates.

The main problem with the plasma–skeleton grain concept is that the distinction between both units is based, by definition, on their stability, a characteristic not objectively detectable in thin sections. Stoops and Jongerius (Paper 7) therefore rejected this criterion and proposed considering only purely morphometric criteria, namely, the relative size that can be determined objectively in thin sections. To avoid any confusion, they did not use the terms *plasma* and *skeleton grains,* but rather referred to *fine* and *coarse* material. The size range of both is left to the user, who nevertheless should state it in any written findings. Brewer and Pawluck (Paper 8) retained the terms *plasma* and *skeleton,* as defined earlier, but gave more emphasis to the terms *matrix* and *framework members* to indicate the finer and the coarser fraction, without considering their stability; they indicated no size limit. Both terms had already been mentioned in Brewer [1964]. Eswaran and Baños (Paper 9) distinguished between plasma and grains, defining plasma as colloidal-sized material and grains as single particles greater than colloidal size without considering their stability. This means that their concept of plasma does not agree with those of Kubiëna or Brewer. An essential feature of Eswaran and Baños's related distribution types is the subdivision of grains in sand and silt, the limit being 50 μm.

The types of related distribution of coarse and fine material, as defined by Stoops and Jongerius (Paper 7), do not only apply to individual particles (e.g., sand and clay particles), but can also be used to express the interrelationship of larger and more complex units (e.g., peds). Stoops and Jongerius distinguished five types of related distribution of coarse and fine material: monic, gefuric, chitonic, enaulic and porphyric. They found it useful to represent these types and intergrades between two or more types in a three-dimensional diagram, a triangular bipyramid from which each corner is occupied by one type. In addition, they represented in the same figure the surface indicating the limit of skeletal function of the coarse grains; beneath this surface the coarse grains have no supporting function as they don't touch each other. In a later work Stoops (1978) changed the position of this surface in such a way that it intersects with the gefuric corner, which implies that the coarse grains may be touching each other in the gefuric type. This makes the system more logical (Brewer, 1976; Stoops and Jongerius, 1977).

Compared to the related distribution patterns defined by Brewer (1964), those proposed by Brewer and Pawluck (Paper 8) cover much more complex types of organizations. This is especially true for the granic and granoidic type. In fact, they can be extended to descriptions of soil structure and pedality, as well as to different types of organic matter. Brewer and Pawluck also considered several possibilities of intergrades and mixtures.

Eswaran and Baños distinguished two types of related distribution patterns: the normal basic related distribution pattern (NRDP) and the specific related distribution pattern (SRDP). The NRDP is the result of random packing of the components (sand, silt, and clay) without any reorganization due to pedogenetic processes. The authors identified four distinct types of the NRDP—granic, phyric, plasmic, and porphyric and their intergrades—which they illustrated on a triangular diagram. The new contribution here is the introduction of the related distribution of three, rather than of two size fractions, which is linked straight away to the texture of the soil. In a statistical comparative study, Brewer (1979) proved that a strong relationship exists between particle size and the related distribution pattern in a number of Australian soils.

Under the influence of pedogenetic processes, the SRDP may be formed by a rearrangement of the components. Eswaran and Baños considered the following types: intertextic, dermatic, congelic, agglutinic, and reticulic. The first three correspond essentially to gefuric, chitonic, and enaulic, respectively. Agglutinic and reticulic types may be considered microstructures.

To describe the related distribution patterns of organic components, Bal (1973) proposed a specific terminology, which has not been extensively used. The related distribution patterns of organic materials were also discussed by De Coninck et al. (1974). In the classification of clastic sedimentary rocks, much attention has been given to the related distribution of coarser and finer particles (e.g., Michot, 1958).

REFERENCES

Bal, L., 1973, *Micromorphological analysis of soils. Lower levels in the organization of organic materials,* Soil Survey Paper No. 6, Netherlands Soil Survey Institute, Wageningen, 175p.

Brewer, R., 1960, The Petrographic Approach to the Study of Soils, *7th Internat. Congr. Soil Sci. Trans.* **1**:1–13.

Brewer, R., 1964, *Fabric and Mineral Analysis of Soils,* J. Wiley & Sons, London, New York, and Sydney, 470p.

Brewer, R., 1973, Some considerations concerning micromorphological terminology, in *Soil Microscopy,* G. K. Rutherford, ed., Limestone Press, Kingston, Ontario, pp. 28–48.

Brewer, R., 1976, Proposal for a micromorphological classification of soil materials. I. A classification of the related distributions of fine and coarse particles. A discussion, *Geoderma* **15:**437-442.

Brewer, R., 1979, Relationships between particle size, fabric and other factors in some Australian soils, *Australian Jour. Soil Res.* **17:**29-41.

De Coninck, F., D. Righi, J. Maucorps, and A. M. Robin, 1974, Origin and micromorphological nomenclature of organic matter in sandy spodosols, in *Soil Microscopy,* G. K. Rutherford, ed., Limestone Press, Kingston, Ontario, pp. 263-280.

Michot, P., 1958, Classification et terminologie des roches lapidifiées de la série psammito-pélitique, *Soc. Géol. Belgique, Annales* **81:**311-342.

Stoops, G. and A. Jongerius, 1977, Proposal for a micromorphological classification of soil materials. I. A classification of the related distribution of fine and coarse particles. A reply. *Geoderma* **19:**247-249.

Stoops, G., 1978, *Provisional Notes on Micropedology,* Geologisch Instituut, R.U.G.

PROPOSAL FOR A MICROMORPHOLOGICAL CLASSIFICATION OF SOIL MATERIALS. I. A CLASSIFICATION OF THE RELATED DISTRIBUTIONS OF FINE AND COARSE PARTICLES

G. STOOPS and A. JONGERIUS

Geological Institute, State University of Ghent, Ghent (Belgium)
Department of Soil Micromorphology, Soil Survey Institute, Wageningen (The Netherlands)

(Received February 2, 1973; revised version accepted October 24, 1974)

ABSTRACT

Stoops, G. and Jongerius, A., 1975. Proposal for a micromorphological classification of soil materials. I. A classification of the related distributions of fine and coarse particles. Geoderma, 13: 189—199.

As a first step towards a new micromorphological classification of soil materials, a classification is made of the mutual spatial distributions of the coarse and fine fabric elements of lowest order: the c/f-related distributions.

The classification comprises five orthotypes: monic — only one size fraction is present; gefuric — bridges of finer material exist between coarser grains; chitonic — finer material coats the coarser grains; enaulic — the finer material occurs as small aggregates in the spaces between the coarser grains; and porphyric — the coarser grains are embedded in a dense mass of finer material.

The relationship between the different types is visualized in a three-dimensional geometric figure: a trigonal bipyramid.

An appropriate terminology makes it possible to express the different combinations of types, the size variations and the main composition of each size fraction.

INTRODUCTION

The types of mutual spatial arrangements of coarse and fine particles are considered to be very important characteristics of soil materials (e.g. Kubiena, 1938; Brewer, 1960), of weathering rocks (Lukashev, 1970) and of sediments (e.g. Ruchin, 1958).

Several of the authors mentioned above have described and named different kinds of such arrangements, creating isolated types. In soil micromorphology these types are restricted to the lowest units of organization and are based on the plasma-skeleton concept, i.e. related to the estimated stability of the constituents (e.g., Kubiena, 1938; Brewer, 1960). The aim of this study is to present a coherent and comprehensive classification (with intergrades and mixtures) of the mutual spatial arrangements (related distributions) of relatively finer and coarser particles on a morphometric base.

BASIC CONCEPTS

Since this is the first paper in a series dealing with proposals for a classification of the microscopic characteristics of soil materials, it might be useful first to explain some basic concepts. This classification system has been elaborated specially with a view to the microscopic description and interpretation of soils and of soil-like materials (e.g., termite mounds, human-made earthy constructions, infillings of drainage tubes); this in contrast to the existing systems which were restricted to natural soil materials. For this reason, the classification criteria should be as objective as possible (e.g. morphometric data), rather than genetic (e.g. stability of materials).

As a consequence of this objective approach, the concepts of coarser and finer material are used. They indicate two groups of individuals which can be distinguished according to their relative size. This means that the size limit between both is not fixed, and will depend upon the requirements of the study (complexity of the material, scale of observations). Compound elements can also be used as units; e.g., in a tropical soil it might be useful to study the relation between coarse sesquioxidic nodules and the finer material (sand, silt and clay) between them. Organic compounds can be treated in the same way as inorganic ones.

Although such a classification of related distributions of coarse and fine materials can be applied to every level of organization of the material, it will be most useful in describing the mutual spatial arrangement of the fabric members of lowest order, i.e. the grains of sand, silt and clay (size definition).

As noted before, several terms exist in soil micromorphology to designate certain related distribution types (Kubiena's elementary fabrics and Brewer's related distribution types in the basic fabric). For the reason that they are restricted to the lowest fabric units of the soil, constituted of the plasma and the skeleton grains (which are genetic units, based on stability, rather than pure morphologic units), they could not be used for the present classification without changing their definitions. Therefore, it was prefered to propose new terms and definitions.

THE c/f-RELATED DISTRIBUTION

The c/f-related distribution (c/f stands for "coarse versus finer") is a specific type of the related distributions as proposed by Brewer (1964). It is defined as follows: "The c/f-related distribution expresses the distribution of individual particles[1] in relation to finer material and associated voids not included in the particles". This definition is kept very broad, so as to include as many situations as possible. There are no restrictions concerning kind of

---

[1] A particle is an individual or compound element delimited by natural boundaries (i.e. not resulting from sample preparation) which behaves or can behave as a unit (with respect to internal or external forces).

material (composition and complexity), absolute size, orientation, granulation and origin. In principle, the system may be used to describe the spatial distribution of elementary particles (e.g. quartz grains) as well as of compound units (e.g. humic micro-aggregates). The coarser particles may be of silt, sand or even gravel size, whereas the finer material may be composed of amorphous material, clay, silt or sand. On a given scale it is not important if the finer material is present as a mass of individual particles, or occurs in a granulated form or as cutans (free-grain cutans are, thus, not excluded in our concept). The last part of the definition ("... finer material and associated voids, *not included in the particles*") has been added to exclude some coarse secondary crystallizations comparable to the poeciloclastic texture observed in some sediments. In some cases only two of the three components (coarser and finer material and voids) are present, namely when only one size fraction occurs (e.g. impure sands), or when, at the scale of observation, all interstitial voids are filled up by finer material (porphyric).

Although independent of absolute sizes, the c/f-related distribution concept may be difficult to apply when using ultramicroscopic enlargements (e.g. scanning electron microscopy) because the morphological characteristics of the material are different at that scale.

As mentioned already in the basic concepts, the size limit in the c/f-related distribution must be fixed for each problem separately. It is possible to indicate the size limit by adding it as a subscript to the lettergroup c/f-, e.g. $c/f_{20\mu}$-related distribution, indicating that the limit between the coarse and the fine material is situated at 20μ. In general one speaks of the c/f_x-related distribution. The ratio between the amount of particles coarser and finer than the given limit is expressed in the c/f_x-ratio; for instance, a porphyric $c/f_{2\mu}$-ratio of 2/3. These ratios can also be expressed in volume percentages when determined quantitatively.

THE BASIC TYPES OF c/f- RELATED DISTRIBUTIONS

As mentioned already before, it has been the aim of the authors to elaborate a classification system of c/f-related distributions, characterized by a gradual transformation of one type into another, rather than a typology of distinct, unrelated types. A logical sequence could be based on the gradual transition from a framework of bare grains of a single size fraction to an entity of grains embedded in a dense groundmass. This sequence is characterized by an increase of fine material and a decrease of interstitial voids. This transition can take place in three essentially different ways: (1) the finer material can occur as bridges between the coarser grains; (2) as coatings around them; or (3) as loose aggregates in the interstitial spaces. All intergrades are possible between two or more of these types.

For the reasons mentioned earlier, new terms are proposed as names for the five basic types in the sequence. These are:

(1) The *monic* c/f-related distribution (from the Greek μονος, meaning one):

only particles of one size group or amorphous material, is present. It must be understood that the associated interstitial voids are always present, even if it seems to be a one-phase system of colloids (ultramicroscopic voids).

(2) The *gefuric* c/f-related distribution (from the Greek γεφυρα, meaning bridge): *the coarser particles are linked by braces of finer material.* This implies that the coarser particles are not in contact with each other, and thus have no skeletal function.

(3) The *chitonic* c/f-related distribution (from the Greek χιτων, meaning tunic, garb, but also peel of an onion; compare with chitine): *a skeleton of coarser particles which are wholly or partly surrounded by a cover of finer material.* The orientation of the finer material is not considered; this means, for instance, that stratified true illuviation clay cutans as well as random clay deposits on grain-surfaces are considered as a chitonic c/f-related distribution.

(4) The *enaulic* c/f-related distribution (from the Greek ἐναυλος, meaning being in something, being in a cavern): *a skeleton of coarser particles with aggregates of finer material in the intergranular spaces.* It must be understood that the aggregates do not completely fill the intergranular spaces, because the c/f-related distribution would then be porphyric.

(5) The *porphyric* c/f-related distribution: *the coarser particles occur in a dense groundmass of finer material.* In the porphyric c/f-related distribution the coarse grains have no skeletal function, except in one case, which will be discussed later.

It must be noted that the above types are not completely mutually exclusive. Only the application of more complicated micromorphometric criteria would allow this.

CLASSIFICATION OF THE BASIC TYPES AND THEIR INTERRELATION

The best way to visualize the relation between the different types is to represent them in a diagram. This can easily be done by plotting the five basic types on the five corners of a trigonal bipyramid, so that the monic type is at the top and the porphyric at the bottom (see Fig. 1).

If we examine the figure, we see that the top is the only place representing the pure material for a given fraction. Moving downward one notes the appearance and subsequent increase of finer material, which may happen according to three different distribution patterns. Under the equatorial plane the amount of finer material can increase continuously (tending ad infinitum), so that only a few grains of the coarser fraction are floating in it (porphyric ortho-type). This is reflected by the dotted lines in the lower part of the figure. In view of the considerable variation of the ratio of coarser to finer material (c/f-ratio) possible in the porphyric type, the following subdivision is proposed: close porphyric (the coarser grains have points of contact); single-spaced porphyric (the distance between the coarser grains is

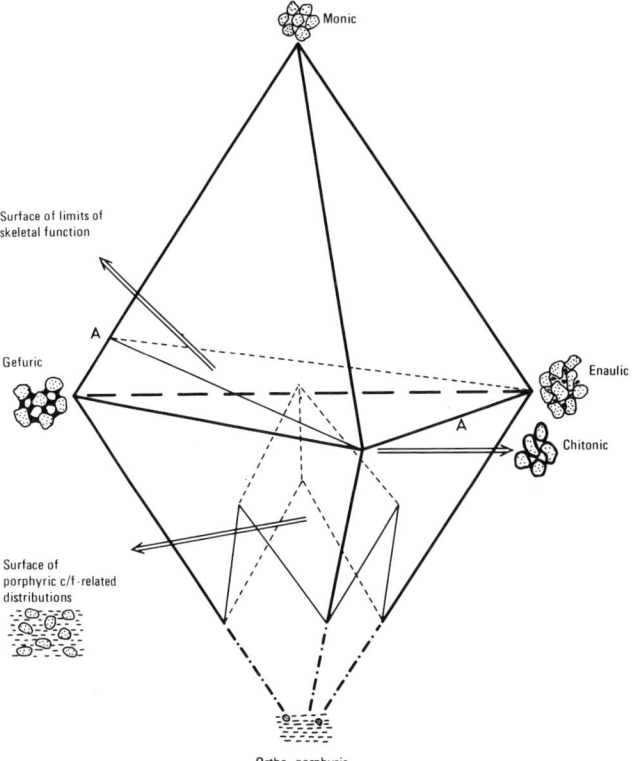

Fig. 1. Space diagram showing the specific characteristics of the five basic types of c/f-related distributions and their interrelationships.

less than their diameters[1]); double-spaced porphyric (the distance is one to two times the diameters); and open porphyric (the distance is more than twice the diameters). In special cases, these adjectives may be used also in connection with other types, e.g. a double-spaced gefuric c/f-related distribution has been noted in humus-rich soil materials. In these cases the coarse material of course no longer has a skeletal function.

Three groups of intergrades are to be distinguished: intergrades located on the edges, on the faces and within the figure. The first types (on the edges) are intergrades between two basic types (e.g., between monic and chitonic: a packing of grains, some of them having a coating of finer material, others being bare). The second types (on the faces and the equatorial plane) are intergrades between three basic types (e.g., between monic, gefuric and chitonic: a packing of bare grains, some however having a coating of finer material, and some grains being united by bridges of finer material). The third and last types (within the figure) are the intergrades between all four basic types (e.g., monic, chitonic, enaulic and gefuric). The exact position of the intergrades is defined by the proportions of each type present.

---

[1] If the coarse grains show an important range of sizes, the estimated mean size is considered.

This simple figure may be extended to some degree with additional surfaces in order to illustrate some specific characteristics. A surface AA cuts the figure in two unequal parts. This surface is the meeting place of all limits of the skeletal function. It means that under this surface no skeletal function exists; above the surface the coarse material has the function of a soil skeleton. This surface goes through two basic types (chitonic and enaulic) but lies above the gefuric corner, as is evident from the definition of the latter.

Even in earthy materials, where the coarser material has a skeletal function, it may show different degrees of packing (cf. Lukashev's unconsolidated and compact inequigranular textures).

Inside the bottom part of the bipyramid, a small trigonal pyramid is inserted. Its top touches the skeletal surface, and its position is defined by a rotation of $60°$ around the trigonal axis of the bipyramid. The surface of this pyramid delimits the c/f-related distributions with interstitial voids (above) and those having all spaces filled up with finer material (below). The top represents the case of a c/f-related distribution just at the limit of the skeletal function, and where all voids are just filled up with finer material. Consequently, all kinds of c/f-related distributions below this pyramidal surface are called porphyric in the broad sense.

Because of the fact that we do not take into account the genesis of the classified c/f-related distributions, the base of the small pyramid may be fixed arbitrarily. If a genetic meaning were given to this pyramid, the position of its base would be kept floating (the small pyramid would then open and close like an umbrella).

THE LOWEST LEVEL OF c/f-RELATED DISTRIBUTION

As already mentioned above, the c/f-related distribution will be most useful to describe the mutual distribution of the fabric members of lowest order, i.e. grains of sand to clay size (this level corresponds more or less to the elementary fabric of Kubiena (1938) and to Brewer's related distributions in the basic fabric (1964).

First of all it is necessary to indicate the essential differences between the plasma/skeleton concept as used in soil micromorphology, and the concept of coarse/fine material as presented here. Plasma and skeleton are differentiated theoretically on a basis of stability, skeleton being the relatively stable individual fabric elements of lowest order in a soil material, and plasma the rather unstable ones. More difficult to evaluate is the chemical and/or physical stability of coarser grains: e.g., under which conditions is a grain of gypsum or calcite stable? The plasma/skeleton concept has been created to describe "normal soils". Once non-soil materials (e.g., termite mounds, earthy cements) or some specific soil horizons (especially the lithified horizons such as the petrogypsic, petrocalcic and petroplinthic) or weathering rocks are concerned, the stability criterium (not easily translocated or transformed by pedogenetic processes) is no longer valid. Only morphometric criteria, e.g. size, seem suited in this case. The essential difference between the plasma/skeleton concept and the coarse/fine concept is, thus, that the former is based only on (partly subjective) genetic criteria (stability), whereas the latter is based on morphometric (thus rather objective) criteria (relative size).

The above-proposed terminology can readily be applied to describe the c/f-related distribution of elementary particles in thin sections. Intergrades and mixtures, the sizes of the different fractions and the kinds of material are then indicated in simple phrases. For compactness of description and to save time, many persons prefer a more elaborate system making it possible to express all these characteristics in one (compound) term. Therefore, a specific terminology will be proposed.

To be a useful tool for micromorphological descriptions, this lowest level must not only give some indications of the real c/f-related distribution of the finer and coarser material, but must give in addition some ideas about absolute sizes of both fractions and kinds of materials. In addition, rules must be set up to distinguish and describe intergrades and mixtures. (See Plates I and II for some examples.)

(a) Size

Since the soil micromorphological descriptions must be related as much as possible to the data from routine soil analysis, the size fractions are delimited roughly in the same way as in usual pedological language, viz., sand, silt and clay. We will not give precise limits, because they must depend, as explained above, in each country, upon the granulometric fractions in use (e.g., I.S.S.S., U.S.D.A., D.I.N. systems). In practice, the size of the finer and the coarser material should be mentioned after the type of related distribution, using following symbols: psef(i) from psefitic for gravel size[1]; psam(mi) from psammitic for sand size; sil(i) from silt for silt size; pel(i) from pelitic for clay size.

It was decided for convenience to mention first the symbol for the coarser fraction, and then the one for the finer fraction, separated by a slanted line. For reasons of euphony the elements are linked by the vowel "-i"; e.g. chitonic sili/pel means that silt-sized grains are surrounded by material of clay-size. If one wishes to say that the finer or the coarser fraction is composed of different-sized material, then more symbols are used; e.g. enaulic psammi/pelisil means that aggregates of clay- and silt-sized material occur between sand grains. A special case would be the monic pel c/f-related distribution, indicating that only a continuous clay mass is present.

If one only wants to express the size limit between the coarser and the finer particles, the expression c/f_x-related distribution can be used, of course.

---

[1] The terms psefitic, psammitic, and pelitic are adopted from sedimentology.

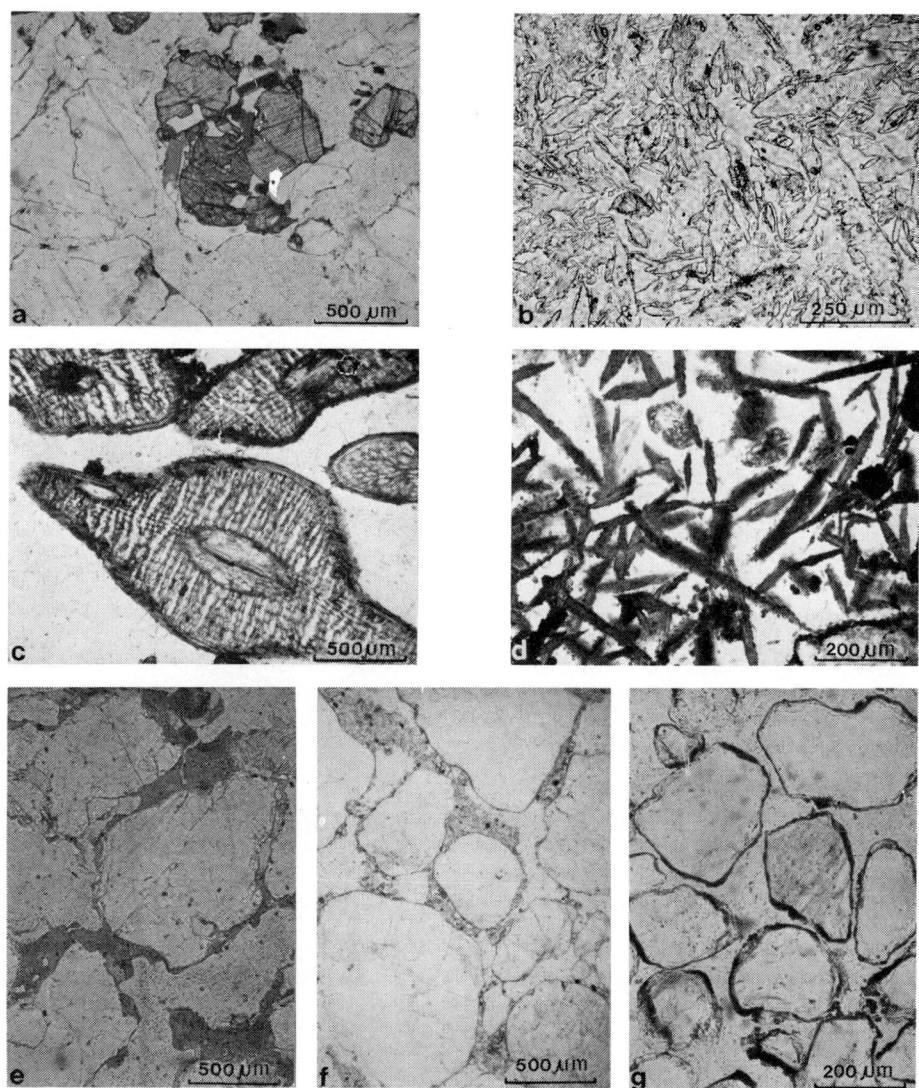

PLATE I

a. Monic lithopsam. Cryorthent. Greenland.
b. Monic gypsopsam. Calciorthid. Iraq.
c. Monic phytopsammisef. Humod. The Netherlands.
d. Monic viviopsam, locally viviopsammi/siderosil. Subsoil of Sphagnofibrist. The Netherlands.
e. Chito-gefuric psammi/argiopel. Haplumbrept. Spain.
f. Gefuric psammi/pelisil. Orthod. The Netherlands.
g. Chitonic psammi/humopel. Humod. The Netherlands.

Morphological Classification

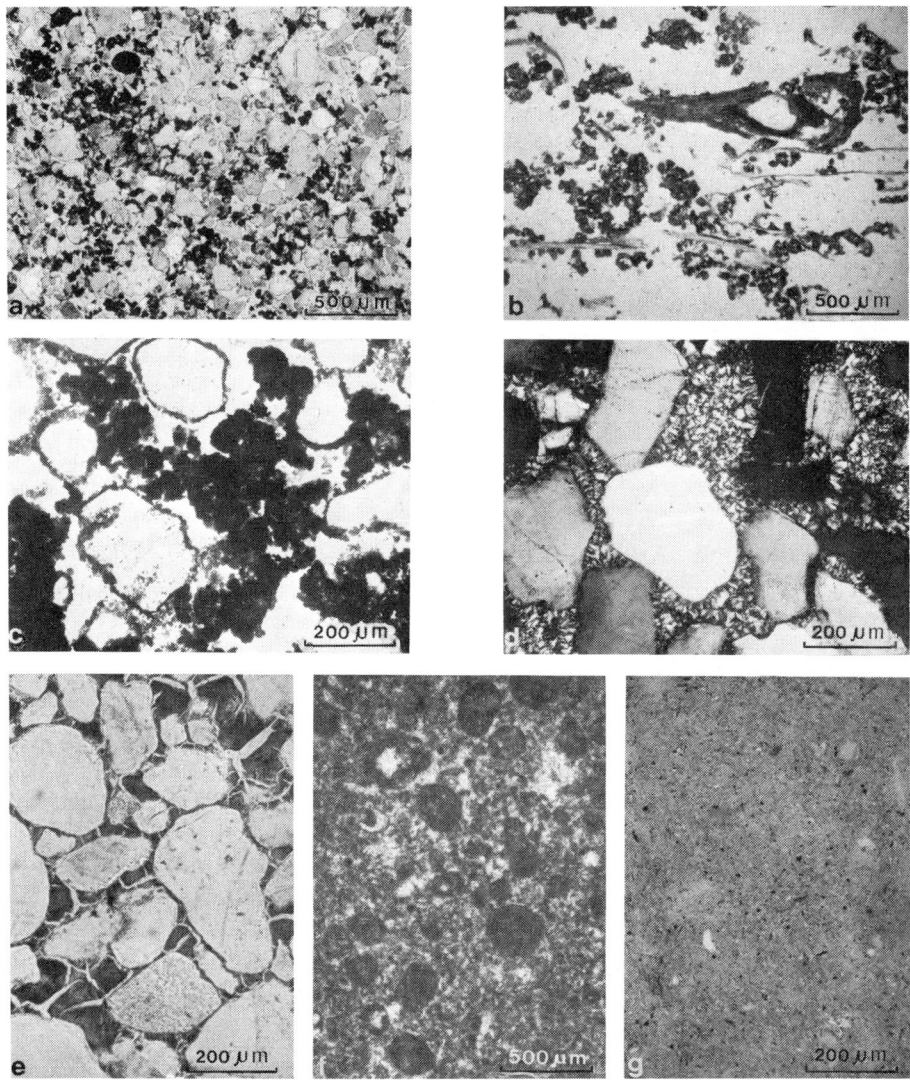

PLATE II

a. Enaulic psammi/humopel and monic psam. Plaggept. The Netherlands.
b. Enaulic phytopsammi/humopelisil. Orthod. The Netherlands.
c. Porphy-chito-enaulic psammi/sideropel. Subsoil of Sphagnofibrist. The Netherlands.
d. Close porphyric psammi/quartzosil. Silicified Tertiary sand. The Netherlands.
e. Close porphyric psammi/humopel (the cracks are artificial, i.e. caused by shrinkage during drying). Aquod. The Netherlands.
f. Double-spaced porphyric calcipsammi/calcisil. Calcitic nodule. Poland.
g. Open porphyric psammi/argiopeliphytosil. Hydraquent. Surinam.

(b) Material

The kind of material generally will be treated in the other parts of the descriptions. However, insofar as materials are easily recognizable, they may be mentioned in the description of the c/f-related distribution. The symbols are coined by adding the suffix "-o" to the name of the material or part of it, e.g., recognizable plant fragments (phyto), humus — i.e. finely comminuted organic matter — (humo), calcite (calco), sesquioxydes (sesquo), rock fragments (litho), clay minerals (argio), gypsum (gypso), quartz (quartzo), silica (silico), etc. These symbols should, by convention, be placed before the name of the relevant size fraction; e.g., viviopsam/siderosil means that silt-sized material composed of siderite occurs together with sand-sized vivianite crystals.

(c) Intergrades

Intergrades are named by joining the two or three basic terms, the most dominant being mentioned last. From the first term, the suffix "-ic" and the last consonant are dropped: chito-gefuric means a c/f-related distribution dominated by bridges of finer material, but with presence of some coatings. If both are present in approximately the same amount, the terms are written out fully: chitonic—gefuric. Each edge of the pyramid can in this way be subdivided into three zones: dominance of one type, more or less equal amounts of both types, and dominance of the other type.

(d) Mixtures

In soil materials not only intergrades are observed, but in several cases it is clear that mixtures of different types of c/f-related distributions are present. In such cases the different c/f-related distributions are indicated by full terms, joined by the word "and": chitonic and gefuric sili/calcipel means the presence of domains with a chitonic and others with a gefuric c/f-related distribution composed of silt-sized grains and fine calcite dust. However, if the c/f-related distributions are dissimilar in size and/or nature of components, the relevant terms should be spelled out completely.

The c/f-related distributions are considered to be homogeneous when they cover more than 95% of the total area. The same rule applies for grain sizes and nature of materials.

ACKNOWLEDGEMENT

The authors wish to acknowledge the members of the Working-Group on Soil Micromorphology and in particular Dr. H.-J. Altemuller (Braunschweig, Germany) for the very valuable discussions on the plasma/skeleton and relative size concepts during the meetings of the Working-Group.

REFERENCES

Brewer, R., 1960. The petrographic approach to the study of soils. Trans. Int. Congr. Soil Sci., 7th, Madison, 1:1—13
Brewer, R., 1964. Fabric and Mineral Analysis of Soils. Wiley, New York, N.Y., 470 pp.
Kubiena, W.L., 1938. Micropedology. Collegiate Press, Ames, Iowa, 242 pp.
Lukashev, K.I., 1970. Lithology and Geochemistry of the Weathering Crust. Israel Program Scientific Translations, Jerusalem, 367 pp.
Ruchin, L.B., 1958. Grundzüge der Lithologie. Akademie-Verlag, Berlin, 806 pp.

8

Copyright © 1975 by the Agricultural Institute of Canada
Reprinted from pages 304–311 of Canadian Jour. Soil Sci. **55**:301–319 (1975)

INVESTIGATIONS OF SOME SOILS DEVELOPED IN HUMMOCKS OF THE CANADIAN SUB-ARCTIC AND SOUTHERN-ARCTIC REGIONS
1. MORPHOLOGY AND MICROMORPHOLOGY

R. BREWER[1] and S. PAWLUK[2]

[1]*Division of Soils, CSIRO, Canberra City, A.C.T., Australia, and* [2]*Department of Soil Science, University of Alberta, Edmonton, Alta. Contribution No. T74-6, Alberta Institute of Pedology, received 16 Sept. 1974, accepted 1 Apr. 1975.*

[*Editors' Note:* In the original, material precedes and follows this excerpt.]

2. MICROMORPHOLOGY
CONCEPTS AND FABRIC TYPES

Fundamental differences between fabrics observed in the soils under investigation were strongly evident. In order to classify fabric types according to their important distinguishing characteristics it was necessary to use different concepts for different arrangements. In the discussion and definition of fabric types, the following concepts are used:

S-MATRIX. S-matrix of a soil material is the material within the simplest (primary) peds, or composing apedal soil materials, in which the pedological features occur; it consists of the plasma, skeleton grains, and voids that do not occur in pedological features other than plasma separations.

MATRIX. In a rock in which certain grains are much larger than others, the grains of smaller size comprise the matrix (American Geological Institute, (AGI) 1966).

FRAMEWORK MEMBERS. The much larger grains (skeleton grains, rock nodules, etc.) associated with the matrix.

PHENOCLAST. A large fragment in a nonuniformly sized sediment such as a pebble in a conglomerate (AGI 1966); in the context of soil fabric, occasional grains (mineral grains, rock nodules, etc.) that are significantly larger than the common framework members.

Some of the fabrics observed are best characterized by the basic distribution pattern (Brewer 1964) of discrete units, and some by the related distribution pattern (Brewer 1964) of like units with regard to other discrete entities. Related morphologically to these are materials composed of soil units that are not entirely discrete but appear to be interconnected through narrow necks. Other distinct fabrics are characterized by the related distribution of framework members and matrix.

Granular and Related Fabrics

Brewer's (1964) definition of granular fabric is based on a concept of packing of discrete units, an important characteristic being that any plasma present occurs in pedological features. Thus, glaebules, faecal pellets, and so on, are considered as units that can have a granular arrangement, either as individual groups or collectively. Logically, too, the arrangement of discrete peds (the secondary fabric of Brewer 1964) can be described as granular; this is not a contradiction but simply an alternative method of description. Similarly the arrangement of plant fragments in various stages of decomposition often conforms to the definition of granular fabric, even though Brewer (1964) specifically excluded such materials in his treatment of fabrics. Since the definition of granular fabric does not include specifications in regard to grainsize distribution, packing or the internal characteristics of the units, granular fabrics can be subdivided on these characteristics, including the degree of accommodation (Brewer 1964) of the faces with adjoining peds which is a parameter of packing.

A useful alternative method of description can also be used for soil materials that appear to be composed of coalesced units, that is, units that are similar in many respects to those that occur in granular fabrics but which are not entirely discrete, appearing to be fused (i.e. united) over a relatively small proportion of their perimeters in thin section. Such fabrics generally fall into the group of porphyroskelic fabrics, although other types are possible, and the presence of incompletely isolated units would be inferred from the description of the voids — interconnected vughs where the units are unaccommodated and skew planes where they are accommodated (Brewer 1964).

The rationale of the following classification and nomenclature is selection of a root for the fabric group, and use of the suffix *ic* for fabrics in which the units are discrete and the suffix *oidic* where the units are not entirely discrete.

GRANIC. ("Granic" was originally suggested by Dr. G. Stoops but subsequently replaced by "monic" (Stoops and Jongerius 1975). The term is used here with Dr. Stoop's permission.) Unaccommodated, typically loosely packed, discrete units without coatings on, or bridges between, units. Although composition of the units is not strictly a parameter of fabric, the materials of many of the horizons (or zones) in the soils examined are characterized by various combinations of recognizably different kinds of units arranged in a granic fabric. Thus, the materials can be described more specifically by using the following terms: *orthogranic*, loosely packed mineral grains and/or rock nodules; *phytogranic* (Fig. 1a), loosely packed, partially decomposed plant fragments, including Bal's (1973) histons and some of his organic skeleton grains; *humigranic* (Fig. 1a), loosely packed, dark, usually isotropic, moder-like organic units; *mullgranic* (Fig. 1b), loosely packed mull-like units consisting of plasma plus skeleton grains

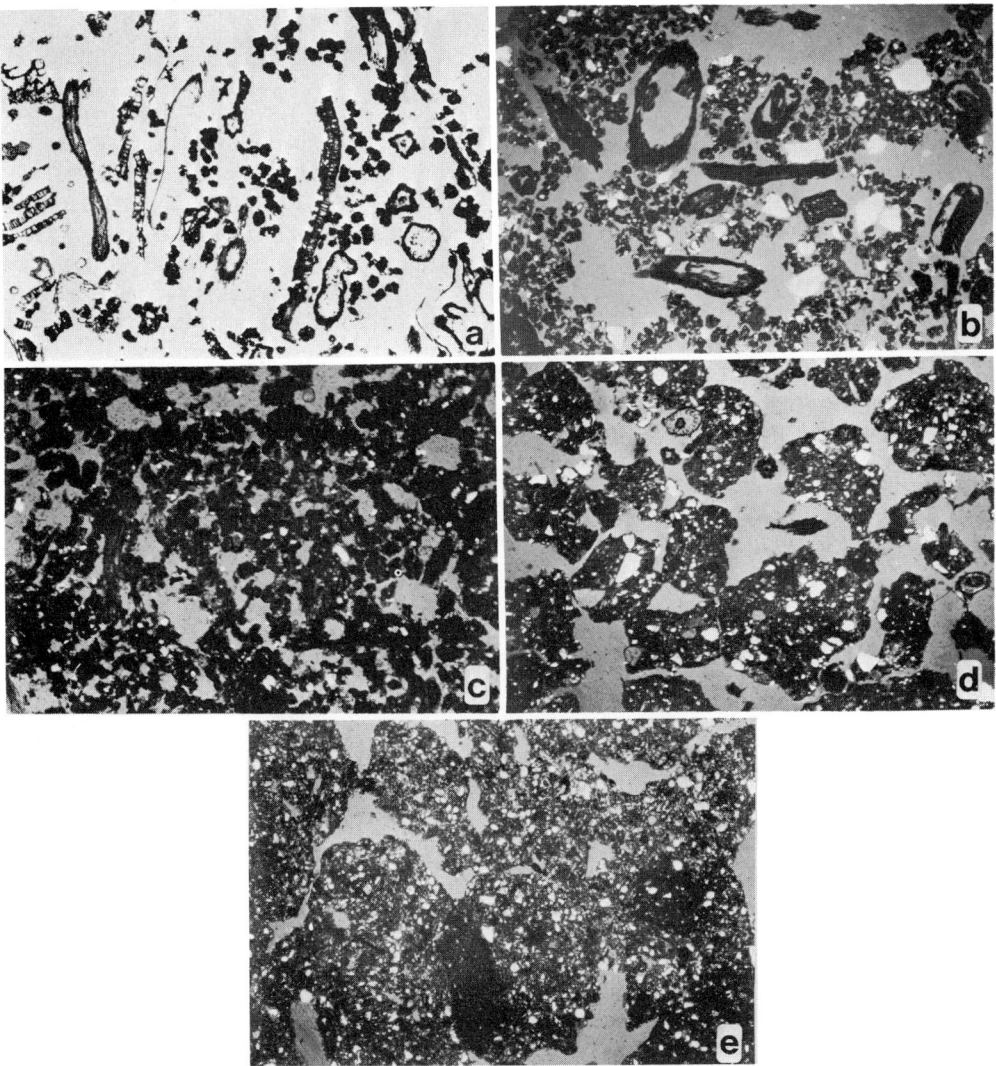

Fig. 1. (a) Humi-phytogranic material; humigranic units 20–200 μm with strong mode. Profile P409, zone I. Plain light. Frame length 3.35 mm. (b) Ortho-humi-phyto-mullgranic (20–200 μm) material. Profile P405, zone Ia. Partially crossed polarizers. Frame length 3.35 mm. (c) Mullgranoidic (20–200 μm) material. Profile P409, zone II. Partially crossed polarizers. Frame length 2.1 mm. (d) Metamatrigranic (400 μm–2 mm) material. Profile P407, zone III. Partially crossed polarizers. Frame length 3.35 mm. (e) Metamatrigranoidic (400 μm–2 mm) material. Profile P407, zone IV. Partially crossed polarizers. Frame length 3.35 mm.

with the birefringence of the plasma masked by finely disseminated, probably colloidal organic matter; *matrigranic* (Fig. 1d), loosely packed units composed of "normal" soil material, that is, plasma plus skelton grains, not uncommonly with included pedological features, but not mull-like material.

GRANOIDIC (Fig. 1c,e). Forms like those in granic fabric but the units are not discrete, appearing in thin section to be fused (i.e. united) at their contacts; the forms of the walls of voids suggest coalescence of units (or "aggregates"). Materials with granoidic fabric can be subdivided as for those with granic fabric except that orthogranoidic materials would not be expected because boundaries of mineral grains are normally quite sharp even when densely packed.

FRAGMIC. Relatively densely packed, accommodated discrete units without coatings on, or bridges between, units. Although fragmic materials could be systematically subdivided as for those with granic fabric, the only type observed in the soils examined is the analogue of matrigranic. In addition, in the materials examined, the degree of accommodation of the units varies, so the term *partially accommodated* is used to qualify fragmic where the units are not fully accommodated.

FRAGMOIDIC. Forms like those in fragmic fabric but the units are not discrete, appearing in thin section to be fused (i.e. united) at their contacts, that is, where the units are fully accommodated there are frequent skewplanes. This fabric would be described by Brewer (1964) according to the internal fabric of the units (usually porphyroskelic) and the visible voids, that is, porphyroskelic with interconnected skewplanes (or craze planes) or skewplanes and vughs.

These fabric groups are quite broad and more specificity can be brought to the descriptions by extended statements of a range of characteristics such as size, shape and packing of the units. In particular, grainsize distribution may be important in interpretation. Absolute sizes and special distributions, such as bimodal, can be recorded by extended statements; a simple abbreviated method is to record the mode (if there is a distinct mode) and/or size range in brackets after the name. An important characteristic in these fabrics seems to be the conformation of the surfaces of the individual units; the prefix *meta* is used where the surfaces are significantly smoothed and rounded (Fig. 1d,e).

Fabrics Defined on Related Distribution of Matrix and Framework Members

A number of these fabrics are defined best by using Kubiena's (1938) concepts of coatings and bridges of finer grained material (matrix) on and between significantly larger grains (framework members). The chlamydic, gefuric and plectic fabric types recognized in the soils examined are defined in these terms. As for granular materials, materials with these fabrics can be subdivided broadly on the composition of the matrix by using prefixes (especially for grainsize distribution). In the soils examined, the matrix consists of "normal" soil material composed dominantly of clay mineral plasma plus skeleton grains; the prefix *matri* is used to designate fabrics with matrixes of this kind; obviously extended descriptions could specify such fabrics more accurately.

CHLAMYDIC. The matrix occurs as complete coatings or uniformly discontinuous coatings (i.e. if incomplete, the partial coatings are not confined to the parts of the grain surfaces close to grain contacts or the point of nearest approach of adjacent grains) on the framework members, otherwise the intergranular spaces are essentially empty. This is a variety of chitonic fabric (Stoops and Jongerius 1975), and Kubiena's (1938) chlamydomorphic fabric is a variety of both chlamydic and chitonic fabric. The coatings in all these fabrics could be regarded as cutans on the framework members (free grain matrans (Brewer 1964), so an alternative description of the arrangement is granular (or granic) with free grain matrans. The variety recognized in the soils examined is matrichlamydic fabric (Fig. 2a) in which the matrans commonly have a moderate striated extinction pattern, so the fabric can be specified as *skelsepic matrichlamydic*.

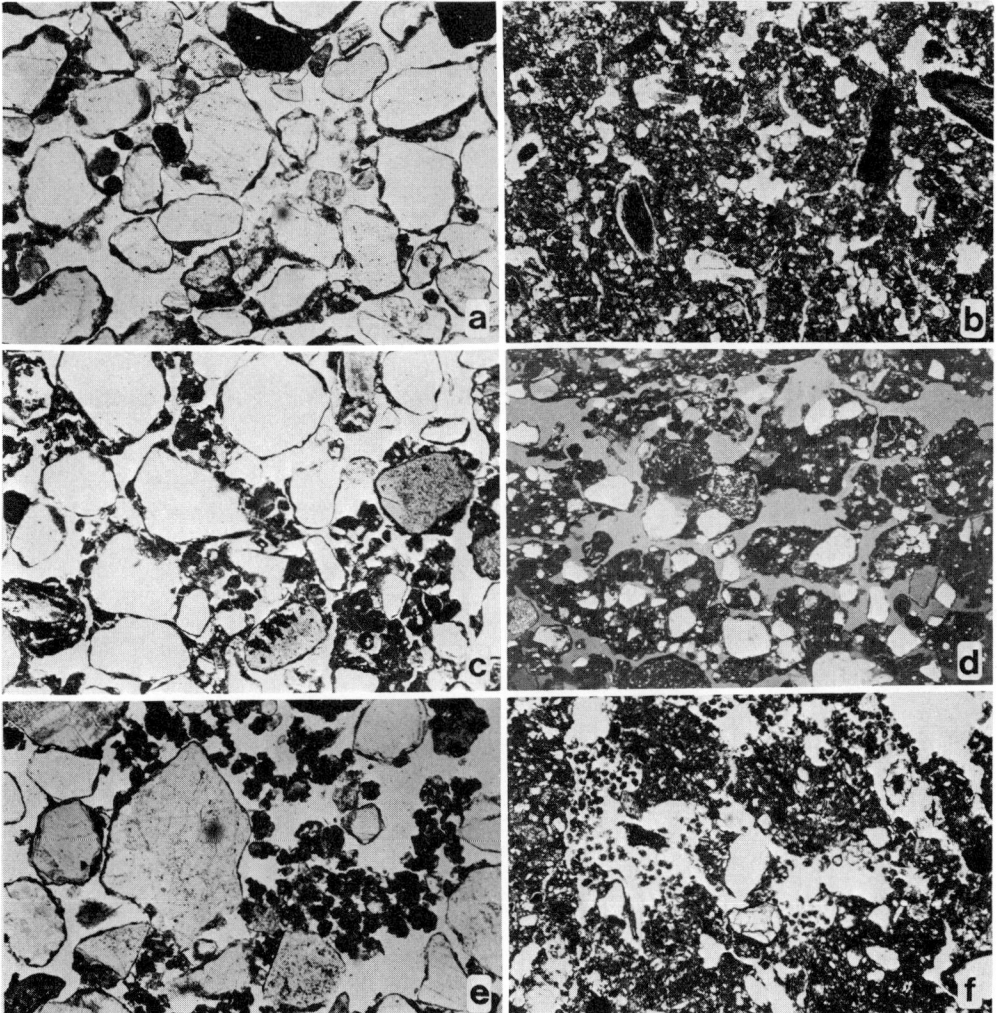

Fig. 2. (a) Matrichlamydic material. Profile P405, zone III. Partially crossed polarizers. Frame length 2.1 mm. (b) Matrigranoidic porphyroskelic material. Profile P409, zone III. Partially crossed polarizers. Frame length 2.1 mm. (c) Matrigranoidic matriplectic material. Profile P405, zone III. Plain light. Frame length 2.1 mm. (d) Banded metagranoidic fabric. Profile P405, zone IVb. Partially crossed polarizers. Frame length 3.35 mm. (e) Matrichlamydic-mullgranic (20–200 µm) material. Profile P405, zone II. Partially crossed polarizers. Frame length 2.1 mm. (f) Matrigranic (20–200 µm) partially accommodated fragmoidic (1-3 mm) material. Profile P407, zone VI. Plain light. Frame length 3.35 mm.

GEFURIC. The coarser particles are linked by braces of finer material (Stoops and Jongerius 1975). Translated into the terms used in this paper the definition would read: the matrix occurs as bridges between the framework members but does not entirely coat them; bare surfaces of framework members commonly form part of the walls of the large voids. This fabric can be subdivided as was chlamydic fabric. Kubiena's (1938) intertextic fabric is a variety of gefuric fabric. The variety recognized in the soils examined is *matrigefuric* fabric.

PLECTIC. The matrix coats the framework members and broadens and extends to form bridges between them; the walls of the large voids always consist of matrix material. This fabric can be subdivided as were chlamydic and gefuric fabrics. Kubiena's (1938) plectoamictic fabric is a variety of plectic fabric. The variety recognized in the soils examined is *matriplectic* fabric.

BANDED. A succession of subhorizontal bands each of which shows a gradation in color and proportion of plasma (matrix) from the top to the bottom; the upper part of each band is relatively darker (and denser) with a sharp boundary between it and the overlying band (after McMillan and Mitchell 1953).

Intergrades

The fabrics defined above are assessed as being modal types. Intergrades are defined as fabrics that have some of the characteristics of more than one modal type so that it is difficult to decide which is the best name to use. Such fabrics are recorded by using adjectives, the last name denoting the most similar modal fabric and the preceding adjectives the kind of modification. If sufficiently important, separate names can be given to specific intergrades since they essentially represent points in a continuum. The intergrades important in the soils examined are:

GEFURIC PORPHYROSKELIC. An intergrade between gefuric (matrigefuric in these soils) and porphyroskelic fabrics. The bridges between framework members are so wide that the larger visible voids are commonly isolated vughs; it is possible to regard the fabric as consisting of thick bridges between otherwise bare framework members, or as a porphyroskelic fabric with frequent vughs in which the occurrence of bare surfaces of framework members adjoining and forming the walls of voids is accidental.

GRANOIDIC PORPHYROSKELIC (Fig. 2b). An intergrade between granoidic and porphyroskelic fabrics. Essentially a variety of porphyroskelic fabric with interconnected vughs whose size, shape and arrangement are such that the material has an overall appearance that suggests it could be formed by strong coalescence of units of the granic (matrigranic in these soils) type, other than orthogranic.

GRANOIDIC MATRIPLECTIC (Fig. 2c). A matriplectic fabric in which the bridges between coated framework members have a granoidic fabric.

BANDED INTERGRADES. These are recorded by using one fabric name as an adjective, for example, banded metamatrigranoidic fabric (Fig. 2d) denotes a metamatrigranoidic fabric in which the units are elongated horizontally giving an overall appearance of banded fabric; the material may be somewhat denser at the top than at the bottom of each unit or band of units.

Complex Fabrics (see Fitzpatrick 1971)

There are two expressions of complex fabrics in the soils examined:

MIXED COMPLEX FABRICS. Complex fabrics in which the component fabrics are inextricably intermixed; in a sense they are intergrades but they have characteristics typical of modal fabrics and not intermediate characteristics. They occur where a modal fabric can be expressed in a discrete unit (e.g. chlamydic fabric) and in a collection of discrete units (e.g. granic fabric); a mixed complex fabric occurs when the units expressing two such fabrics occur in a random distribution pattern. Such fabrics are designated by hyphenating the two component fabric names, the last name designating the dominant (by area in thin section) component fabric; if two component fabrics are equally important areally, the hyphen is replaced by an equals sign (=). The variety observed in the present soils is *matrichlamydic-mullgranic* fabric (Fig. 2e) in which a granic fabric contains a significant proportion of randomly distributed framework members that are coated with matrix composed of plasma plus skeleton grains up to about 20 μm.

Although composition is not a true characteristic of fabric, granic fabrics in which the units have different compositions occur in arrangements analogous to mixed complex fabrics. To characterize specific kinds of soil materials it is advantageous to record such arrangements by hyphenating the compositional prefixes, or using equals signs, as for the true mixed complex fabrics. The following are examples of the types that occur in the soils examined: *humi-phytogranic* (Fig. 1a) in which a phytogranic material contains a significant proportion of randomly distributed humigranic units; *ortho-humi–phyto-mullgranic* (Fig. 1b) in which a mullgranic material contains a significant proportion of randomly distributed, discrete, partially decomposed plant fragments, and some discrete mineral grains and organic aggregates.

SEPARATED COMPLEX FABRICS. Complex fabrics that consist of repeating areas of two or more component fabrics. They are designated by using the names of the component fabrics separated by one slash (/) where the boundaries between the areas of each fabric are sharp, and two slashes (//) where the fabrics grade into each other; the dominant fabric (areally) is named last. An example from the soils examined is matrigranic/matrigranoidic material (Fig. 2f) in which the dominant fabric is granoidic but there are distinct areas with granic fabric. The pattern can be described more specifically in terms of the related distribution pattern of areas of the various component fabric types. Some of the materials examined have the same fabric expressed in two size ranges that constitute a separated complex; these are recorded by using the one fabric name with the two size ranges separated by one or two slashes in brackets after the name, for example, mullgranic (20–200 μm/400–600 μm).

REFERENCES

[*Editors' Note:* Only the references cited in the preceding excerpt are reproduced here.]

AMERICAN GEOLOGICAL INSTITUTE. 1966. Glossary of geology and related sciences. Washington, D.C.
BAL, L. 1973. Micromorphological analysis of soils. Soil Survey Inst., Wageningen, The Netherlands. 174 pp.
BAVER, L. D. 1948. Soil physics. John Wiley & Sons, Inc., New York, N.Y., 398 pp.
BREWER, R. 1964. Fabric and mineral analysis of soils. John Wiley & Sons, Inc., Sydney. 470 pp.
FITZPATRICK, E. A. 1971. Pedology — A systematic approach to soil science. Oliver & Boyd, Edinburgh. 306 p.
JUNG, E. 1931. Untersuchungen uber die Einworkung des Frostio auf den Erdboden. Kolloidchem. Beihefte. 32: 320–373.
KUBIENA, W. L. 1938. Micropedology. Collegiate Press. Ames, Iowa. 243 pp.
McMILLAN, N. J. and MITCHELL, J. 1953. Microscopic study of platy and concretionary structures in certain Saskatchewan soils. Sci. Agric. 33: 178–183.
STOOPS, G. and JONGERIUS, A. 1975. Proposal for a morphological classification of soil materials. 1. A classification of the related distributions of fine and coarse particles. Geoderma 13: 189–200.

RELATED DISTRIBUTION PATTERNS IN SOILS AND THEIR SIGNIFICANCE

by

H. ESWARAN * and C. BAÑOS **

RESUMEN

MODELOS DE DISTRIBUCION RELACIONADA EN SUELOS Y SU SIGNIFICADO

Los modelos de distribución relacionada son parámetros micromorfológicos de gran importancia en los estudios edafogenéticos. Se consideran dos grupos: uno constituido por *modelos de distribución relacionada normal* (básica), y otro por *modelos de distribución relacionada específica*, dependiendo de la función del plasma. Dentro de cada grupo se han determinado varios tipos, definiéndose cada uno de ellos. Se establece, en diagrama triangular, la correlación entre granulometría y distribución relacionada básica (normal, «NRDP»). Se presentan ejemplos de suelos donde ocurren ambos grupos, ilustrándose algunos de ellos.

INTRODUCTION

The concept of the related distribution pattern (RDP) in Soil Micromorphology was introduced by Brewer (1964) who considered it as the «distribution pattern of like individuals with regard to the distribution of individuals of a different kind». In this concept both the ratio of plasma to skeleton grains (e. g. granular and porphyroskelic) and specific arrangement patterns of the plasma with respect to the skeleton grains (e. g. agglomeroplasmic and intertextic) are considered, following Kubiëna (1938). Since the contribution of Brewer, more RDP's are recognised, e. g. phyric (Eswaran et al., 1969) and argillamatric (Eswaran, 1972; Bellinfante et al., 1974). These latter ad hoc suggestions are the result of the study of a wider range of soil materials. Reviewing the situation, the authors felt the need for a more systematic terminology and this forms the objetive of this paper.

(*) Geological Institute, Ghent (Belgium).
(**) Centro de Edafología y Biología Aplicada del Cuarto. C. S. I. C., Sevilla (Spain).

Stoops et al. (1974), confronted with a similar situation proposed the «c/f related distribution» and defined it as «the c/f related distribution expresses the distribution of individuals particles with relation to finer material and associated voids not included in the particles». In this concept, size limits of the particles is left to decision of the user.

The concepts and terminology developed here are purely for pedological purposes and not for other uses of the Micromorphological techniques as envisaged by Stoops et al. (1974). It is the intention here to relate micromorphological properties of the soil to the pedological properties. Related distribution patterns are one of the micromorphological parameters that can be employed to evaluate soils and soil genesis.

Concepts

The pedon (USDA, 1974) is the smallest volume that is recognised as soil individual. The vertical section of the pedon is the soil profile which is composed of horizons. A thin-section is a sample of a horizon. The basic descriptive unit is the s-matrix which is «the material within the simplest (primary) peds, or composing apedal soil materials, in which the pedological features occur; it consists of the plasma, skeleton grains and voids that do not occur in pedological features other than plasma separations» (Brewer, 1964).

The plasma, grains and voids are the three basic components of the s-matrix. The pedological features are a result of a specific combination or arrangement of one or more of the components. The concept of the plasma and grains differ from that of Brewer's and so is given in detail.

The plasma

The plasma is the colloidal fraction of the soil which may be mineral or organic. This is the most active component of the soil material and is capable of reorganisation, translocation and neoformation. In Soil Taxonomy, soils are classified based on the activity of the plasma or on the characteristics given to the soil by a specific behavior of the plasma. The oxic horizon is one where the chemical activity is low (type of plasma); in the argillic horizon plasma has accumulated by translocation; in the albic horizon plasma has been removed and in the spodic horizon a specific kind of plasma is accumulated. These diagnostic horizons characterise certain groups of soils. So in the choice of parameters for micromorphological indicators of pedogenesis (Eswaran, 1972) study on plasma has important bearing.

Although a colloidal size is specified, individual plasma cannot be seen with the petrographic microscope and even with the scanning electron microscope (SEM) a magnification of more than 10,000 is generally necessary. However, plasma domains are readily discernable. Presence or absence of domains and the size and arrangement of domains are important micromorphological characteristics which will be evaluated in a later contribution.

The grain

The grain is a single particle greater than colloidal size. Fragments of plant remains larger than colloidal size are frequently present in soils especially in the surface horizons. These are recognisable entities and are described as such. Consequently grains will only include the mineral materials.

Grains comprise a range of minerals which are primary or secondary, which vary in solubility, which are or may be present in all stages of transformation and which are present in all size grades. A classification of grains based on these parameters will be useful and may be made when necessary but will not add any additional information than simple descriptive terms. For example, gypsum relative to quartz is a soluble mineral and so a subdivision into soluble and non-soluble minerals or restricting the grains to their resistent properties is an attractive proposition. However, in the context of the Aridisols, where gypsum is most frequent, it is a stable mineral and so the division looses its relevance.

For these reasons the classification of grains is reduced to a minimum. This will not preclude the use of comparative or descriptive terms based on interpretation neoformed quartz, biotite pseudomorph, plasmified feldspar or secondary gypsum. One useful division of grains is into sand and silt size, the limit being 50 microns. This division is petrographically possible and will correspond to textural analysis.

The voids

The voids are empty spaces in thin-sections. A few may be artifacts caused by the preparation of the thin-section. Others show some regularity of shape and configuration that they can be grouped together. Brewer's nomenclature of voids is adhered to.

Related distribution patterns

The components of direct interest here are the plasma and grains; voids being a consequence of the arrangement of the other components.

The two basic aspects of RDP are the proportions of plasma, silt and sand in the s-matrix, and the specific arrangement of the plasma with respect to the other two. The arrangement does not lead to formation of distinct entities in which case they are pedological features. However, the arrangement gives a specific aspect to the s-matrix.

What is the cause of the arrangement patterns? In most cases this is due to pedogenesis; some sedimentary features can also attain such forms. In the absence of pedogenetic influence, there is a random distribution of the plasma, silt and sand; random in the sense that the plasma does not play any specific role in the arrangement of the silt and sand. This is considered as the normal-basic situation and the RDP is termed the Normal Basic Related Distribution Pattern (NRDP). The NRDP does not exclude such features as banded or clustered arrangement of the silt and sand grains as in some sediments. With pedogenesis, the plasma attains a different role. It bridges or coats grains, it aggregates silt and sand or it forms clusters. What the plasma does depends on the type of plasma, the NRDP and the formation stage of the soil. One will argue that this is a genetic division. It is genetic but the resulting micromorphological features are distinct enough to group them as Specific Related Distribution Patterns (SRDP). They are in fact extragrades to NRDP.

The normal related distribution patterns

The NRDP's are differentiated on the proportion of plasma to sand to silt. The textural triangle (USDA) is employed for this purpose. This diagram, fig. 1, attempts to relate the texture of the soil to micromorphology.

Granic

Granic NRDP is characterised by a dominance of sand-sized particles with small amounts of plasma and silt. The associated voids are those resulting from a close packing of sand grains — intergranular voids or simple packing voids. The field textures are commonly sands and loamy sands.

This is a typical related distribution pattern of the albic horizon in some Alfisols and in Psamments. When necessary the name of the mineral is used as a prefix. e. g. A Psamment has a quartzi-granic NRDP; some gypsic horizons have a gypsi-granic NRDP.

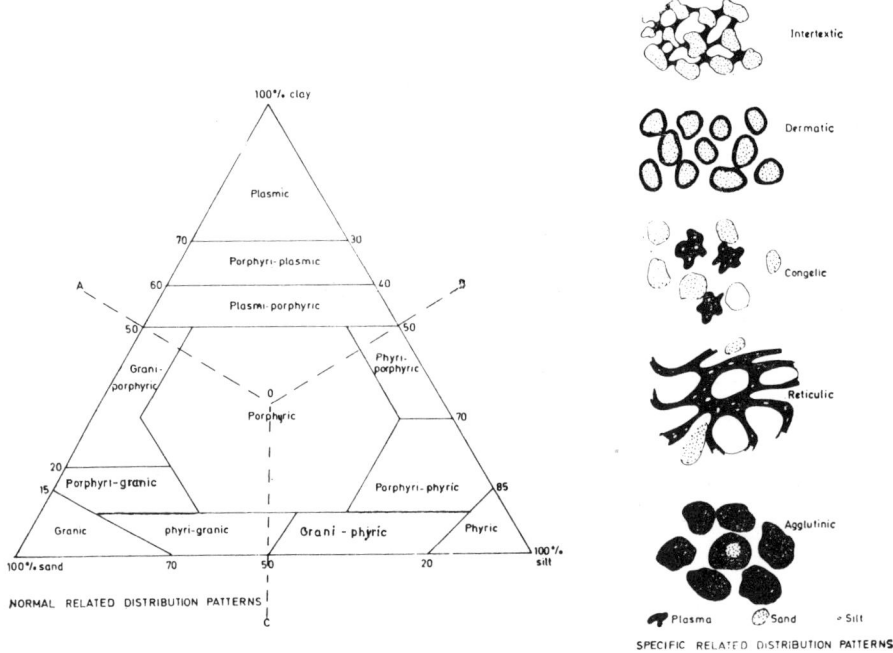

Fig. 1.—Normal (basic) and specific related distribution in soils.

Phyric

Phyric NRDP is characterised by a dominance of silt-sized grains with small amounts of plasma and sand. This is a much closer packing than granic; the voids are also intergranular. The field texture is silty and the consistence is compact and firm.

The albic horizon of some fine textured podsols (Eswaran et al., 1969) typify this NRDP. Sediments composed of silt have this NRDP. Some calcic horizons or calcareous soil materials have a calci-phyric NRDP. Some gibbsic horizons have a gibbsi-phyric NRDP (Plate 1, c).

Plasmic

Plasmic NRDP is characterised by a dominance of plasma with small amounts of silt and sand. Voids that are present include vughs, channels, vesicles and planes. The field texture is heavy clay and this NRDP characterises clayey sediments and highly weathered soils on basic and ultrabasic rocks — Ultisols and Oxisols (Plate 1, a).

This term supercedes the original term — argillamatric — of Eswaran (1972).

Porphyric

Phorphyric NRDP is characterised by an balanced amount of plasma, silt and sand. This NRDP grades to plasmic, phyric or granic when the amount of one of the component exceeds the sum of the other two. The sand frequently appears more prominent and is generally embedded in the s-matrix. A clay loam and loam texture characterise this NRDP. All types of voids may be present.

Intergrades

As shown in fig. 1, eight intergrades are recognised. These are characterised by the dominance of one component over the other. The intergrades are:

1. Phorphyri-plasmic.
2. Plasmi-porphyric.
3. Phyri-porphyric.
4. Porphyri-phyric.
5. Grani-porphyric.
6. Porphyri-granic.
7. Phyri-granic.
8. Grani-phyric.

PLATE 1

a) *Plasmic* normal related distribution pattern. Dominance of plasma without tendency to aggregate is the characteristic feature. Oxisols with agglutinic SRDP in the upper part of the profile have a plasmic NRDP in the lower.

b) *Agglutinic* specific related distribution pattern. The soil is an Acrorthox. It is a highly weathered soil and the plasma is dominantly sesquioxides with some kaolinite. Grains are few. Aggregation of the plasma to give a highly porous material is a comun feature.

c) *Phyric* normal related distribution pattern. The soil is a Gibbsiorthox. Silt-sized grains of gibbsite form the s-matrix of the gibbsic horizon. There is little or no plasma. The RDP is infact gibbsi-phyric.

d) *Congelic* specific related distribution pattern. The soil is a Andept. The plasmic, clusters are distributed randomly between the grains.

(The white bar on the photos has a length of 0.02 mm.)

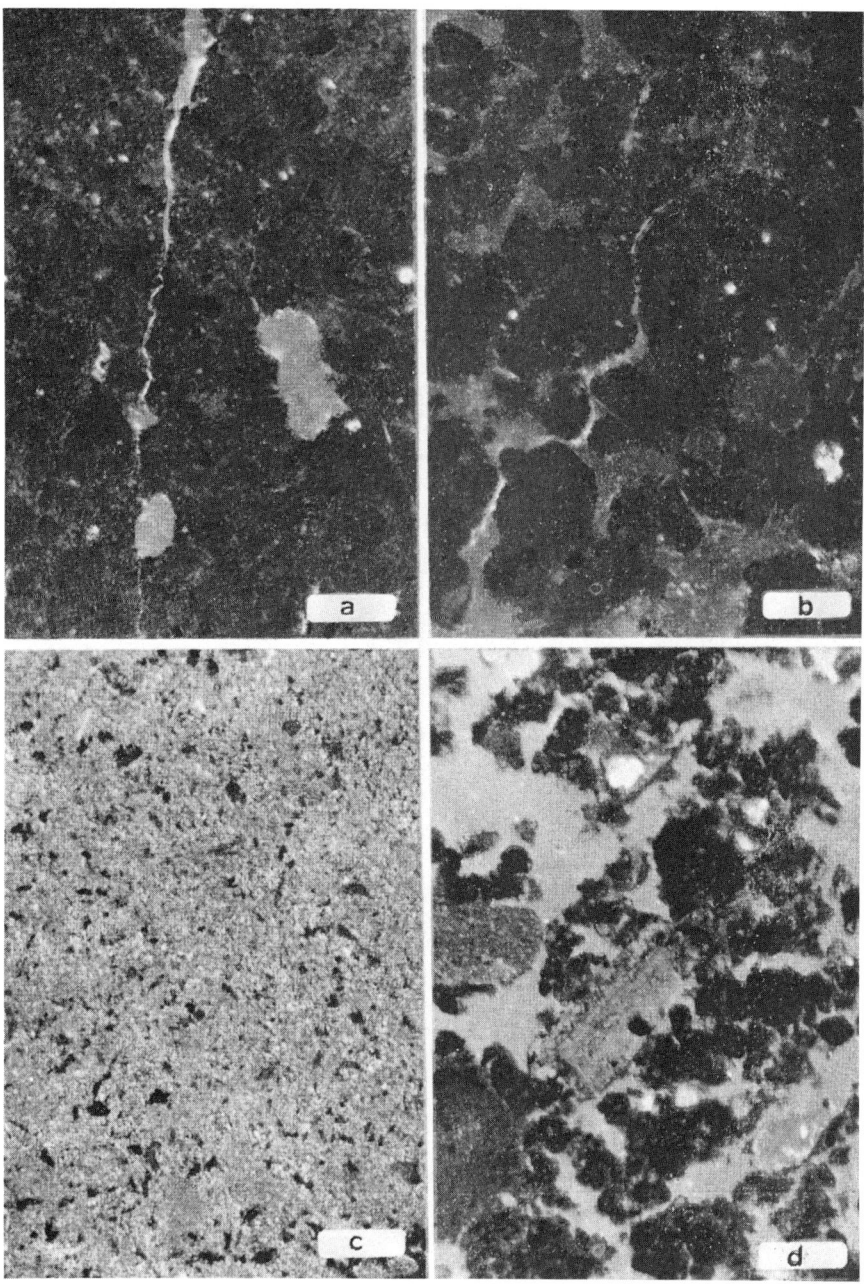

PLATE 1

[*Editors' Note:* Plates 1 and 2 are reproduced in color in the original.]

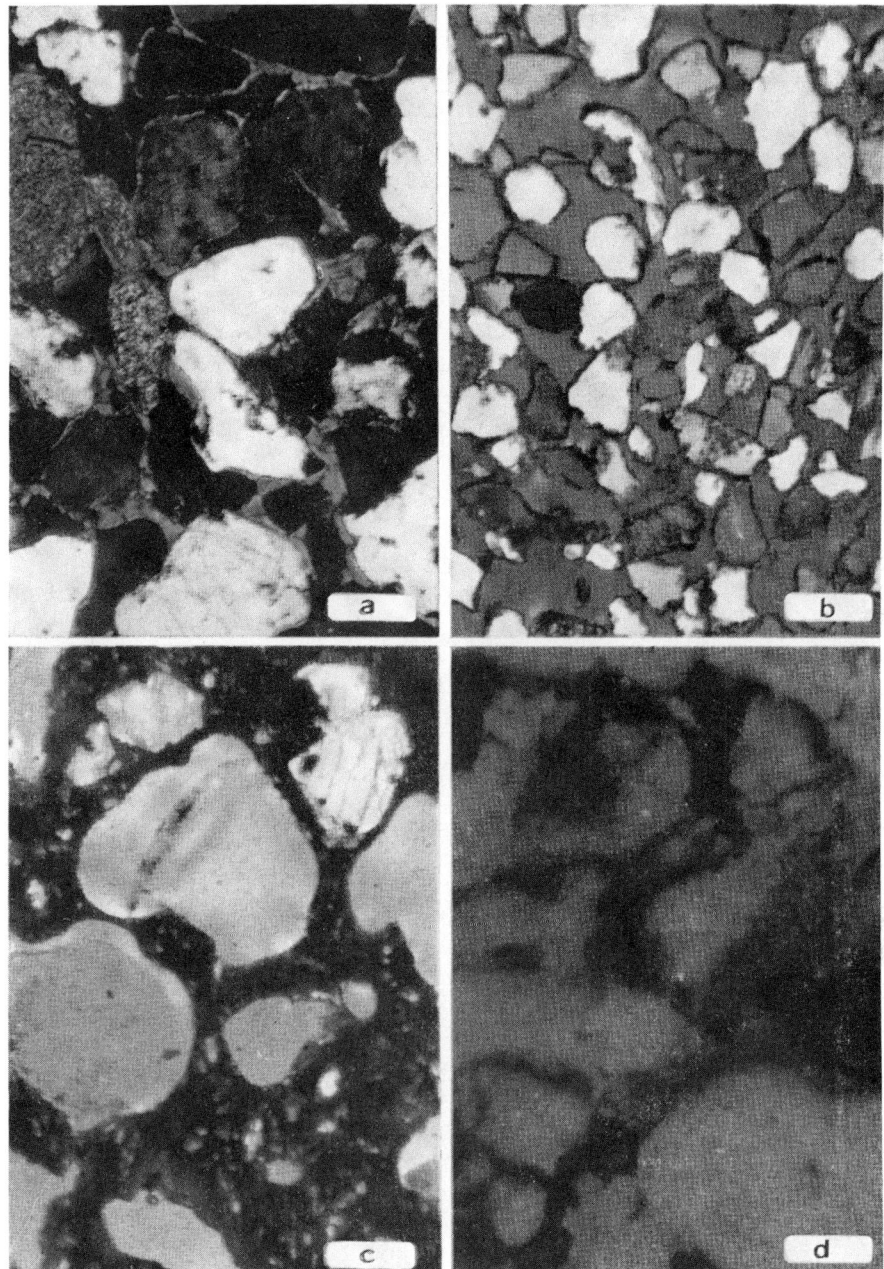

PLATE 2

a) *Intertextic* specific related distribution pattern. This is a typical feature of the argillic horizon on sands. The translocated plasma forms bridges between the grains.

b) *Dermatic* specific related distribution pattern. The spodic horizon in sands shows this. The translocated plasma coats the grains.

c and d) *Reticulic* specific related distribution pattern. Two types are shown. c) is an instance where a vesicular basalt is plasmified without volume changes.

In d), the sesquioxidic plasma attains a reticulic pattern in a plasmic kaolinite mass. This is the normal morphology of plinthite.

(The white bar on the photos has a lenght of 0.02 mm.)

A second group of intergrades are those where a single s-matrix shows more than one NRDP. For these combinations, the most dominant is indicated last:

«a granic and porphyric» NRDP characterises a s-matrix which is dominantly porphyric but where there are patches of granic.

The texture of the soil as determined by granulometric analysis, gives a first idea of the NRDP. The actual NRDP may not exactly coincide due to localised differentiation in the s-matrix — presence of intergrades — or due to pedogenesis which gives rise to the SRDPs.

Specific related distribution patterns

Due to the properties of the plasma, consequent to pedogenesis, specific related distribution patterns result. In Soil Taxonomy (USDA, 1974), soils which are due to special processes which result in specific morphological traits are grouped together. The macromorphological features have their counterparts in thin-sections and SRDPs are one of them.

Specific related distribution patterns (SRDP) are those which cannot be attributed to the random arrangement of plasma with respect to the silt and sand.

Five SRDPs are recognised for the moment; more will inevitably be added as distinct types are studied.

Intertextic

Intertextic SRDP is one where the plasma forms bridges connecting the sand grains (Brewer ss.). The SRDP is confined to materials with granic NRDP. The field texture is sandy to sandy loam and the consistence is fluffy.

The plasma is generally translocated and this SRDP is usually an early stage of the next type dermatic. The SRDP is present in some spodic and argillic horizons. Many banded textural B horizons on sandy materials show this SRDP (Plate 2, a).

Dermatic

Dermatic SRDP is one where the plasma forms a complete coating around the grains which are usually sand sized (Plate 2, b).

This SRDP is typically associated with sandy materials. In fig. 1, it is confined to the region delimited by AOC and the apex of the triangle but expression becomes indistinct in materials other than granic. For all practical purposes, dermatic is considered to be derived from granic.

The genesis of dermatic SRDP is due to plasma accumulation by translocation in sandy materials. The plasma is both mineral and organic. These transformations are possible in Spodosols and in sandy Alfisols and Ultisols. The micro-pedomorphosis of the s-matrix is examined in greater detail in several soils. Fig. 2, shows the sequence of evolution of the soil and the concomitant changes in the related distribution patterns.

Congelic

Congelic SRDP is one where the plasma aggregates silt-sized materials; the larger grains do not generally participate in the process and so the resulting morphology consists of silt-sized particles aggregated together by plasma and present in between coarser sand-sized particles (Plate 1, d).

The process described above takes place only if the plasma is or was in an amorphous state or has a significant amount of amorphous colloids and the NRDP must be phyric. If only sand and amorphous plasma are present, a dermatic SRDP results. The previous conditions are present in groups of soils such as:

> Spodosols on fine textured materials and in Andosols or Andepts.

The NRDP of the parent materials is generally in the region BOC (fig. 1) but is best expressed in soils with phyric NRDP-intergrades.

In Spodosols on silty parent materials, the translocated amorphous plasma aggregates the silt particles leading to the congelic SRDP. This was studied by Eswaran et al. (1969). In the field, the B_{2ir} horizon which shows this SRDP is spongy and fluffy whilst the overlying A_2 which has a phyri-granic NRDP is compact. Roots of plants pass trough the A_2 and proliferate in the more porous B_{2ir}.

In Andepts, the allophane behaves similarly and the field consistence is similar. The high porosity and rapid internal permeability of these soils or horizons is explained by the formation of the SRDP.

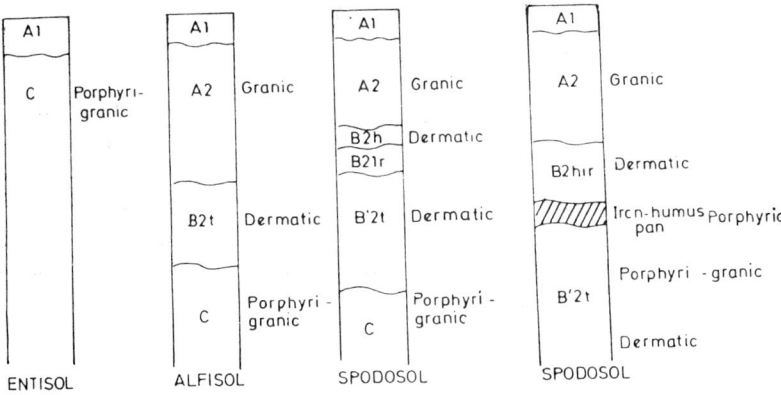

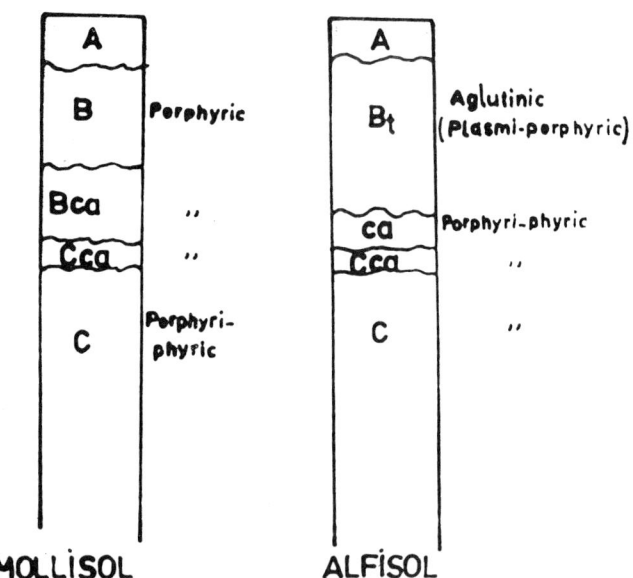

Fig. 2.—Changes in related distribution patterns as a function of evolution of the soils: a) developed on sandy materials. b) Developed on Quaternary calcareus sediments.

Agglutinic

Agglutinic SRDP is one where the plasma is aggregated together into sand or silt sized aggregates incorporating any silt or sand grains (Plate 1, b).

The processes involved are similar to the ones leading to the congelic SRDP; the difference is that they act on plasmic materials. Agglutinic SRDPs have been observed in clayey Oxisols. Certain Oxisols, the Acrorthox, have in the field a very weak subangular blocky structure which breaks into fine crumb. These crumbs are water stable aggregates and are alumino-silicate clays cemented by sesquioxides. The related distribution of such soil materials is agglutinic. The aggregates — there is practically no silt or sand — sometimes have a rim of oriented clay (ooidsepic plasmic fabric), subcutanic to the aggregate.

In fig. 1, this SRDP is confined to the region AOB but is best expressed in materials with plasmic NRDP or intergrades. The degree of expression of this SRDP is a function of the nature of the plasma — amount of sesquioxides — and the dessication of the soil. In an Acrorthox, agglutunic SRDP is present in the upper part of the profile; in the middle part, the aggregates are present but are coalesed whilst in the lower part the RDP is plasmic.

Reticulic

The reticulic SRDP is one where the plasma is arranged in a reticulate pattern. The plasma may incorporate silt and fine sand grains. This SRDP is rare.

Complete plasmafication of a vesicular basalt or granite without collapse of the material results in this SRDP. Formation of this SRDP is confined to the saprolite zone of weathering (Plate 2, c).

In some cases, two distinct kinds of plasma — kaolinite and sesquioxides — in a plasmic to porphyric NRDP may show this. The sesquioxidic plasma attains a reticulate pattern (Plate 2, d). This feature is common in the mottled zone of deep weathering profiles in the humic tropics. In this case a reticulic SRDP is superimposed on a plasmic NRDP.

Acknowledgements

We wish to acknowledge the cooperation of Drs. G. Stoops and G. Paneque. We are also gratefull to Geological Institute, State University, Ghent, Belgium and Patronato «Alonso de Herrera» del

Consejo Superior de Investigaciones Científicas (C. S. I. C.), Spain, which enabled us to undertake this study.

Summary

Related distribution patterns are important micromorphological parameters for pedogenetical studies. Two groups are recognised, a normal (basic) related distribution pattern and a specific related distribution pattern, depending on the role of plasma. Within each group several types are evaluated. Relation between granulometric data and basic related distribution patterns is studied in triangle diagram. Each of the types is defined; some are illustrated and examples of the soils where they occur is given.

References

Bellinfante, N.; Paneque, G. y Baños, C. 1974. Estudio micromorfológico de un suelo sobre sedimentos del Trías del Viar. An. Edaf. Agrob., C. S. I. C., Madrid.

Brewer, R. 1964. Fabric and Mineral Analysis of Soils. John Wiley and Sons. London, 470.

Eswaran, H.; De Conick, F. and Conry, M. J. 1969. A comparative micromorphological studie of light and medium textured podzols. Proc. 3rd Int. Work, Meet. Soil Micromorphology. Wroclaw, Poland, 269-286.

Eswaran, H. 1972. Micromorphological indicators of pedogenesis in some soils derived from basalts from Nicaragua. Geoderma, 7, 15-31.

Kubiena, W. L. 1938. Micropedology. Collegiate Press. Ames. Iowa.

Stoops, G. and Jongerius, A. 1975. Proposals for a morphological classification of soils materials. I. A classification of the related distributions of fine and coarse particles. Geoderma, 13 (3), 189-199.

Recibido para publicación: 21-XII-74

Part IV

ORGANIC MATTER

Editors' Comments
on Papers 10 Through 13

10 JONGERIUS and SCHELLING
Micromorphology of Organic Matter Formed under the Influence of Soil Organisms, Especially Soil Fauna

11 BABEL
Humuschemische Untersuchung eines Buchen-Rohhumus mittels mikroskopischer Methoden

12 BARRATT
A Revised Classification and Nomenclature of Microscopic Soil Materials with Particular Reference to Organic Components

13 DE CONINCK et al.
Excerpt from *Origin and Micromorphological Nomenclature of Organic Matter in Sandy Spodosols*

Kubiëna (1953) was the first to provide an extensive description of organic matter in the soil. As his work of that period is characterized by a morphogenetic approach, we should not be surprised that he paid little attention to individual organic particles, whereas he emphasized the total fabric, or humus form. Accordingly, he focused mainly on the degree of granulation of the organic residues, their mixing and binding with inorganic soil components, the color of the fine mass, and the microfabric. Most micromorphologists followed the same approach for many years. This general trend is clearly illustrated in Paper 10, in which Jongerius and Schelling consider the terrestrial and semiterrestrial humus forms (raw humus, moder, mull, and anmoor) of the Netherlands. They interweave both descriptive and genetic considerations, relying on Kubiëna's (1953) concepts of humus forms but giving more attention to the origins of those forms. These authors were among the first to extend their knowledge of the micromorphological characteristics of natural humus layers to arable land, as an attempt to apply microscopic investigations to soil-management techniques.

This morphogenetic approach is contemporaneous to U. Babel's morphoanalytical approach (Paper 11). Instead of emphasizing the general aspect of the different humus layers, Babel stressed the morphology of the individual organic remnants. A thorough knowledge of plant anatomy is therefore required for this type of research. Babel was without doubt one of the most active researchers in this domain. Applying microchemical tests, staining techniques, and ultraviolet fluorescence on uncovered thin sections, he acquired a better insight in the morphological aspects of humus formation and the different stages of decomposition of plant remains (Babel, 1964a, 1964b, 1972, 1975). In his 1975 paper, he gave one of the best reviews of the general knowledge of the micromorphology of soil organic matter. The concepts, as discussed in Paper 11, are the basis of a new classification of organic materials, as proposed in Bullock et al. (in press). Babel distinguished between relatively coarse plant remains and finer material, which does not appear as recognizable units under the microscope. Two main types of plant residues may be distinguished: organ residues (e.g., leaf, root) consisting of more than one tissue type, and tissue residues (e.g., xyleme, sklerenchyme). The fine material is composed of microscopic amorphous materials and individual fragments of cell walls. Using different staining methods to indentify the different components (lignin, cellulose, cutin, etc.), Babel described the transformation of beech leaves through the L, F, and H layers.

As mentioned earlier (p. 41), organic components, whether fresh or decomposing, are practically excluded from Brewer's (1964) system. This near omission has been seen as a shortcoming by many researchers involved in the study of soils with well developed humus layers. Two attempts to supplement Brewer's system by adding a section of organic matter have been published: one by Barratt (1968 and 1969) and a second by Bal (1970 and 1973).

Barratt (Paper 12) tried to apply Brewer's analytical approach to the traditional humus forms; the system is, however, sufficiently open to be extended to other forms. It is based primarily on the distinction between plasmic and skeletal materials, which are further subdivided according to their chemical, mineralogical, and histological composition. This is clearly summarized in the table shown here. In the third column of the third row, however, the term *sphericity* should be replaced by *roundness* (B. Barratt, pers. communic. 1984). Unlike Brewer's system, Barratt gave a great deal of attention to microstructure, which she subdivided according to the shape of the aggregates, their size, and their arrangement.

Bal (1970; 1973) proposed a very detailed terminology to describe the composition and fabric of organic components in the soil. The introduction of a large number of newly coined terms was probably one of the reasons why most micromorphologists never accepted his system. Moreover, at the time of its publication, some of the concepts on which it was based (e.g., plasma-skeleton) were being seriously questioned (see also p. 99).

Paper 13, by F. De Coninck et al., deals specifically with the organic matter in spodosols. In contrast to the papers mentioned before, this paper gives special attention to the description of amorphous organic matter. The distinction made between polymorphic and monomorphic (amorphous) organic matter is fundamentally important, both for the description and genetic interpretation of spodosols. This new nomenclature has been readily accepted by many scientists.

Classification of Microscopic Soil Materials*

Criteria of Classification	Category	Name of Class (and Description)					
Composition Physical	Category I	Skeletal Materials (non-colloidal)			Plasmic Materials (colloidal)		
Chemical	Category II (main classes)	humiskel (organic)	lithiskel (inorganic)	humicol (organic)	argillicol (inorganic)	mullicol (intimate organic-inorganic association)	
Mineralogical or histological (plant tissues)	Category III	lignic (woody) parenchymal (soft)	calcitic (of calcite) silicic (of quartz) feldspathic (of feldspar) †	not named†	Montmorillonitic (of montmorillonite) illitic (of illite) allophanic (of allophane) †		
Microstructure Shape of particles	Category IV	very angular angular subangular subrounded rounded well-rounded	sphericity of single grain fabrics		massive blocky pelleted platy spongy expanded †		
Size of simple particles or aggregates (based on U.S.D.A. sand and silt sizes)	Category V		*Size classes* very coarse coarse medium fine very fine extremely fine	*Diameter (mm)* 1·0 –2·0 0·5 –1·0 0·25 –0·5 0·1 –0·25 0·05 –0·1 0·002–0·05			
Arrangement (or distribution) of particles ‡	Category VI		random clustered banded radial concentric †				
Orientation of particles ‡	Category VII		strongly oriented moderately oriented weakly oriented unoriented indeterminate				

* The classes of Category II are the main classes. The full name of the soil material is derived from the classification in the sequence Category VII to Category II, e.g. moderately oriented, radial, coarse, subrounded, calcitic, lithiskel.
† New classes may be added when required.
‡ The distribution and orientation patterns of particles and associated voids together comprise the "fabric" of Brewer (1964).

Source: B. C. Barratt, 1971, Micromorphology of some zonal soils of New Zealand, *New Zealand Jour. Sci.* 12(3):651-697.

REFERENCES

Babel, U., 1964a, Opake organische Gemengteile in Auflage-humusformen, Scheffer-Festschr. Inst. f. Bodenkunde, Göttingen, pp. 95-110.

Babel, U., 1964b, Chemische Reaktionen an Bodendünnschliffen, *Leitz-Mitt. Wiss. u. Techn.* **3**(1):12-14.

Babel, U., 1972, Fluoreszenzmikroskopie in der Humusmikromorphologie, in St. Kowalinski ed., *Soil Micromorphology*, Paristwowe Wydawnictus Naukowe, Warsaw, pp. 111-127.

Babel, U., 1975, Micromorphology of soil organic matter, in *Soil Components*, vol. 1, John Gieseking, ed., Spinger, Berlin, Heidelberg, New York, pp. 369-473.

Bal, L., 1970, Morphological investigation in two moder-humus profiles and the role of the soil fauna in their genesis, *Geoderma* **4**:5-36.

Bal, L., 1973, *Micromorphological analysis of soils. Lower levels in the organization of organic materials,* Soil Survey Paper No. 6, Netherlands Soil Survey Institute, Wageningen, 175p.

Barratt, B. C., 1968, Micromorphological observations on the effects of land use differences on some New Zealand soils, *New Zeland Jour. Agric. Res.* **11**(1):101-130.

Barratt, B. C., 1969, A revised classification and nomenclature of microscopic soil materials with particular reference to organic components, *Geoderma* **2**:257-271.

Barratt, B. C., 1971, Micromorphology of some zonal soils of New Zealand, *New Zealand Jour. Sci.* **12**(3):651-697.

Brewer, R., 1964, *Fabric and Mineral Analysis of Soils,* J. Wiley and Sons, London, New York and Sydney, 470p.

Bullock, P., N. Fedoroff, A. Jongerius, G. Stoops and T. Tursina, *Handbook for Soil Thin Section Description,* Waine Research Publications, Wolverhampton (in press).

Kubiëna, W. L., 1953, *The Soils of Europe,* Thomas Murby & Co., London, 314p.

MICROMORPHOLOGY OF ORGANIC MATTER FORMED UNDER THE INFLUENCE OF SOIL ORGANISMS, ESPECIALLY SOIL FAUNA

by

A. Jongerius[*], and J. Schelling[**]

The processing of dead organic matter on and in the soil can occur in various ways and is determined by such different factors as the parent material, the type of vegetation, fauna and flora of the profile, the hydrological conditions, etc. The humus profile resulting from this process, including its causal factors, is termed the humus form (10). Since living creatures in particular are such an important factor in the genesis of the humus form, this term, which was introduced by Müller, was defined by Kubiëna (7) as follows:

'the 'humus form' is a concept of a formation in nature: that is to say, a complex consisting of the biotic community plus its biotope.'

Over the years many research workers have helped to increase our knowledge of the humus forms and their more detailed classification (e.g. 1, 2, 3, 4, 6, 7). Some workers occasionally overshot the mark (e.g. 11) by failing to remember that the essential feature of the humus forms is their vertical succession of humus horizons, i.e. a particular natural combination of layers of organic matter which may have entirely different configurations. In micromorphological studies in particular, in which microscopic patterns are described and an attempt made to analyse them, there is a considerable risk of the real nature of the humus forms being forgotten. To prevent this danger we prefer the term humus types for designating micromorphologically distinct units. Thus a certain genetically determined vertical succession of humus types constitutes a humus form.

In our contribution we start from the micromorphological concepts of the humus forms as defined by Kubiëna (6). We will confine ourselves to the humus forms of the mineral soils which are of most importance in Holland and have hitherto received most attention from research workers, viz. the terrestric and semi-terrestric half bog soils. The humus forms of the organic soils are discussed in another contribution (v. Heuveln, Jongerius and Pons: Soil formation in organic soils). We by no means propose to offer a new classification of humus forms but only to make a contribution towards knowledge of these forms. But there is one respect in which we differ from the view generally held, as in our opinion the existing terms, albeit in a somewhat modified form, are just as suitable for the nomenclature of the humus forms of the cultivated soils. Hence we employ such terms as mull and moder in this connection as well.

Raw Humus

This form, which is characterized by an almost complete absence of conversions under animal influence and in which the microflora and microfauna are entirely responsible for the decomposition and reformation of organic materials, is subdivided by us into a number of variants in practically the same manner as that adopted by Kubiëna (6), Ehwald (1) and others. The criteria are the degree of development of the organic matter horizons, the degree of granulation of the plant remains and the percentages of

[*] Micromorphologist of the Netherlands Soil Survey Institute.
[**] Director Soil Survey Investigations of the Netherlands Soil Survey Institute.

opaque amorphous organic matter. We should also point out that such raw humus forms occur both below natural vegetation and grass. This is especially the case on organic soils where peat formation or irreversible drying out occurs (cf. van Heuveln, Jongerius and Pons, this congress). In this environment, which is practically inhabitable for animal organisms, there is also an accumulation of dead surface and underground remains of the grass vegetation. The final outcome is a felt-like mat of plant remains only attacked by fungi and bacteria and of which the structure is still quite visible. Variants of this kind are also found on heavy clay soils below grass.

MODER

By moder formation is meant the process in which the organic matter on and in the mineral soil is wholly or largely processed by small soil fauna (chiefly microarthropodes) without this being accompanied by an intense mixing and binding of the mineral particles.

The process invariably commences as follows (8, 9): dead surface and underground vegetable matter is attacked by what are known as the primary consumers. A large part of the soft vegetable parts prone to attack are soon consumed, a large amount of excretions being produced. But this effects little change in the composition of the organic matter since most of these fauna find it difficult to digest hemicellulose, cellulose, lignine, etc. Hence the effect of the primary consumers is mostly confined to mechanical granulation of the material and intense saturation with moisture. But these two changes in the organic matter greatly promote the activity of the micro-organisms, so that humification begins, viz. the reforming of the organic matter in which 'humic' substances are created and the plant structures disappear. The resultant substratum (aggregated and weakly humified organic matter) becomes in turn a good nutrient medium for what are known as secondary consumers, viz. mostly smaller fauna which in most cases granulate the organic matter still further in an extensive succession a process which is accompanied by constantly progressing humification by the micro-organisms. In this way a dark earthy 'humus' mass may be gradually formed, substantially consisting of dark excretions usually 25 to 60 μ in diameter, these excretions being an entirely or largely structureless mass. They may vary from dark brown and — in thin sections — semi-transparent to jet black and opaque; viewed from outside they are matt to highly glossy parallel thereto. These differences may be due to chemical differences in the starting material, viz. the nature of the parent material, the types of vegetation and any dressings of fertilizer. Good moder humus types are dark brown and matt, and poor ones black and highly glossy.

This process may, however, be continued in different ways, viz.:

1. *Mechanical illuviation, aggregation and possible deliquescence of the excretions* (5).

Below natural vegetation moder formation chiefly takes place at the surface, forming the Ao- horizon. From here in the pedologically higher profiles mechanical influences (chiefly precipitation) transport excretions into the mineral soil. As the depth increases there can be seen an increasingly concentrated conglomeration of excretions to spongy micro-aggregates generally some hundreds of μ in diameter. Under eutrophic conditions nothing further is seen to cocur, but under oligotrophic conditions there is a marked compression of the micro-aggregates, viz. the excretions lose their shape and run together into a more or less formless clot owing to the fact that much of the organic matter is converted into dispersed humus. The organic matter often becomes entirely mobile so that the micro-aggregates completely 'dissolve' and the dispersed humus is deposited around the sand grains at a certain level (humus B formation). Similar phenomena also take place in organic soils in which the moder formation is otherwise not confined to a thin surface layer but may extend over a depth of tens of centimetres (cf.: van Heuveln, Jongerius and Pons).

But these moder types of humus are found both below natural vegetation and sandy grassland and arable land. The following cases occur.

Black, ancient sandy arable lands have been made by man as a result of centuries of fertilization with farmyard manure which in this instance consisted of a fermented mixture of heath sods and animal manure. These humous layers may be over 1 metre thick. Below the tilth the organic matter in the covers entirely consists of loose excretions or, more usually, micro-aggregates which have hardly if at all deliquesced and are generally jet black and extremely glossy. In this case, therefore, we are dealing with very distinct moder types of humus, and we believe that these forms, which owing to arable tillage obviously show no enrichment of organic matter on the surface, should also be unhesitatingly included in the moder group. Moreover it is noticeable that in contra-distinction to its behaviour below natural vegetation the glossy black organic matter does not substantially turn into ooze.

Moder types of humus are also frequently encountered in recent sandy arable lands. Below the tilth they are the original types in the form in which they developed below natural vegetation. In the tilth the moder patterns described are only found provided there is a reasonably good level of fertilization, good tillage, etc. and then probably only at certain periods of the year, or in other words when the environmental conditions are such that a reasonably good biotic community can be maintained in the soil. The excretions of this arable moder are generally dark brown and rich in vegetable fragments. This is because they usually have a much shorter life than those existing below natural vegetation (a lower biological level, more extreme climatic influences, which means that both mechanical granulation and humification do not proceed so far). In addition to the recent moder formations the remains of the original moder types of humus are often found in the tilths. See also under 2.

The same conversion processes are generally found in the sandy grasslands as well as in the soils under natural vegetation. Owing to fertilization good moder formation is nearly always found in these grasslands, viz. formation of porous zoögenic micro-aggregates. Often there is a very thin Ao- horizon. Such grasslands may occasionally contain fairly large amounts of earthworms, even of the typical mull-forming species such as A.longa and L. terrestris. In this case the Ao may be entirely absent, although not necessarily so. Thus some of these cases correspond to the sandy mull of Hoover and Lunt (4). Despite this we also include this humus form among the moders; the arguments in favour of this being as follows. The bond between the organic matter and the sand granules is very weak. Consequently the soil micro-aggregates only have slight water stability, so that micro-erosion may occur and hence scouring. Extremely humous sandy soils are known which have a very dense population of deep-burrowing worms and in which even horizons of the A_2 types are developed! These phenomena cannot be seen to occur until there is at least some few percent of lutum in the soil and it is possible to speak of sandy mull.

2. *The disintegration of the excretions.*

The phenomena dealt with are typical of the higher sandy soils. On moist sandy soils a somewhat different course of events may be observed. Here as well large amounts of excretions are generally produced, but these tend to disintegrate rather than to form aggregates, although this process may also eventually lead to the formation of fairly large amounts of dispersed humus. This phenomenon (the formation of micro-aggregates of organic matter with a typical debris structure, i.e. a nondescript blending into a shapeless aggregate of plant remains, often only slightly humified, and very fine earthy matter), is no doubt partly due to the nature and activity of the microflora in these damp soils. For a great part the disintegration is caused purely mechanically, influenced by hydrological environmental conditions. These types of humus may occur both below natural vegetation and in moist grasslands and the tilths of most arable land. In the latter case humification is usually far less extensive.

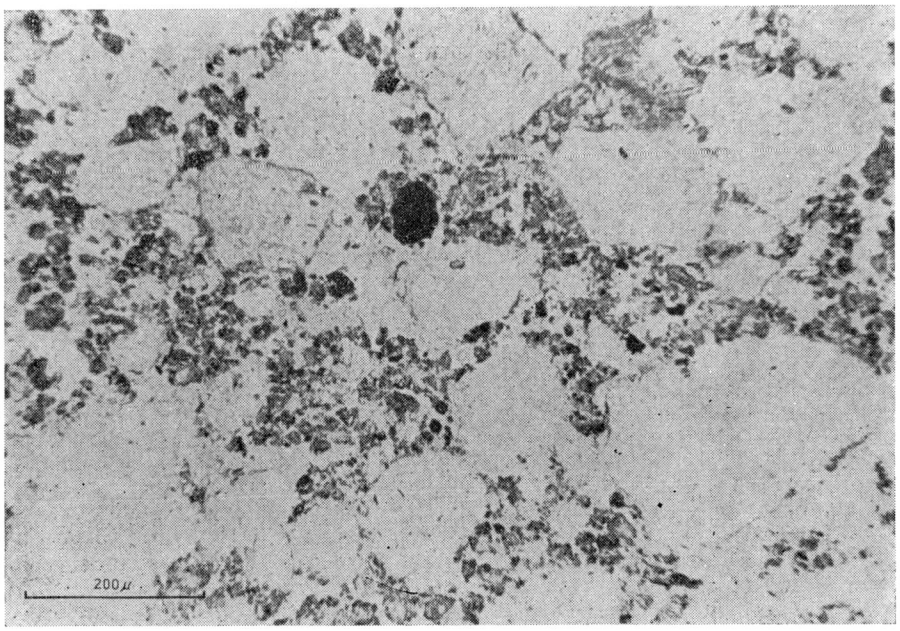

Photo 1. Vertical thin section of a favourable moder debris structure in the Ap-horizon of a brown podzolic soil.

The debris structure are also found in the tilths of the higher sandy soils (photo 1), but in this case they are the result of extreme environmental conditions and a low biological level. When there is a very low level of fertilization, so that the C/N-quotient of the organic matter produced is high, or when, as in the oligotrophic high moor soils, the parent material is a very poor organic material, one often only finds in the tilth small, black, opaque, woody fragments and possibly coatings of dark, opaque, amorphous humus on the sand grains. Most of the organic matter is illuviated in the form of dispersed humus and occasionally forms humus-B horizons.

Moder formation is also found in clay soils, viz. the conversion to moder of withered roots, dung, leaf mould, etc., but this moder is soon worked up further by the mull formers. For the moder types of the organic soils cf. van Heuveln, Jongerius and Pons. The conversion to moder on loess and lime decomposition soils has not yet been adequately studied by us.

Mull Formation

Mull in the micromorphological sense (6) means the zoögenically formed, mechanically inseparable mixtures of organic matter and mineral particles, particularly a certain percentage of lutum (the exact amount is not known). It may contain a certain percentage of sand, but in this case, unlike the sand in moder, it is very highly embedded and cemented with the clay-humus complex. Typical mull is dark in colour.

It is generally assumed that various oligochaetae are primarily responsible for mull formation, and termites in the tropics. In western Europe mull formation depends on various lumbricidae and enchytrae. They are capable of intensively blending clay and organic matter (so that an Ao-horizon is not found on mull profiles), the organic matter undergoing various changes in the animal intestine. Thus the mull-forming worms have various ferments in the intestinal system, these including cellulase and chitinase which, together with the microbial changes which subsequently also occur in the excretions, result in a very stable clay-humus complex. Typical mull has a very porous structure.

This concept, which is particularly related to soils under natural vegetation, and was moreover only valid when the environmental conditions were very favourable, is in our opinion too limited, especially when soils under conditions of cultivation are also taken into consideration. Our reasons are as follows:

a. Preliminary stages are known which are characterized by large intact plant remains in a mass of clay with only a small percentage of finely divided and bound organic matter. They especially occur on soils in which the C/N-quotient of the organic matter is very high. A good example is constituted by recently highly dehydrated peat soils with a clay cover below arable cultivation. Polygon formation occurs in the peat as a result of the dehydration. Clay from the overlying cover falls into the cracks caused by shrinkage, and the earthworms in the cover enter the cracks where they blend the clay and the organic matter from the peat. In this way a mull is very gradually formed, but it will always continue to be of the heterogeneous type referred to under b.

b. But in cases in which we are undoubtedly dealing with 'ripe' mull (clay or peaty clay soils with intense worm activity and a good level of fertility), micromorphological study shows that frequently the humus form does not answer the conventional definition. This is because we often find in the clay-humus complex a further smaller or greater percentage of woody plant remains, sometimes even fragments of amorphous humus (for example, moulded peaty clay topsoils (photo 2) below grass, certain calcium-rich alluvial levee soils below grass, various wetter mull formations). We therefore think several heterogeneous mull forms are to be distinguished in addition to the conventional mull which we would term homogeneous mull.

c. Finally there are the stages of decay of the mull. On true mull soils we can generally observe the individual excretions coalescing to form porous aggregates (photo 2). In this process the individual excretion forms are usually entirely lost. They are sometimes even difficult to trace in thin

Micromorphology of Organic Matter

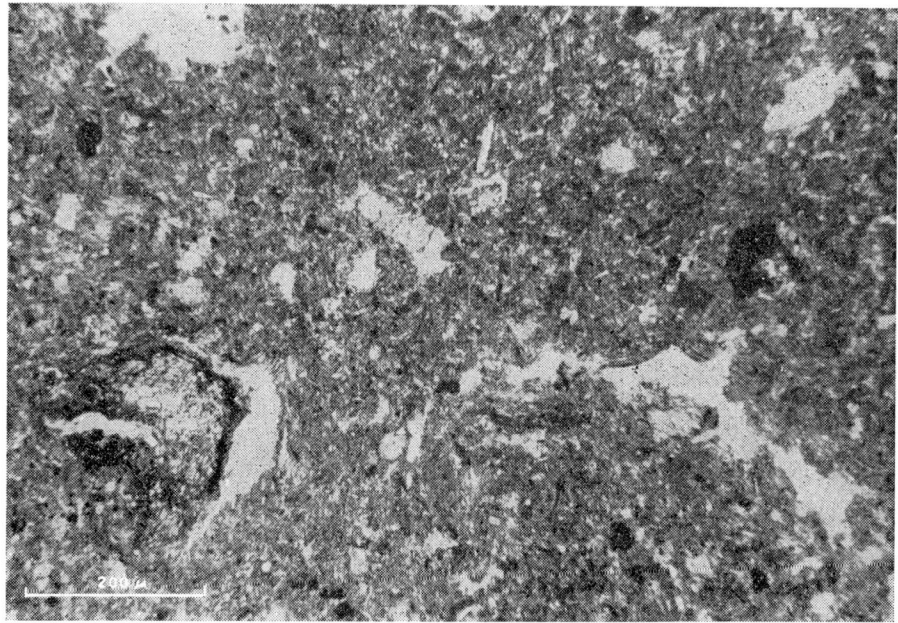

Photo 2. Thin section of a moderately heterogeneous mull formation in the topsoil of a clayey organic soil under grassland. The individual worm excrements are coalesced to a porous mass.

sections. This stage is characteristic of mull in its optimum development, whether it is homogeneous or heterogeneous. This pattern is only found below woodland, grassland or in the best garden soils. Should arable farming be practised on such soils the mull will decay. Owing to the marked lowering of the biological level that usually accompanies arable farming there is no longer that regular passage through the intestines which is required for maintaining the mull state. Much of the organic matter is lost owing to mineralisation processes. We can observe a gradual structural decay, the porosity is reduced and the bonds between the mineral particles and the organic matter are broken down, the colours become lighter. These stages of decay, for which we have not yet been able to compile any satisfactory classification, we combine under the term post-mull.

Current research has shown that julidae (e.g. Cylindro Nitidus and C. Silvarum) are capable of forming very intense clay-humus mixtures both below grass and on arable land. Since they are also able to burrow for at least some tens of centimetres we are justified in speaking of mull formation in this instance.

Semi-terrestric Zoögenic Humus (half bog soil)

This is a mineral rich humus form found in soils which are continuously or periodically saturated with water. The form shows no peat formation, viz. no accumulation of distinct intact plant remains, but chiefly consists of a black to dark grey blend of mineral particles and mechanically degraded, finely divided well-humified coprogenous particles produced by aqueous fauna and which are continually enriched (6).

When this humus is studied in thin sections the most surprising feature about this material is that the contours are often very hazy, whereas in the organic matter numerous very gradual transitions occur from very dark, opaque amorphous material to very light brown, transparent material. The bond with the mineral soil particles is fairly weak.

This humus form was found to occur fairly frequently in Holland, for example in many AC profiles and in various profiles with an impermeable subsoil (e.g. till), while muck should also be regarded as belonging to this category. Many of the soils referred to originally were more wet then they are nowadays. As a result a different biotic community has arisen in soils which are generally used as grassland, and in particular many worms and enchytrae may occur. It is remarkable that this has not been accompanied by any essential micromorphological change in the humus form. Dehydration has, however, led to illuviation of colloidal organic matter which can then be identified in the subsoil as coatings around sand grains, or as microscopically fine fibres. Under permeating light this accumulated humus is generally honey-brown and often somewhat cloudy, viz. flocculated. For the behaviour of muck below arable cultivation, cf. van Heuveln, Jongerius and Pons.

Literature

1. EHWALD, E., 1958. Die Einteilung der Waldhumusformen. Forstliche Standortsaufnahme. Begriffe und Fachausdrücke, 23—30.
2. HARTMAN, F., 1951, Der Waldboden. Humus-, Boden- und Wurzeltypen als Standortsanzeiger. Wien.
3. HEIBERG, S. O. and R. F. CHANDLER, 1941, A revised nomenclature of forest humus layers. Soil Science 52: 87—99.
4. HOOVER, M. D. and H. A. LUNT, 1952, A key for the classification of forest humus types. Proc. Soil Sci. of Am. 16: 368—370.
5. JONGERIUS, A., 1956, Etude micromorphologique des sols sableux secs des bois et bruyères aux Pays Bas. VIe Congrès Intern. de la Science du Sol. Paris. Rapp. E, 353—357.
6. KUBIËNA, W. L., 1953, Bestimmungsbuch und Systematik der Böden Europas. Stuttgart.
7. KUBIËNA, W. L., 1955. Animal activity in soils as a dicision factor in establishment of humus forms. Soil Zoology. Proc. of the Univ. of Nottingham Second Easter School in Agr. Sci., 73—82.
8. KÜHNELT, W., 1957. Die Tierwelt der Landböden in ökologischer Betrachtung. Verh. der Deutschen Zoologische Gesellsch. in Graz, 39—103.
9. KÜHNELT, W., 1958, Zoogene Krümelbildung in ungestörten Böden. Tagungsber. 13 der Deutsche Akademie der Landwsch. zu Berlin, 193—199.
10. MÜLLER, P. E., 1887, Studien über die naturlichen Humusformen. Berlin.
11. RAMANN, E., 1911., Bodenkunde, Berlin.

Summary

Humus forms are genetically determined by the following factors: vegetal and faunal life in the soil profile, nature of the parent material, organic matter and environmental conditions.

Important humus forms are raw humus (Duff, Rohhumus), moder and mull. Moder is characterized by surface attack of organic matter by non- or little burrowing soil animals. Down in the profile organic matter originates from illuviation. Moder has various appearances. Mull is understood to be zoogenetically formed, mechanically unseparable admixtures of clay and organic matter; larger mineral components, if any, are firmly cemented in this clay-humus complex. This conception, however, is too restricted, as authors discuss.

Anmoor (half-bog soil) is another, mineral-rich, humus form found in soils, which are continuously or periodically saturated with water. It consists of disintegrated, strongly humificated coprogeneous parts produced by aquate animals.

Other humus forms occur in peat soils.

ZUSAMMENFASSUNG

Humus erhält genetisch seine Formen im Boden durch die Wirkung folgender Faktoren: das vegetabile und faunistische Leben im Bodenprofil, die Art des Muttermaterials, der vorhandenen organischen Substanz und der Verhältnisse in der Umgebung.

Wichtige Humusformen sind Rohhumus, Moder und Mull. Moder wird gekennzeichnet durch oberflächlichen Angriff der organischen Substanzen durch nicht oder nur wenig wühlende Bodentiere. Organische Substanz tiefer im Profil ist von Einspülung herkünftig. Moder hat verschiedene Erscheinungsformen. Unter Mull versteht man zoögenetisch gebildete, mechanisch untrennbare Mischungen von Ton und organischer Substanz; wenn grössere Mineralkomponenten dabei bezogen, zo sind sie in dem Ton-Humus-Komplex fest zementiert. Diese Auffassung ist jedoch, wie die Verfasser erörtern, zu kurz gehalten. Anmoor ist noch eine andere Humusform, mineralreich, zu finden in Böden, welche durchlaufend oder periodisch mit Wasser übersättigt sind. Sie besteht aus desintegrierten, stark humifizierten, koprogenen Resten, durch in Wasser lebende Tiere ausgeschieden.

In Torfböden kommen noch andere Humusformen vor.

DISCUSSION

W. KUBIENA: Remarks: The transformation product of peat we have seen in the last thin section shows high desintegration, disappearance of recognizable plant residues and a formation of a more or less homogeneous organic decomposition product which is highly peptized and shrinks highly — as you have noticed by the numerous cracks — when drying. The colour of this substance or rather substances is reddish-brown, reddish to yellow. These unknown substances (which have been called humo-lignins occasionally) are not humic substances, but products which are found under unfavourable conditions for humus formation and from unfavourable organic parent material. They never occur in mull. Therefore I would suggest not to include this humus form in the mull group (I called it 'Pechtorf' in my book). It is a dystrophic formation. The environment is not even fit for the life of soil fungi which we have seen present only in the form of sclerotia in the last preparation.

C. O. TAMM: May I ask Dr. Schelling whether he has considered the possibilities of using entirely different terms than those once introduced by P. E. Müller as characteristics for soil horizons, not for substances. There is certainly a need for good names of the humus types, so well described by Dr. Schelling, but I find it highly desirable to avoid a confusion with the old, but still used terminology.

J. SCHELLING: I certainly agree that better names have to be found. We wanted first to find more followers for these ideas. Afterwards we can try to find agreement on the names.

G. REUTER: Verzichten Sie auf die von Kubiena als besonders wichtig herausgestellte Humusform 'mullartiger Moder'?

J. SCHELLING: Diese kurze MiHeilung ist beschränkt auf die meist typischen Humusformen. Übergangsformen wie ,,mullartiger Moder" sind deshalb nicht erwähnt.

Humuschemische Untersuchung eines Buchen-Rohhumus mittels mikroskopischer Methoden

von Ulrich Babel

(Aus der Bundesforschungsanstalt für Forst- u. Holzwirtschaft Reinbek — Abt. Bodenkunde [Leiter: Prof. Dr. W. Kubiena])

Einleitung

Ausgehend von der Mikromorphologie der Humusformen läßt sich eine Arbeitsweise entwickeln, die zur Untersuchung chemischer Fragen der Humusbildung dient. Sie benutzt in erster Linie morphologische Methoden. Sie geht vom makromorphologischen Aufbau des Humusprofils aus und wechselt dann zu mikroskopischen Untersuchungen über. Im Grunde handelt es sich um die Ausnutzung botanisch-histologischer Kenntnisse für die Humusforschung.

Es gibt bisher merkwürdigerweise noch fast keine Arbeiten, die systematisch in diesem Sinne vorgehen. Eigentlich sind nur zwei Autoren zu nennen; beide sind bezeichnenderweise durch das forstliche Problem der Rohhumusdecken zu dieser Arbeitsweise geführt worden, denn dort liegt die prinzipielle Untersuchung der im Profil von oben nach unten allmählich immer stärker zersetzten Pflanzenreste besonders nahe. Großkopf (4) hat schon Ende der zwanziger Jahre morphologische Untersuchungen über die Zersetzung von Fichtennadeln gemacht, als er die Frage der Rohhumusbildung in Fichtenreinbeständen bearbeitete. Handley (5) hat die prinzipiellen Unterschiede der Mullbildung und der Auflagehumusbildung im Walde im wesentlichen mit Hilfe mikroskopischer Methoden zu erkennen versucht.

Sonst wurden mikroskopische Humusuntersuchungen mehr zur Klärung humusbiologischer als humuschemischer Fragen benutzt. Kubiena (7) hat auf diese Weise die große Rolle der Bodenkleintiere in den meisten Humusformen nachgewiesen. Kononowa (6) berichtet über mikroskopische Beobachtungen von mikrobiologischen Vorgängen bei der Humusbildung.

Einige Autoren haben morphologische Untersuchungen als Ergänzung zu anders angelegten Arbeiten ausgeführt; so hat Wittich (11) zu seinen Studien über den Verlauf der Streuzersetzung morphologische Beobachtungen gemacht und Kullmann (8) hat zu Verrottungsversuchen Dünnschnittuntersuchungen herangezogen.

Herrn Prof. Dr. W. K u b i e n a, der mir die Anregung zu dieser Arbeit gegeben hat, bin ich zu großem Dank verpflichtet. Ich danke auch der D e u t s c h e n F o r s c h u n g s g e m e i n s c h a f t, die die Arbeit durch Sachbeihilfen unterstützt hat.

Objekt, Fragestellung und Grundlegung der Methode

Im Folgenden sollen am Beispiel eines Buchen-Rohhumus die Möglichkeiten der mikroskopischen Arbeitsweise gezeigt werden. Dabei wird nur e i n e Frage behandelt: Aus welchen Gruppen von Stoffen ist die H u m u s s t o f f s c h i c h t in der Hauptsache aufgebaut? Aus der Vielfalt der mikroskopischen Bilder und der Unzahl der Einzelbeobachtungen, die an den Bauteilen des Auflagehumus gemacht werden können, wird für die vorliegende Besprechung nur das herausgegriffen, was zur Klärung dieser Frage dient. Auf Einzelheiten der Methoden und der morphologischen Untersuchungsergebnisse wird nicht eingegangen. Veröffentlichungen dazu sollen folgen.

Die von oben nach unten einander folgenden Schichten des Rohhumusprofils (L = Streu; F = Vermoderungsschicht; H = Humusstoffschicht) sind die zeitlich einander folgenden Umwandlungsstadien des Bestandesabfalls. Das heißt, daß zunehmende Tiefe im Profil zunehmendem Alter entspricht. (Tiefen- und zugehörige Altersangaben finden sich in Tab. 1.)

Statt Modellversuche — etwa zur Verrottung von Blättern — im Labor anzusetzen, kann also der Verlauf der Streuzersetzung und Humusbildung aus dem Studium der einzelnen Rohhumusschichten und deren Einzelbestandteile erarbeitet werden — angefangen von den kaum veränderten Pflanzenresten der Streu bis hin zu den schon Jahrzehnte im Boden liegenden stark umgewandelten Materialien der Humusstoffschicht. Auf diese Weise arbeitet man an einem Material, das unter natürlichen Bedingungen zersetzt worden ist. Der Nachteil einer solchen Arbeitsweise ist zwar die außerordentliche Kompliziertheit eines Systems, wie es eine natürliche Rohhumusauflage im Vergleich zu einem Verrottungsmodell ist. Aber diese Kompliziertheit wird bei Benutzung des Mikroskopes im Auge behalten. Man sieht die Wirkung neuer in der Streu nicht vorhandener Kräfte — man sieht, wie die Blattreste von Tieren zerkleinert werden, wie Wurzeln zwischen die Blattpakete hineinwachsen, wie in großer Menge Pilzmyzel gebildet wird und so fort — und kann das alles bei Versuchsansätzen und Deutungen berücksichtigen.

Das Mikroskop löst außerdem die makroskopisch fast homogene Masse der Humusstoffschicht (H) in morphologisch deutlich unterschiedene Einzelgemengteile auf (Abb. 1, s. Farbtafel). Was diese morphologischen Unterschiede chemisch bedeuten, darüber erhält man in erster Näherung eine Aussage, wenn man feststellen kann, aus welchen Gewebe- oder Zellteilen die Gemengteile der Humusstoffschicht entstanden sind. Denn der Chemismus der Gewebe und Zellen ist in großen Zügen von der botanischen Histologie her bekannt oder er läßt sich doch mit Hilfe bekannter mikrochemischer Reaktionen feststellen.

Um die Herkunft der Gemengteile der Humusstoffschicht zu klären, müssen die frischen Ausgangsmaterialien und die früheren Zersetzungsstadien, wie sie in der Streu (L) und Vermoderungsschicht (F) vorliegen, in ihrem Gewebeaufbau und in den Veränderungen der Gewebe bei der Zersetzung systematisch untersucht werden.

Für diese Untersuchungen wurden zur Orientierung Boden-Dünnschliffe (Vestopal-Technik) verwendet; diese geben einen Überblick über a l l e Bestandteile der Rohhumusschichten und über deren natürliche räumliche Beziehungen zueinander, aus denen oft auf die genetischen Beziehungen geschlossen werden. kann. Die morphologischen Detailuntersuchungen und die Mikroreaktionen wurden an Mikrotomschnitten von isolierten Pflanzenresten oder an Streupräparaten der Feinsubstanz ausgeführt. [Chemische Reaktionen lassen sich in beschränktem Umfange auch an Vestopalschliffen ausführen; s. Babel (2).]

A n g a b e n z u m S t a n d o r t: Hausbruch bei Hamburg (Harburger Berge); Buchenalthoz mit einzelnen Eichen; Strauch-, Kraut- und Moosschicht fast fehlend; 10—15 cm mächtige Rohhumusdecke (Mächtigkeit etwas durch Einwehung von Blättern mitbedingt) auf Podsol (mittelstarke Ausbildung) aus schwachlehmigem Sand der Saale-Grundmoräne. Nähere Angaben, besonders zur Vegetation, siehe bei Meyer (9); dort auch einige chemische Analysen und vor allem mikrobiologische Untersuchungen.

Tabelle 1

Abnahme der Pflanzenreste und Zunahme der Feinsubstanz mit zunehmendem Alter der Rohhumusschichten

Horizont		Tiefe	Alter Jahre	Anteil (in Vol.% der organischen Substanz) Pflanzenreste	Feinsubstanz
Streu	L 1	0 - 2 cm	1	99 %	1 %
	L 2	2 - 3 cm	2 (bis 3)	95 %	5 %
Vermoderungs-	F 1	3 - 4 cm	3 bis 7	85 %	15 %
schicht	F 2	4 - 6 cm		55 %	45 %
	F/H	6 - 7,5 cm		35 %	65 %
Humusstoff-	H 1	7,5 - 10 cm	40*	25 %	75 %
schicht	H 2	10 - 13 cm	76*	20 %	80 %

* Mindestwerte von Meyer (10) für die Zeit bis zur vollständigen Veratmung der betreffenden Schichten, hier unter der Annahme eingesetzt, daß ein dynamisches Gleichgewicht herrscht. Von Zachariae (12) wurden für die untere Grenze des F/H durch eine direkte Methode 40 - 50 Jahre ermittelt.

Spezielle Untersuchungen und Diskussion

Bevor die Untersuchungen an den am stärksten zersetzten und daher humuschemisch am meisten interessierenden Teilen besprochen werden, muß nach dem oben Gesagten wenigstens ein kurzer Überblick über die morphologische Zusammensetzung der Schichten des Rohhumusprofils gegeben werden.

Auf den ersten Blick können Pflanzenreste (mikroskopisch gut erkennbare Organ- und Gewebereste) und organische Feinsubstanz (mikroskopisch amorphes Material und Zellwandbruchstücke außerhalb eines Geweberverbandes) unterschieden werden (Abb. 7). Das sind die beiden für diese Betrachtung wichtigsten genetischen Hauptgruppen.

Die Feinsubstanz ist (von wenigen Ausnahmen abgesehen) durch Zerkleinerung der Pflanzenreste entstanden, muß also aus den mehr oder weniger umgewandelten Bruchstücken der Pflanzenreste bestehen. Freilich nicht aus allen, denn es werden Teile aus den Pflanzenresten in Lösung fortgeführt und ein sehr hoher Anteil wird von den Mikroorganismen veratmet — nach Meyer (10) in der Streu und der Vermoderungsschicht zusammen ca. 60—80% der gesamten organischen Substanz.

Der Anteil der Pflanzenreste nimmt mit zunehmender Tiefe im Profil zugunsten der Feinsubstanz ab. Die Tabelle 1 gibt dazu Werte, die mikroskopisch mit dem Punktzähler (Firma Zeiss) und teilweise durch Schlämmsiebung ermittelt wurden.

Organ- und Gewebereste

In der Streu (L) sind an Organ- und Geweberesten fast nur Blätter vorhanden. Deren Zellwände sind meist unverändert; aus den Zellinhaltsstoffen der Mesophyll- und der Epidermiszellen haben sich Bräunungsstoffe gebildet (Abb. 2).

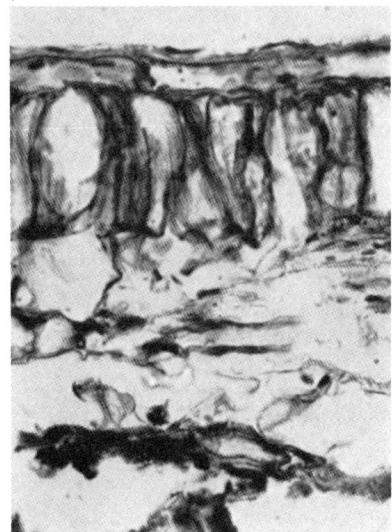

Abb. 2
Buchenblatt aus der Vermoderungsschicht (F 1)
(Querschnitt; nat. Bildhöhe 0,16 mm)
Die Zellulosewände sind vor allem im Palisadenparenchym noch unverändert, aber durch dünne Lagen gelber bis brauner Umwandlungsprodukte der Zellinhaltsstoffe völlig verkrustet.
(Aufnahme mit Leitz-Spektralfilter 50 (blaugrün):
die gelben und braunen Teile erscheinen dunkel)

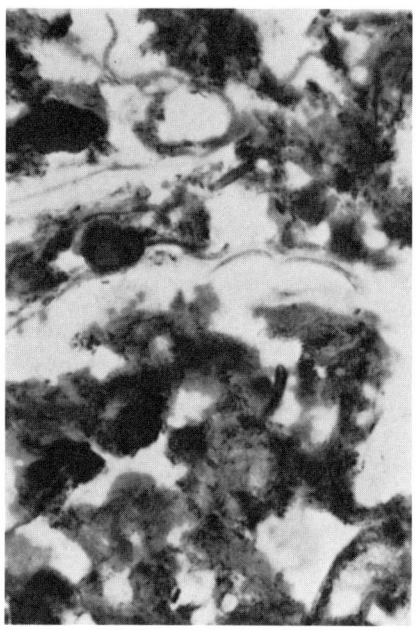

Abb. 1 zu Babel

Humusstoffschicht H_1 des Buchenrohhumus (Dünnschliff, nat. Bildhöhe 0,2 mm).

Starke Differenzierung der Gemengteile in Form und Farbe; direkt ansprechbar ist nur eine Außenwand der Blattepidermis, die horizontal etwas über der Mitte durch das Bild geht, und direkt daneben in der linken Bildhälfte eine Pilzspore sowie zwei Hyphenstücke (Mitte oben und etwa Mitte).

[*Editors' Note:* These photographs are reproduced in color in the original.]

Aufnahmematerial: Agfacolor-Umkehrfilm für Tageslicht. 18° DIN, Kleinbildformat, mit Lifa-Farbtemperaturfiltern blau CB 12 und CB 1,5.

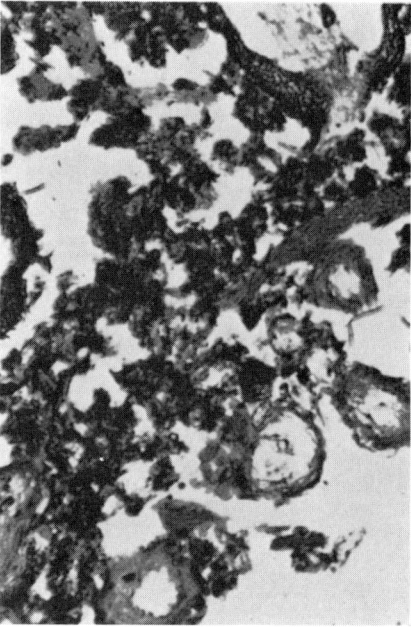

Abb. 3 zu Babel

Wurzelreste in der Humusstoffschicht H_2 (Dünnschliff, nat. Bildhöhe 0,6 mm).

Die Zentralzylinder der Wurzeln sind weitgehend verschwunden, während die Rinden noch erhalten sind und zwar besonders die durch Phlobaphene rotbraun gefärbten äußeren Teile, die sich stellenweise ablösen und in die Feinsubstanz eingehen. Weitere Erläuterungen siehe Abb. 6.

In der Vermoderungsschicht (F) stammen etwa 90% der erkennbaren Gewebereste von Blättern. Die Zellwände in den Blattadern sind meist stark zersetzt und in braune Stoffe umgewandelt (Abb. 4). Die Wände der Mesophyll- und Epidermiszellen sind nur wenig zersetzt; die Bräunungsstoffe aus den Zellinhaltsstoffen sind meist noch erhalten. Etwa 10% der Gewebereste in der F-Schicht sind schwach oder mittelmäßig zersetzte Wurzelreste.

Innerhalb der Humusstoffschicht (H 1 bis H 2) nehmen die Blattreste von 30% der Gewebereste auf fast 0% ab; es sind meist stark zersetzte Blattadern. Korkartige Reste (von Ast- und vor allem Wurzel-Borken) nehmen auf etwa die Hälfte zu; sie sind ziemlich wenig zersetzt. Etwa die Hälfte (also nach Tabelle 1 etwa 15% der gesamten organischen Substanz) sind Wurzelreste. Diese sind mittelmäßig bis stark zersetzt; oft sind die gerbstoffhaltigen Rindengewebe noch erhalten, während der zentrale Holzteil fehlt (Abb. 3, s. Farbtafel, und Abb. 6).

Feinsubstanz

Die Feinsubstanz besteht — was aber erst bei eingehenderen Untersuchungen klar erkennbar wird — aus einer großen Menge von Wandfetzen, welche in der Vermoderungsschicht mit hellbraunen und gelbrötlichen, in der Humusstoffschicht mit matt- und dunkelbraunen, morphologisch amorphen Stoffen verkrustet und vermengt sind.

Farblose Wandreste in der Feinsubstanz

Zur genaueren Untersuchung der Feinsubstanz wurden Streupräparate von durch Schlämmsiebung gewonnenen Fraktionen unter 200 μ hergestellt; die meisten Teile waren sehr viel kleiner als 200 μ, meist zwischen 20 und 60 μ. Mit Natriumhypochlorit ließen sich die gefärbten Stoffe entfernen; die farblosen Wandreste blieben zurück und waren jetzt für die morphologische Bearbeitung und auch für Mikroreaktionen zugänglich.

Zur Orientierung, welchen Anteil diese offenbar chemisch noch wenig veränderten Wandsubstanzen an der gesamten organischen Substanz der Streu (L), Vermoderungsschicht (F) und Humusstoffschicht (H) haben, wurde die Behandlung mit Natriumhypochlorit auch quantitativ ausgeführt. In L 1 waren 43%, in F 2 33% und in H 2 23% unlöslich. Der Anteil von chemisch höchstens schwach veränderten Wandresten nimmt mit zunehmendem Alter also erheblich ab; jedoch ist auch noch in der unteren H-Schicht dieser Anteil mit etwa 1/4 der organischen Feinsubstanz mengenmäßig von Bedeutung.

An den Wandresten wurden Mikroreaktionen auf Cutin (Sudan III), Lignin (Phloroglucin) und Zellulose (Chlorzinkjod) ausgeführt und zusätzlich ihre Primärfluoreszenz beobachtet. Zur Diskussion der Ergebnisse dieser Reaktionen war ihre Prüfung an frischen und an wenig zersetzten Blättern nötig, worauf hier aber nicht eingegangen werden soll. Die Menge der auf die einzelnen Reagenzien ansprechenden Teile wurde zu schätzen versucht (Tab. 2).

Der Anteil an cutin-, lignin- und zellulosehaltigen Wandresten wird aus einer zusammenfassenden Interpretation aller Reaktionen

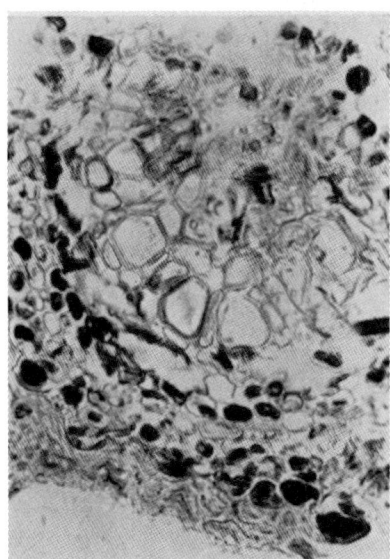

Abb. 4
Rest einer Buchenblattader aus der Humusstoffschicht (H 1)
(Querschnitt; nat. Bildhöhe 0,2 mm)

Besonders in den lignifizierten Geweben (Xylem und Sklerenchym) ist eine Umwandlung der Zellwände in gelbe und braune Substanzen deutlich erkennbar. Histologische Erläuterungen vgl. Abbildung 5. Näheres zur Blattaderzersetzung vgl. Babel in 13.

(Aufnahme mit Leitz-Spektralfilter 50 (blaugrün): die gelben und braunen Teile erscheinen dunkel)

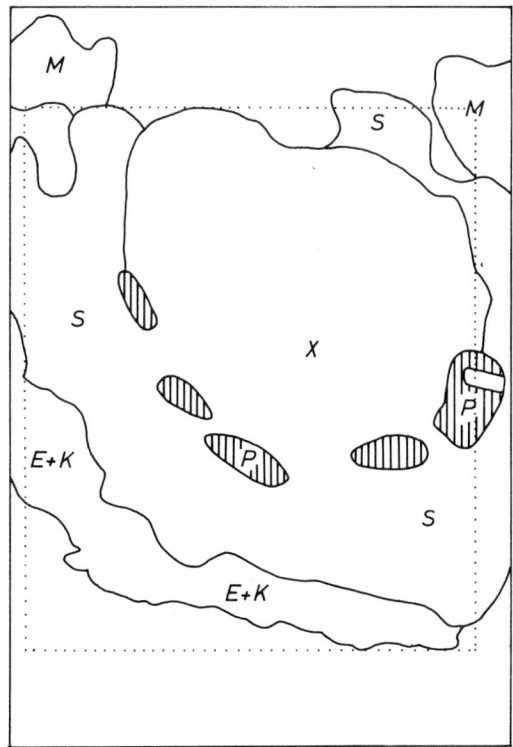

Abb. 5
Histologische Erläuterungen der Reste einer Buchenblattader aus der Humusstoffschicht H 1 (zu Abbildung 4)

E + K:	Reste von Epidermis und Kollenchym
S:	Reste im wesentlichen von Sklerenchym
X:	Reste von Xylem
M:	Reste im wesentlichen von Mesophyll
P:	ursprüngliche Lage des ganz verschwundenen Phloems (dort rechts eine Pilzhyphe)
...:	Ausschnitt in Abbildung 4

erhalten, da die Reste jeweils nicht nur auf das genannte Hauptreagens ansprechen.

Zum Beispiel werden Cuticulen durch Sudan III rosa gefärbt, durch Chlorzinkjod gelb — wie aber auch ein Teil der verholzten Zellwände; sie fluoreszieren meist gelb, z. T. aber so schwach, daß ihre Fluoreszenz bei Betrachtung der Teile in der Aufsicht nicht erkennbar ist.

Weil die verschiedenen Reaktionen mehr oder weniger deutlich sind, und weil die Mengenschätzungen an den verschieden großen und verschieden dicken Teilen ohnehin schwer sind, werden in Tab. 2 die Mengen nur nach einer Schätzskala von 1—5 (1: wenig, 5: sehr viel) angegeben.

Die Mengenverhältnisse der einzelnen Wandreste zueinander entsprechen in der Feinsubstanz der Vermoderungsschicht (F) am ehesten den Verhältnissen im Blatt. Das rührt daher, daß in der F-Schicht so gut wie alle Blätter zerkleinert werden — und zwar ebenso Epidermen (cutin-reich) wie Mesophyll (zellulose-reich) wie Blattadern (lignin-reich). Nur die cutinhaltigen Wände sind in der Vermoderungsschicht schon etwas angereichert.

In der Humusstoffschicht (H) ist die Anreicherung der cutinhaltigen Wände erheblich verstärkt. Sie sind dort fast so häufig wie Zellulosewandreste. Cutin zeigt sich also als relativ zersetzungsresistent. Dagegen ist Lignin in der H-Schicht nur noch in geringer Menge zu finden. Der Anteil der Zellulose geht ebenfalls zugunsten von Cutin zurück; Zellulose ist in der H-Schicht jedoch noch häufig. Die Zahlen zeigen, daß Lignin mindestens gleich schnell wie Zellulose abgebaut wird.

Histochemische Untersuchungen von lignifizierten Zellwänden aus dem hier bearbeiteten Rohhumus zeigten simultanen Abbau von Lignin und Zellulose, siehe Babel (13).

Die Schätzwerte für die Feinsubstanz der Streuschicht (L) sind für die Zersetzungsgeschwindigkeit der Zellwandreste nicht weiter aufschlußreich: sie besagen nur, daß die mechanische Zerkleinerung der Cuticula und in viel geringerem Maße auch der Epidermen und des Mesophylls (Zellulosewände) schon in der unteren Streuschicht (L 2) einsetzt, während die Blattadern (Ligninwände) dort noch kaum zerkleinert werden. Das ist auch bei der direkten Beobachtung der Blätter der Streuschicht mit Lupe und Mikroskop festzustellen.

Amorphe, gefärbte Stoffe in der Feinsubstanz

Für die zweite Gruppe der Feinsubstanz, die mikroskopisch amorphen, gefärbten Stoffe, die durch Natriumhypochlorit gelöst werden und die nach den oben angegebenen Zahlen die Hauptmenge der gesamten Feinsubstanz ausmachen — in der Humusstoffschicht etwa $^3/_4$ —, können bis jetzt noch keine direkten Kennreaktionen wie für die Zellwandreste benutzt werden.

Abb. 6
Wurzelreste in der Humusstoffschicht H 2 (zu Abbildung 3)

Horizontal schraffiert:	Reste von Feinwurzeln, vor allem deren Rinden (rechts oben längs getroffen, darunter in mehreren Fällen quer)
Horizontal gestrichelt:	Morphologisch stärker umgewandelte Teile von Wurzelrinden
Punktiert:	Rest des Holzkörpers einer größeren Wurzel
Weite Kreuzschraffur:	Mykorrhizamäntel
Dichte Kreuzschraffur:	Dunkelbraune, hyphenreiche lockere Feinsubstanz mit einzelnen Quarzen

Tabelle 2

Ungefähre Mengen der verschiedenen Wandreste im natriumhypochlorit-unlöslichen Teil der Feinsubstanz (unter 200 μ)

(Schätzskala von 1 bis 5; 1 = wenig; 5 = sehr viel)

	cutinhaltige Zellwandreste	ligninhaltige Zellwandreste	zellulosehaltige Zellwandreste
Blätter aus der Streu	1	2 - 3	5
Streu (L)	4	1	4
Vermoderungsschicht (F)	2	2	5
Humusstoffschicht (H)	3	1	4

Auf Grund der oben kurz besprochenen morphologischen Untersuchungen ergibt sich aber das Folgende:

Die Hauptmenge der gefärbten Stoffe sind Bräunungsstoffe aus dem Blattmesophyll (s. Abb. 2); das ist das, was von Handley (5) als Gerbstoff-Eiweiß-Komplexe bezeichnet wurde. Ihre Grundsubstanz sind Oxydations- und Polymerisationsprodukte von Gerbstoffen aus der Vakuole, in die Eiweiße aus dem Protoplasten eingebaut werden können; vgl. die Untersuchungen zur Verbräunung von Tabakblättern von Bäbler (3).

Eine zweite Komponente wird gestellt durch die braunen Umwandlungsprodukte der lignifizierten Blattadern-Zell-

Zusammenfassung

Am Beispiel eines Buchenrohhumus wird gezeigt, wie die Kenntnisse und Methoden der botanischen Morphologie und Mikrotechnik sich bei der Bearbeitung humuschemischer Fragen anwenden lassen. Mit Hilfe des Mikroskopes lassen sich, wenn man von Bau und Zusammensetzung des frischen Ausgangsmaterials ausgeht, dessen allmähliche Umwandlungen im Laufe der Humifizierung verfolgen. Dazu müssen keine Modellversuche herangezogen werden, es werden vielmehr die verschieden alten Schichten des Humusprofils und dessen Einzelbestandteile untersucht. Es wird mit Vestopal-Schliffen, Paraffinschnitten und Streupräparaten gearbeitet.

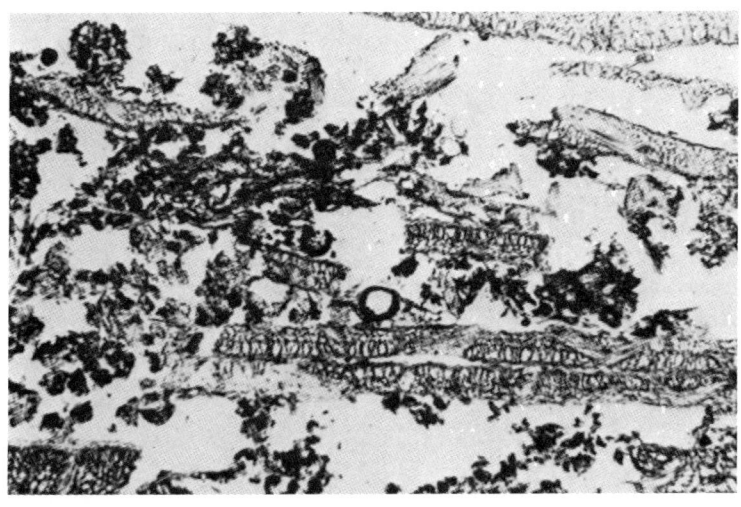

Abb. 7
Vermoderungsschicht (F 2)
(Vertikaler Dünnschliff; nat. Bildbreite 2 mm)
Recht gut erhaltene Reste von Buchenblättern, quer getroffen. Dazwischen Feinsubstanz, vorwiegend aus zerfallenen Kleintierlosungen bestehend.

wände (Abb. 4). Sie bestehen also zu einem großen Teil aus Ligninabbau- und -umbauprodukten.

Für diese beiden ersten Komponenten konnte morphologisch gezeigt werden, daß sie bis in die untere Vermoderungsschicht (F 2) zum großen Teil erhalten bleiben und bei der Zerkleinerung der Gewebe in die Feinsubstanz eingehen.

Es kommt noch eine dritte Komponente hinzu, nämlich gefärbte Stoffe aus den Wurzelrinden, die in der Botanik als Phlobaphene bezeichnet werden und als Gerbstoffpolymerisate bekannt sind (s. Abb. 3). Auch diese Stoffe können noch nach der Zerkleinerung der Wurzelrinden stellenweise in der Feinsubstanz einwandfrei angesprochen werden, ehe sie durch weitere stärkere Vermengung morphologisch nicht mehr von der Gesamtmenge der braunen Stoffe der Feinsubstanz unterschieden werden können.

Auf den Chemismus der ersten und der dritten Gruppe dieser gefärbten Stoffe wurde bei einer früheren Gelegenheit näher eingegangen (vgl. Babel in 1).

Das makroskopisch amorphe Material der Humusstoffschicht des untersuchten Buchen-Rohhumus besteht also zu etwa einem Viertel aus kaum veränderten Pflanzenstoffen (Zellulose und Cutin), zum Hauptteil aus stärker veränderten Stoffen, deren ungefähre chemische Zusammensetzung auf Grund der Kenntnis ihrer histologischen Herkunft angegeben werden kann.

Für die Zusammensetzung der dem bloßen Auge fast homogenen Humusstoffschicht ergab sich, daß 80% des Materials feinzerteilte organische Reste ohne Gewebestrukturen waren (20% waren Kork- und Wurzelgewebereste). Von diesen 80 Prozent waren etwa ein Viertel wenig veränderte Zellwandbruchstücke, vor allem aus Cutin und Zellulose. Drei Viertel waren mikroskopisch amorphe, gelb, braun und rotbraun gefärbte (selten schwarze, opake) Substanzen, die im wesentlichen aus Gerbstoff-Eiweiß-Polymerisaten (aus dem Blattmesophyll), aus Lignin-Umwandlungsprodukten (aus Blattadern) und aus Gerbstoffpolymerisaten (aus Wurzelrinden) bestanden.

Literatur

(1) Babel, U.: Mikroskopische Humusuntersuchungen. Vortrag auf der Tagung der Dtsch. Bodenkdl. Ges., Wien 1961.
(2) Babel, U.: Chemische Reaktionen an Bodendünnschliffen. Leitz-Mitt. Wiss. u. Techn. III, S. 12–14, Wetzlar 1964
(3) Bäbler, S.: Über die Verbräunung des Tabaks. Diss. ETH Zürich 1957.
(4) Großkopf, W.: Stoffliche und morphologische Untersuchungen forstlich ungünstiger Humusformen. Tharandt. Forstl. Jb. 86, S. 48–111, 1935.
(5) Handley, W. R. C.: Mull and Mor Formation in Relation to Forest Soils. Forestry Comm., Bull. 23, London 1954.

(6) Kononowa, M. M.: Die Humusstoffe des Bodens. Dtsch. Verl. d. Wissensch., Berlin 1958.

(7) Kubiena, W.: Die mikroskopische Humusuntersuchung. Z. Weltforstwirtsch. 10, S. 387–410, 1943.

(8) Kullmann, A.: Über die Verrottung von Futterpflanzenwurzeln an Hand von Modellversuchen. Z. Pfl. Ern., Düng., Bodenkde. 84, S. 127–132, 1958.

(9) Meyer, F. H.: Untersuchungen über die Aktivität der Mikroorganismen in Mull, Moder und Rohhumus. Arch. Mikrobiol. 33, S. 149–169, 1959.

(10) Meyer, F. H.: Vergleich des mikrobiellen Abbaus von Fichten- und Buchenstreu auf verschiedenen Bodentypen. Arch. Mikrobiol. 35, S. 340–360, 1960.

(11) Wittich, W.: Untersuchungen über den Verlauf der Streuzersetzung auf einem Boden mit Mullzustand, I. Forstarch. 15, S. 96–111, 1939.

(12) Zachariae, G.: Zur Methodik bei Geländeuntersuchungen in der Bodenzoologie. Z. Pfl. Ern., Düng., Bodenkde. 97, S. 224–233, 1962.

(13) Babel, U.: Dünnschnittuntersuchungen über den Abbau lignifizierter Gewebe im Boden. In: A. Jongerius (editor), Soil Micromorphology, Proc. 2nd Int. Working Meeting Arnhem, Amsterdam 1964, S. 15–22.

SUMMARY

Title: *Chemical investigations of beech raw humus by microscopic methods*

Qualitative information on the chemical composition of the humified (H) layer of raw humus formed under a beech forest was obtained through microscopic investigations. Knowledge of plant morphology and microtechnique were particularly valuable. Preliminary studies were conducted using thin, polished sections of soil ("Vestopal" technique, Fig. 7). These sections provided a survey of all constituents in the raw humus layers and their spatial organization. The latter often gave an indication of the origin and development of these constituents.

Detailed investigations were conducted by preparing microtome sections of the successive stages of decomposed plant remains in each humus layer. The fine substance, microscopically amorphous material and cell wall fragments, was studied using dispersed preparations because oriented sections could not be made. Microscopic studies of the morphological changes during the decay process also gave clues to the composition of the fine substance.

The chemical constituents of the raw humus layers were determined through comparative morphological studies and micro-reaction tests of the cell wall materials. Birefringence, fluorescence, Sudan III, and zinc chloroiodide were used for determination of cutin; fluorescence, phloroglucide, and zinc chloroiodide were used to identify lignin; birefringence and zinc chloroiodide were used to identify cellulose.

In the raw humus profile the proportion of easily recognizable organ and tissue remains decreased from 99% in the litter (L 1 in Table 1) to 20% in the lower H layer (H 2 in Table 1). Root residue accompanied that of leaf remains in the fermentation layer (F) and below. In the lower H layer very little leaf residue was still microscopically visible.

In the H layer 80% of the materials were transformed into organic fine substance (Fig. 1), 25% of which was composed of colorless cell wall fragments. They were composed chiefly of cutin or cellulose. In Table 2 the amount of cell wall fragments in the leaves of the litter *(Blätter aus der Streu)*, in the F layer, and in the H layer was compared. Lignin and cellulose decomposed at approximately the same rate while cutin was relatively resistant to decay.

Three-fourths of the fine substance in the H layer was amorphous and dark yellow to brown in color. In contrast to cell wall fragments, the amorphous material was soluble in sodium hypochlrorite. This portion of the fine substance was composed mainly of the reaction products of leaf browning (polymarisates of tannins and proteins) (Fig. 2), the transformed products of lignified cell walls of leaf veins (Fig. 4), and phlobaphenes of the bark of roots (polymerisates of tannins) (Fig. 3).

(Barnes)

12

Copyright © 1969 by Elsevier Scientific Publishing Co.
Reprinted from pages 257-261 and 264-271 of Geoderma **2**:257-271 (1969)

A REVISED CLASSIFICATION AND NOMENCLATURE OF MICROSCOPIC SOIL MATERIALS WITH PARTICULAR REFERENCE TO ORGANIC COMPONENTS

BERYL C. BARRATT

Soil Bureau, Department of Scientific and Industrial Research, Lower Hutt
(New Zealand)

(Received December 20, 1967)
(Resubmitted October 22, 1968)

SUMMARY

The nomenclature of microscopic soil materials in soil organic horizons, as it stands, is unsuitable for modern terminology and requires revision.

Soil materials are reclassified according to their mineral and organic components into five main classes : humiskels, lithiskels, humicols, mullicols and argillicols.

They are further subdivided according to the kinds of mineral grains or organic particles they contain into calcitic, feldspathic, lignic, mycetic etc., and according to structure into very coarse to extremely fine, massive, blocky, single grain, pelleted, spongy and expanded subclasses, and into fabric subclasses.

Soil materials encountered in the literature are reclassified according to the revised nomenclature.

INTRODUCTION

Humus forms and "microfabrics", now termed soil materials (Brewer, 1964), of grassland soils were reclassified in a previous paper (Barratt, 1964) with due regard to the rule of precedence. For humus forms the original terms, mull and mor, were retained with structural qualifications and these terms have proved satisfactory in use. Considerable confusion remains, however, in the nomenclature of soil materials of organic horizons, which closely follows the terminology of Kubiena (1953).

The main reasons for this confusion are:
(1) Use for soil materials of terms (such as mull humus and silica moder) that are closely similar to terms (such as mull and moder), still in use for humus forms.
(2) Use of similar terms for soil materials that in many respects are quite dissimilar. For example, soil materials that contain abundant discrete faecal pellets are all called moders although the pellets may be of different kinds, consisting of comminuted plant remains, strongly decomposed organic matter, or clay—organic matter complexes.

Some attempt has been made to separate classes of pelleted soil materials, but the use of similar terms such as mull-like moder for a pelleted soil material with clay–humus complexes, and mull-like rendzina moder for a coarsely pelleted but dominantly organic soil material, is very confusing.

(3) Use of genetic and site information in classifying soil materials such as leached rendzina moders. On a morphological basis these moders are generally indistinguishable from the quartz-free subclass of so-called silica moders.

(4) The failure to distinguish category levels in the classification with the result that the relationships of intergrades (such as weak mull humus) and complexes (such as raw soil humus) to the main classes of soil materials are obscured.

Important additions to the nomenclature of soil materials have been made by Brewer (1964) but the soil materials of organic horizons are touched upon only under the heading of faecal pellets.

RECLASSIFICATION OF SOIL MATERIALS

The following classification is based entirely upon micromorphological observations and eliminates the need for genetic interpretations. The classification is not claimed to be complete, but is considered sufficiently flexible to allow the addition of new terms for naming other soil materials as they are encountered. However, it seems likely that most other soil materials will be subclasses of, or intergrades between, the main classes that are named.

The soil materials are divided into two broad groups, skeletal and plasmic. Skeletal soil materials consist essentially of the relatively undecomposed mineral soil skeleton or organic matter. Plasmic soil materials consist of the decomposed, colloidal plasma matrix.

The skeletal and plasmic soil materials, excluding intergrades, are divided into five main classes according to their mineral and organic components as follows:

Main classes

Skeletal materials
(1) *Humiskel* (humus skeleton). Organic residues that are essentially undecomposed or chemically preserved.
(2) *Lithiskel* (lithic skeleton). Mineral grains and rock fragments.

Plasmic materials
(3) *Humicol* (humus colloid). Strongly decomposed organic residues of colloidal size.
(4) *Mullicol* (mull colloid). Strongly decomposed organic residues of colloidal size and clay colloids intimately mixed or associated.
(5) *Argillicol* (argillaceous colloid). Clay.

Subdivision of main classes of soil materials

Subdivision according to kinds of constituents

The main classes are subdivided where necessary according to the kinds of rock particles, mineral grains, plant materials and colloids that are present, e.g., calcitic lithiskel (of calcite grains), dolomitic (dolomite), silicic (quartz), feldspathic (feldspars), lignic humiskel (of woody fragments), parenchymal (of soft leaf or stem residues), mycetic (of fungal hyphae), and sclerotial (of fungal resting bodies). Subdivision according to the kinds of clay and their optical properties is introduced at this level if required, e.g., montmorillonitic argillicol (of montmorillonite), following the scheme for optical identification of clay minerals set out by Deer et al. (1967). Although clay minerals in soils generally have to be identified by other than optical methods, they are thought to be capable of optical identification where sufficiently pure and sufficiently aggregated and oriented, to behave, optically, like single crystals of sand or silt size.

Subdivision according to microstructure

Soil materials are further subdivided according to their microstructure under the headings of shape, size, and arrangement (fabric) of solid particles and voids as follows:

According to shape

(1) *Massive.* Without apparent aggregation, and without appreciable numbers of pores of silt size or larger. Occur in plasmic soil materials.

(2) *Single grain.* Of discrete particles without apparent aggregation. Occur in skeletal soil materials. Although, strictly speaking, single-grain "microstructure" is a fabric, it is included here for convenience (see "granular" fabric of Brewer, 1964). Roundness and sphericity of grains may be described using the illustrative chart of Hatch and Rastall (1965), who recognise the following sphericity categories: very angular, angular, sub-angular, sub-rounded, rounded and well-rounded.

(3) *Blocky, platy etc.* With ordinary soil structures that are recognisable by characteristic shape of aggregates, which becomes evident during air-drying prior to thin section preparation. Occur in plasmic soil materials.

(4) *Pelleted.* Of discrete pellets, generally faecal, (Brewer's, 1964, "single faecal pellets"). Where necessary, pellets may also be subdivided into shape subclasses, e.g., rugose, cylindrical, obovate etc. Occur in both skeletal and plasmic soil materials.

(5) *Spongy.* Aggregated, with irregular cavities that interconnect by irregular fissures. In this respect, cavities differ from Brewer's (1964: "vughs", p.189). Aggregates are commonly produced by worm casts and include Brewer's (1964) "welded faecal pellets". Occur in plasmic soil materials.

(6) *Expanded.* Of organic materials with swellings (including swollen pellets) formed presumably by bacterial action. Occur in humicols and humiskels.

According to size

The skeletal soil materials are subdivided according to the size of their (unaggregated) particles, and the plasmic soil materials according to

TABLE I

Particle-size classification for sand and silt

Size classes	diameter (mm)	
Very coarse	1.0–2.0	sand
Coarse	0.5–1.0	sand
Medium	0.25–0.5	sand
Fine	0.1–0.25	sand
Very fine	0.05–0.1	sand
Extremely fine	0.002–0.05	silt

the mean width of faecal pellets and other aggregates, using the particle size classification for sand and silt (Soil Survey Staff, 1951), with the addition of an extremely fine class as in Table I.

According to fabric or arrangement

Soil fabric is described by Brewer (1964) as "the physical constitution of a soil material as expressed by the spatial arrangement of the solid particles and associated voids". Fabrics of subsoils are named and described in detail by Brewer, but less attention is paid to the fabrics of topsoils and other organic horizons.

In comparison with the fabrics of mineral soils, the fabrics of organic horizons are generally uncomplicated and of rather small importance for nomenclatural purposes, though useful for description. Brewer's (1964, pp.168–174) scheme for describing the basic distribution and orientation patterns seems satisfactory for describing the main classes of soil materials in organic horizons. His fabric names for the s-matrix (groundmass), however, refer mostly to soil material complexes and seem more applicable to mineral than to organic horizons.

PLATE I

A. Parenchymal humiskel. Comprising stem residues with resinous, decay-resistant cell inclusions; from F-horizon of a laminated mor produced by topdressings of ammonium sulphate. Cockle Park Farm, Northumberland, Great Britain.
B. Single-grain silicic lithiskel. Consisting predominantly of quartz grains; from C-horizon, Foxton black sands, New Zealand.
C. Finely pelleted humicol. Faecal pellets containing decayed, very finely comminuted organic matter; H-horizon of a podzolic soil, Waldridge Fell, Durham, Great Britain.
D. Argillicol. Clay-dominant soil material from B-horizon of Naike clay, New Zealand.
E. Spongy mullicol. Clay-organic matter soil material with abundant pores (and skeletal grains); from A_1-horizon of a strongly granular mull produced by topdressings of lime and basic slag, Waikiwi silt loam, New Zealand.
F. Humiskel—humicol complex. Parts of undecayed plant residues, parts of residues decayed beyond recognition; from Awakeri peat profile, New Zealand.

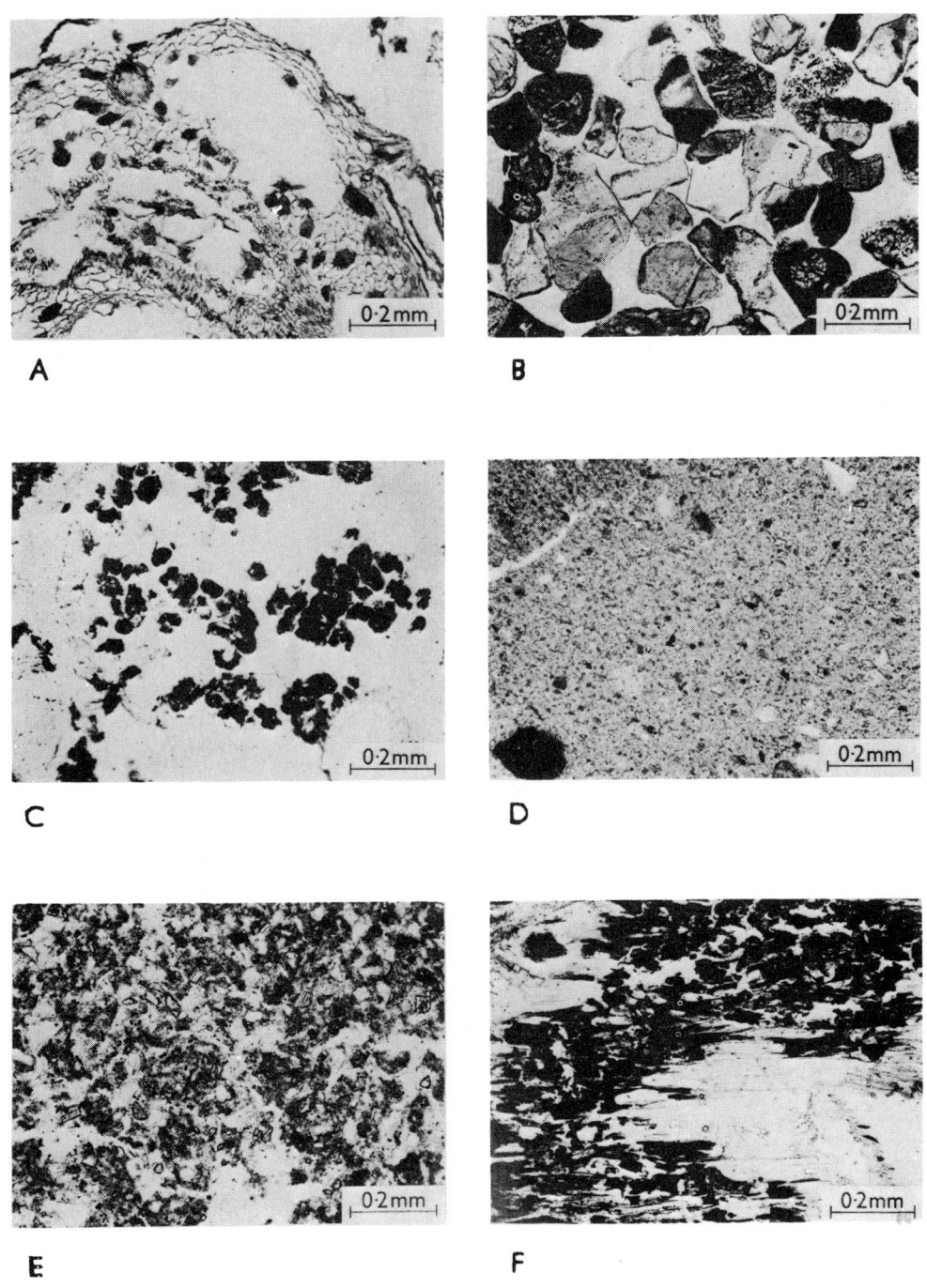

[*Editors' Note:* Table II has been omitted, but see page 136.]

Nomenclature of the main classes

With the aid of this classification, soil materials are named according to their main classes, e.g., humiskel, or with more precision as required, e.g., with kind of material added - lignic humiskel; with shape of aggregates added - pelleted lignic humiskel; with size of aggregates added - coarsely pelleted lignic humiskel; with fabric of aggregates added - random coarsely pelleted lignic humiskel.

In practice the full terminology is rarely required except for descriptive purposes. The revised classification is set out in Table II.

Intergrades and complexes

Soil materials are encountered that cannot be fitted directly into the main classes described above. These include soil materials that are intergrades between two main classes, and soil material complexes or mixtures, in which parts belong to one main class and parts belong to another (when examined using the selected magnification and field of view).

Thus a soil material consisting of strongly decomposed but still coarsely fragmented plant tissues with still recognisable cell structure is a humicol–humiskel intergrade, and similarly a colloidal soil material that contains a recognisable but low quantity of organic matter is termed a mullicol–argillicol intergrade. By comparison, a humicol–humiskel complex, at the given magnification, shows interpenetrating areas of undecomposed plant tissue and of humus colloid, and a mullicol–argillicol complex shows interpenetrating areas of humus–clay complex and clay colloid.

In all examples given above the dominant soil material is named last. The boundary between a main class and a complex is provisionally set at about 15% of subordinate soil material components. Where necessary, however, the percentage of subordinate components should be stated, e.g., 20% humicol–humiskel complex.

In Plate I the five main classes of soil material, together with one example of a complex are illustrated.

Phases

Phases are used where necessary, such as where a small amount of impurity is genetically important, to show soil material differences of lower rank than those which differentiate the main classes. For example, a pelleted humicol (main class *humicol*) with a few, less than 15%, calcite grains (main class *lithiskel*) is distinguished as a pelleted humicol,

---

Fig.1. A–L. Soil materials of organic horizons as seen in thin-section (previous names of microfabric, from Barratt, 1964, in parenthesis). A = air or waterspace; C = calcite or dolomite grains; D = dropping or small cast; F = fungal hypha; I = insect-bitten cavity in plant residue; L = resistant cell inclusion, "lignohumin" of Kubiena; O = oriented, poorly flocculated plasma; P = plant fragment; Q = quartz grain; R = root, rhizoid; S = fungal sclerotium (resting body).

Revised Classification and Nomenclature 265

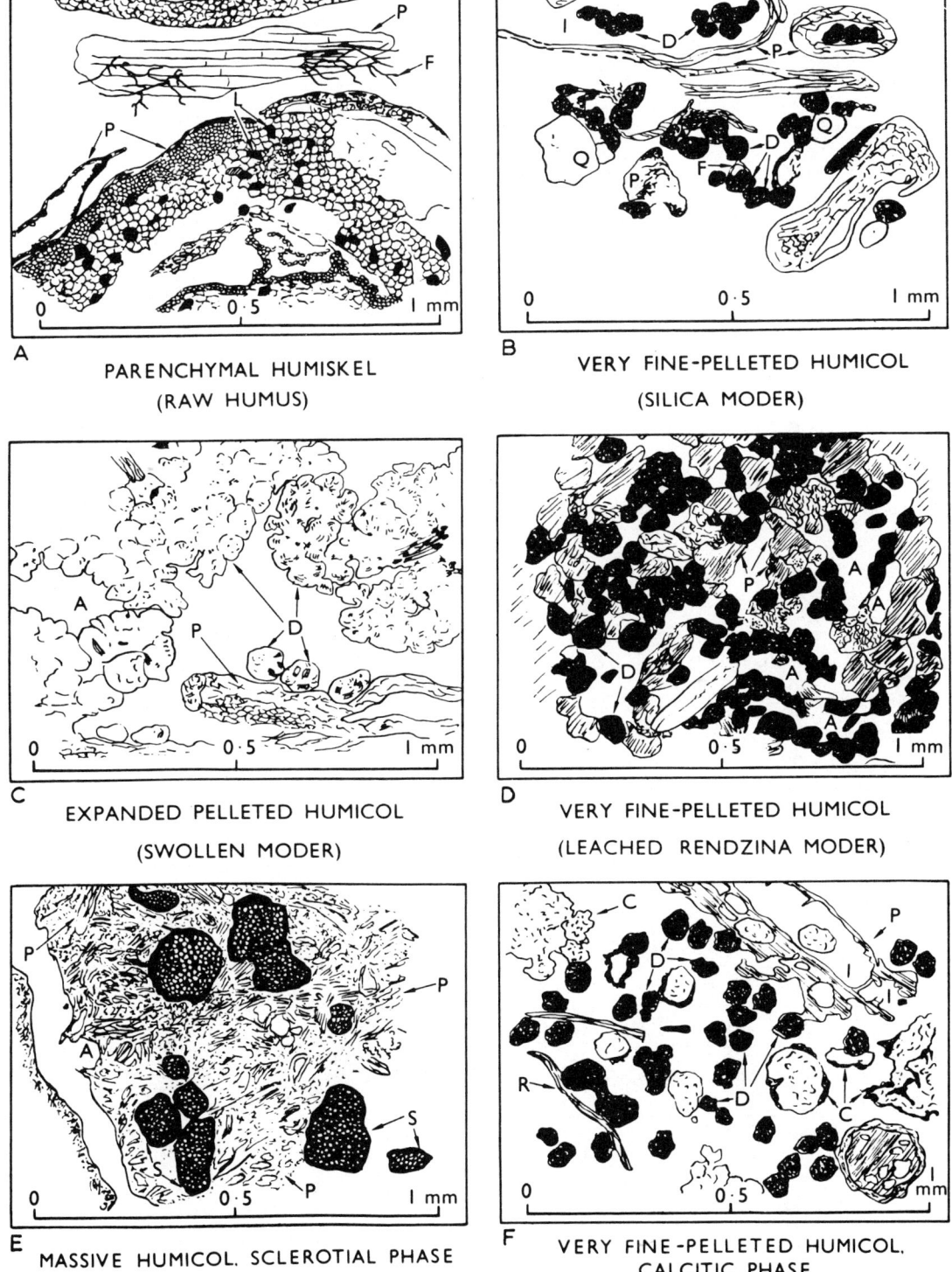

A PARENCHYMAL HUMISKEL
   (RAW HUMUS)

B VERY FINE-PELLETED HUMICOL
   (SILICA MODER)

C EXPANDED PELLETED HUMICOL
   (SWOLLEN MODER)

D VERY FINE-PELLETED HUMICOL
   (LEACHED RENDZINA MODER)

E MASSIVE HUMICOL. SCLEROTIAL PHASE
   (SCLEROTIAL HUMUS)

F VERY FINE-PELLETED HUMICOL,
   CALCITIC PHASE
   (RENDZINA MODER)

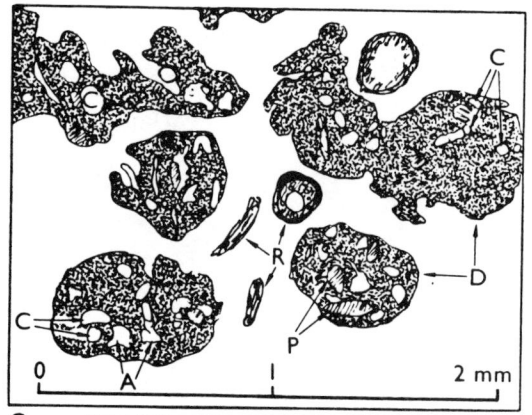

G COARSE-PELLETED HUMICOL,
CALCITIC PHASE
(MULL-LIKE RENDZINA MODER)

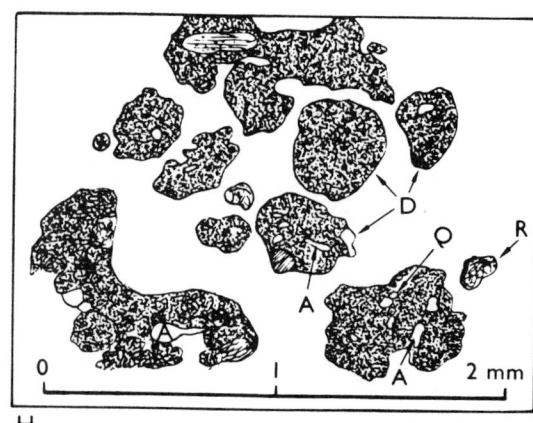

H COARSE-PELLETED HUMICOL
(LEACHED MULL-LIKE RENDZINA MODER)

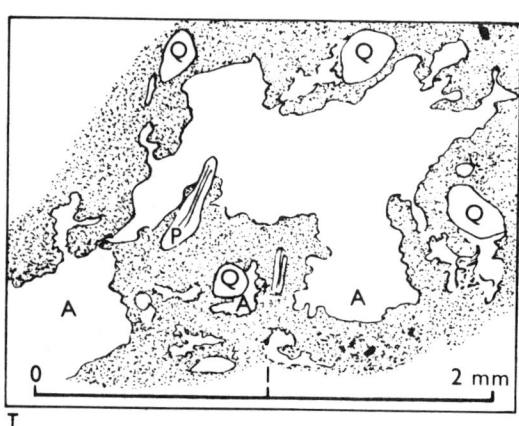

I SPONGY MULLICOL
(STRONG MULL HUMUS)

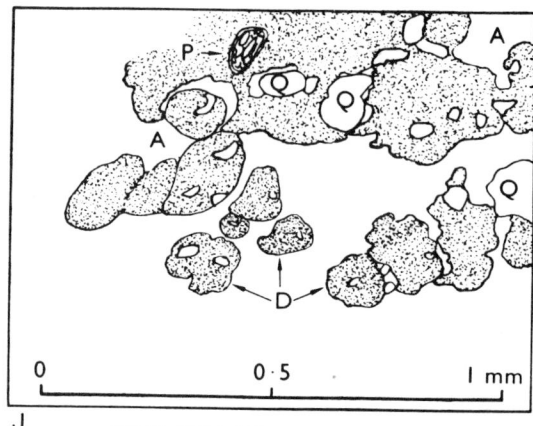

J FINE-PELLETED MULLICOL
(MULL-LIKE MODER)

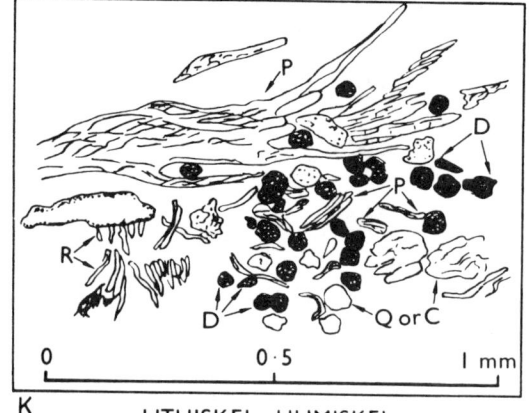

K LITHISKEL–HUMISKEL–
PELLETED HUMICOL COMPLEX
(RAW SOIL HUMUS)

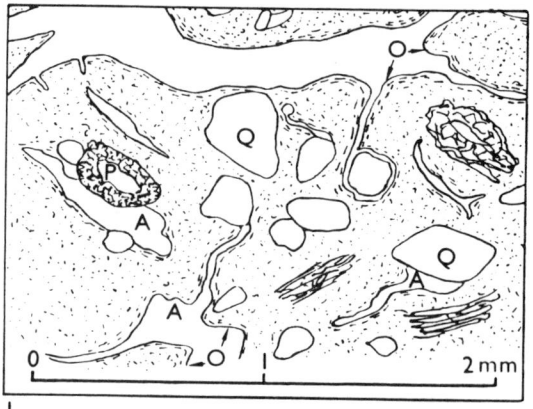

L MULLICOL–ARGILLICOL INTERGRADE
(WEAK MULL HUMUS)

160

calcitic phase. The actual percentage of the subordinate component is stated where necessary, e.g., pelleted humicol, 5% calcitic phase. If necessary, texture can be similarly emphasised in a class other than a lithiskel, e.g., pelleted humicol, coarse sandy phase. However, such precision in nomenclature is rarely required.

In Table III the proposed nomenclature is compared with previous nomenclatures after Kubiena and in Fig.1 illustrations from Barratt (1964) are reprinted with slight alteration and using the revised nomenclature.

DISCUSSION

The new classification of soil materials deals effectively with the four main problems that have caused confusion in older classifications, as noted in the Introduction.

(1) It removes the confusion between the terminology of soil materials and that of humus forms.

(2) It differentiates the wide range of soil materials included originally in the class "moder" as demonstrated in Table III.

(3) It removes the need to distinguish soil materials (such as leached rendzina moders) by genetic information. If genetic interpretations are required any additional information must be consciously acquired from other sources.

(4) With the establishment of main classes and subclasses it is possible to classify many soil materials at the selected magnification as either intergrades between one main class and another, complexes of one class of soil material and another, or phases.

Using the classification it is now possible to name soil materials that were formerly unclassifiable. Thus, for example, a weakly decomposed peat with some accumulating "amorphous" humus (Jongerius and Pons, 1962, fig.4) is a humicol–humiskel complex soil material whereas a strongly decomposed, compressed and platy peat (*ibid.*, fig.5) is a platy to massive humicol. At the other extreme, many B horizons such as Tekapo fine sandy loam contain faecal pellets in which there is little organic matter (Barratt, 1965, fig.4). Such a soil material, which was difficult to classify using the previous systems, is now called a finely pelleted argillicol.

Several of the soil materials illustrated by Kubiena (1953, plates I, II) are now shown to be intergrades or complexes, and all the soil materials he illustrates can be named using the new classification (assuming that they have been magnified about 50 times linear), as in Table IV.

The revised classification can be applied to practical problems, such as identifying the changes in topsoils that result from different systems of land use. Topsoils have been shown to be changed by replacement of the native vegetation by pasture, by fertiliser topdressing and by animal treading (Barratt, 1968), by the introduction of certain species of earthworms (Stockdill, 1966) and by the use of herbicides (Bulfin, 1966, note that captions to illustrations should be reversed). Using the revised classification the microfabrics of these altered topsoils can now be named to facilitate their comparison with the required degree of precision.

The present classification also provides a method for correlating soils classified according to other systems, either national or international.

TABLE III

Soil materials according to older classifications and re-named according to the revised classification

Previous name	Tentative name according to the revised classification			
	microstructure		kind of mineral or organic particles	main class, complex, intergrade or phase
	size of aggregates or unaggregated particles	shape and arrangement (fabric) of particles or aggregates and of voids		
Bleached sand[1]	fine	single grain	silicic	lithiskel
Bleached "clay"[4]	extremely fine	single grain	silicic	lithiskel
Raw humus[2,3]	very coarse	single grain	parenchymal	humiskel
	very coarse	single grain	lignic	humiskel
	very coarse	single grain	mycetic	humiskel, or humiskel, mycetic phase
Swollen raw humus[3]	very coarse	expanded, single grain	parenchymal	humiskel
Silica moders of many forest soils[3]	fine	pelleted	lignic	humiskel
Rendzina moder[2,3]	fine or very fine	pelleted		humicol, calcitic phase, or humicol–calcitic lithiskel complex
Pech or leached rendzina moder[2,3]	fine or very fine	pelleted		humicol
Silica or silicate moder[2,3]	fine or very fine	pelleted		humicol, or humicol, silicic phase
Swollen moder[3]	fine or very fine	expanded, pelleted		humicol
"Mull-like" rendzina moder[2,3]	coarse	pelleted		humicol, calcitic phase, or humicol–calcitic lithiskel complex
Pech or leached "mull-like" rendzina moder[2,3]	coarse	pelleted		humicol
Coarse silica moder[5]	coarse	pelleted		humicol, or humicol, silicic phase
Sclerotial humus[3]		massive		humicol, sclerotical phase, or humicol, sclerotical humiskel complex

Mull-like moder[2,3]	very fine to very coarse	pelleted	mullicol, or mullicol, silicic phase
Mull, strong mull humus[2,3]	very fine to very coarse	spongy	mullicol, or mullicol, silicic phase
Weak mull humus[3]	very coarse	blocky	mullicol–argillicol intergrade, silicic phase
Braunerde[4]	very fine to very coarse	spongy or pelleted	argillicol, or argillicol, silicic phase
Sandy braunerde[4]	very fine to very coarse	spongy or pelleted	argillicol–silicic lithiskel complex
Braunlehm[4]	very fine to very coarse	massive or blocky	argillicol, or argillicol, silicic phase
Sandy braunlehm[4]	very fine to very coarse	massive, blocky or spongy	argillicol–silicic lithiskel complex

[1] Kubiena (1938).
[2] Kubiena (1953).
[3] Barratt (1964).
[4] Barratt (1965).
[5] Barratt (1968).

TABLE IV

Revised nomenclature for soil materials previously named by Kubiena (1953)

Illustrations in Kubiena (1953)	Kubiena's terminology: explanation and revised name of soil material
Plate I_1	Gyttja: illustrates a waterdrop preparation, not a soil material s. str. However, the soil material from which it has been extracted is probably a humicol, silicic (diatomaceous) phase.
Plate $I_{2,2,4,5}$	Turf, sedge, *Hypnum* moss and *Sphagnum* moss peat moors: unprepared but appear to be parenchymal humiskels.
Plate I_6	Old *Sphagnum* moss peat moor: a humiskel-humicol intergrade.
Plate I_7	Syrosem humus, raw soil humus: a silicic lithiskel-humiskel complex with some very finely pelleted humicol within the humiskel fragments.
Plate I_8	Raw humus: a humiskel.
Plate II_1	Tangek humus: a mullicol–humiskel complex.
Plate II_2	Moder, silicate moder: essentially a finely pelleted humicol. (Some silicic lithiskel and some humiskel etc. are also shown.)
Plate II_3	Pitch moder: a very finely pelleted humicol.
Plate II_4	Mull-like rendzina moder: a very coarsely pelleted humiskel–humicol intergrade, calcitic phase.
Plate II_5	Mull: a spongy mullicol.
Plate II_6	Pitch peat anmoor: a massive to blocky humicol.
Plate $II_{7,8}$	Humus ortstein: a humicol–silicic humiskel complex.

The Seventh Approximation (Soil Survey Staff, 1960) together with its supplement (Soil Survey Staff, 1967) is an important system developed in the United States for international soil classification, and micromorphological examination of its diagnostic soil horizons should be undertaken.

The micromorphology of New Zealand reference profiles is being studied and this will facilitate correlation with the United States system.

ACKNOWLEDGEMENTS

I am very grateful to Mr. I.J. Pohlen of the Soil Bureau for advice on the terminology used in this paper, and to Mr. A.V. Weatherhead who prepared the photomicrographs.

REFERENCES

Barratt, B.C., 1964. A classification of humus forms and microfabrics of temperate grasslands. J. Soil Sci., 15: 342–356.

Barratt, B.C., 1965. Micromorphology of some yellow-brown earths and podzols of New Zealand. New Zealand J. Agr. Res., 8: 997–1042.

Barratt, B.C., 1968. Micromorphological observations on the effects of land use differences on some New Zealand soils. New Zealand J. Agr. Res., 11: 101–130.
Brewer, R., 1964. Fabric and Mineral Analysis of Soils. Wiley, New York, N.Y., 470 pp.
Brown, R.W., 1954. Composition of Scientific Words. U.S. National Museum, Washington, D.C., 882 pp.
Bulfin, M., 1966. Micromorphology studies in the soils division. Farm Res. News, 7: 79–81.
Deer, W.A., Howie, R.A. and Zussman, J., 1967. An Introduction to the Rock-Forming Minerals. Longmans, London, 528 pp.
Hatch, F.H., and Rastall, R.H., 1965. Petrology of the Sedimentary Rocks. 4th ed. Revised by J. Trevor Greensmith, Murby, London, 408 pp.
Jongerius, A. and Pons, L.J., 1962. Soil genesis in organic soils. Boor en Spade, 12: 156–168.
Kubiena, W.L., 1938. Micropedology. Collegiate Press, Ames, Iowa, 243 pp.
Kubiena, W.L., 1953. The Soils of Europe. Murby, London, 318 pp.
Little, W., Fowler, H.W. and Coulson, J., 1959. The Shorter Oxford English Dictionary, 3rd ed., vol. I,II. Oxford.
Simpson, D.P., 1964. Cassell's New Latin–English - English–Latin Dictionary. 3rd. ed. Cassell, London.
Soil Survey Staff, 1951. Soil Survey Manual. U.S. Dept. Agr., Handbook, 18: 1–503.
Soil Survey Staff, 1960. Soil Classification - A Comprehensive System, 7th Approximation. U.S. Dept. Agr., Soil Conserv. Serv., Washington, D.C., 265 pp.
Soil Survey Staff, 1967. Supplement to Soil Classification System (7th Approx.). U.S. Dept. Agr., Soil Conserv. Serv., Washington, D.C., 207 pp.
Stockdill, S.M.J., 1966. The effect of earthworms on pastures. Proc. New Zealand Ecol. Soc., 13: 68–75.

ORIGIN AND MICROMORPHOLOGICAL NOMENCLATURE OF ORGANIC MATTER IN SANDY SPODOSOLS

F. DE CONINCK, Geologisch Instituut, State University, Ghent, Belgium.

D. RIGHI, C.N.R.S., laboratoire de Pedologie, Faculté des Sciences, Université de Poitiers, France.

J. MAUCORPS, I.N.R.A., Service de cartographie des Sols, Aisne, France.

A. M. ROBIN, Laboratoire de Géologie Dynamique, Faculté des Sciences, Université de Paris, France.

INTRODUCTION

During our study of sandy spodosols from Belgium (Antwerp campine) and France (forêts de Rambouillet and Fontainebleau: West of Paris, Les Landes: Bordeaux and Aisne: Northern France), we were confronted with the fact that the terms, currently used for the micromorphological description of the organic features, did not permit an accurate description. Moreover, the meaning of the terms is neither accurately defined nor generally accepted.

For this reason we have tried to define a nomenclature, based exclusively on micromorphologically distinguishable characteristics without any reference to chemical composition. This nomenclature will be used to explain the micromorphological aspects of the podzolization in the sandy soils we have studied, especially of the organic matter in the spodic B horizon.

NOMENCLATURE

Terms have been proposed for the description of the different characteristics of soil units such as boundary, porosity, shape, degree of covering and density. These terms are:

Sharpness of the boundary. *Sharp:* knife-edge boundaries between a unit and its surroundings, especially visualized by the difference in color. *Diffuse:* gradual color transition between a unit and its surroundings.

Porosity. *Very porous:* the distance between the units is

NOMENCLATURE OF ORGANIC MATTER

larger than the size of the units themselves. *Porous:* the difference between the units is smaller than the size of the units themselves. *Non-porous:* the units stick together completely.

Shape. *Regular:* globular; ovoid; elongate. *Irregular:* rounded (without marked angles); subangular (with rounded angles); angular (with sharp angles).

Degree of covering (for coatings). *Partial: complete.*

Density of degree of opacity. *Opaque, weakly transparent; transparent.*

Levels of the system

The nomenclature proposed is a system with three levels. *The lowest level* denotes the different kinds of organic matter.

Plant remains. Fragments of plant tissues, directly recognizable by their shape and/or their structure (Fig. 1).

(a) *Non Transformed.* The original cellular structure can be recognized, and some birefringence remains. (b) *Transformed.* The original cellular structure cannot be recognized and no birefringence is present.

The plans residues either compose isolated units in the form of fragments of large size, or are included in different units of the organic matter such as fecal pellets, aggregates and accumulations.

Elements of fragments from the fungal microflora includes hyphae, sclerotiae and spores.

Amorphous organic matter. (a) *Polymorphic organic matter* (Figs. 2, 7). Organic matter without recognizable vegetal or fungal structures. It forms a discontinuous mass formed by the juxtaposition of polymorphic elements that have a globular form, with sharp or diffuse boundaries caused by differences in color and density. (b) *Monomorphic organic matter* (Figs. 3,6).

Organic matter without recognizable vegetal or fungal structure. It forms a continuous mass with relatively uniform color and density. It is present in the form of coatings or irregular, angular or subangular units with sharp boundaries.

The second level contains the units in which the different organic materials can be assembled.

Fecal pellets (Fig. 1). Units with a regular, ovoid or spherical shape and sharp boundaries. The size varies about 25 μ to about 150 μ.

(a) *Fresh*. The fecal pellets are composed of fragments of non-transformed plant remains, still showing their structure and their original birefringence. The color is yellowish brown to brown. Their shape is perfectly regular.

(b) *Transformed*. The fecal pellets composed of transformed organic compounds and/or polymorphic organic matter. The color is brown to very dark brown.

Pellets (Figs. 1, 2, 4). Units that are essentially composed of organic matter with brown to very dark brown color. They form irregular, rounded to subangular units, with sharp to diffuse boundaries. They can be isolated, but mostly form aggregates or accumulations. The size varies from about 25 μ to 150 μ.

Aggregates (Figs. 2, 4). Units which have a more or less porous stacking of one or more organic compounds: non-transformed or transformed plant remains, fragments of fungal microflora, fresh or transformed fecal pellets and pellets and have clear or distinct boundaries. They may or may not be associated with fine or silt-size mineral particles. *Simple aggregates* are composed only of organic materials whilst *complex aggregates* are composed of organic materials and mineral particles.

Accumulations (Figs. 5, 9) have the same composition as aggregates, but they lack distinct boundaries, being bounded mostly by skeleton grains.

Coatings. Layers of organic matter (monomorphic or polymorphic organic matter, plant remains, fragments of the fungal microflora) that may be in combination with fine mineral particles. They cover coarse mineral compounds or other units of organic matter in such a way that their outer boundary is roughly parallel with the outer boundary of the units they coat.

1. Uniform coating. Composed of a single kind of organic matter.

(a) *Uniform monomorphic coatings* which are composed of monomorphic organic matter (Figs. 3, 6).

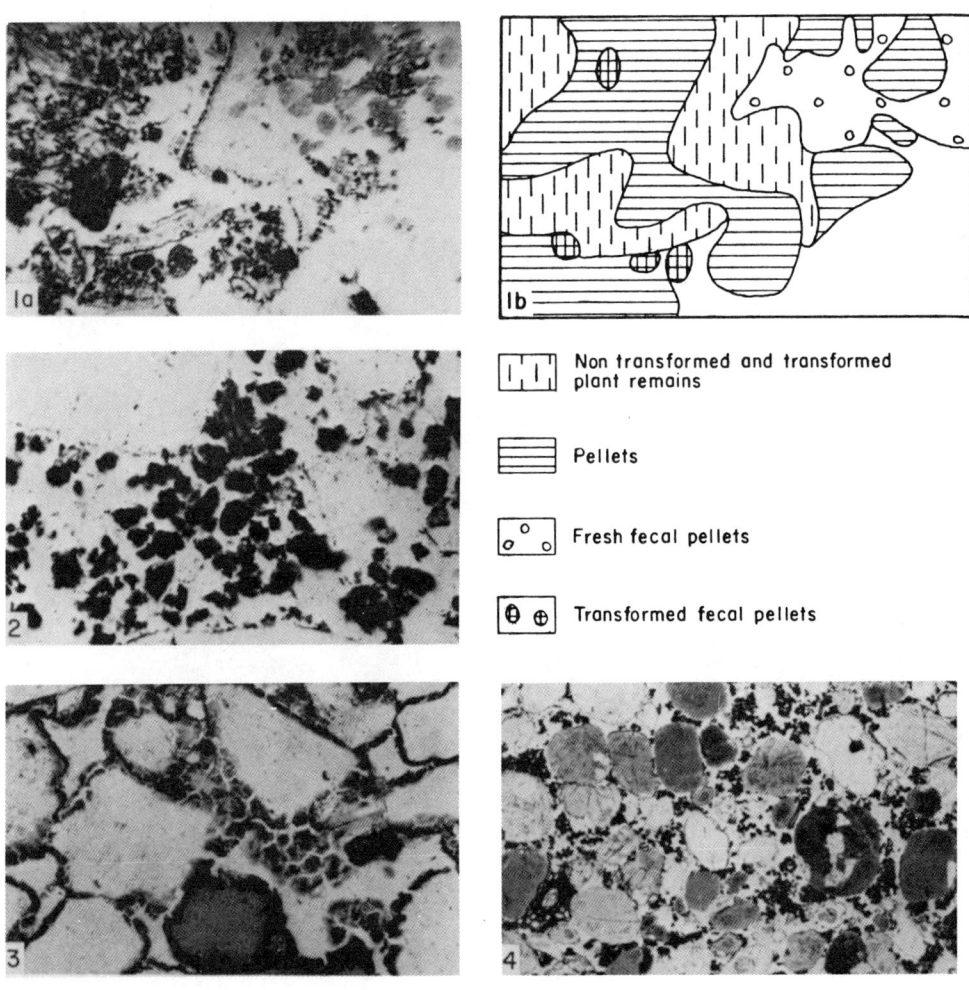

Figure 1. O_1 horizon; x 40. Fresh and transformed fecal pellets and
a and b: plant remains; pellets.

Figure 2. $B_{21}h$ horizon; x 350. Polymorphic organic matter forming pellets.

Figure 3. $B_{22}hir$ horizon; x 100. Monomorphic organic matter forming uniform coatings and concentrations. (Coated related distribution).

Figure 4. $B_{21}h$ horizon; x 30. Pellets and transformed plant remains. (Juxtaposed related distribution).

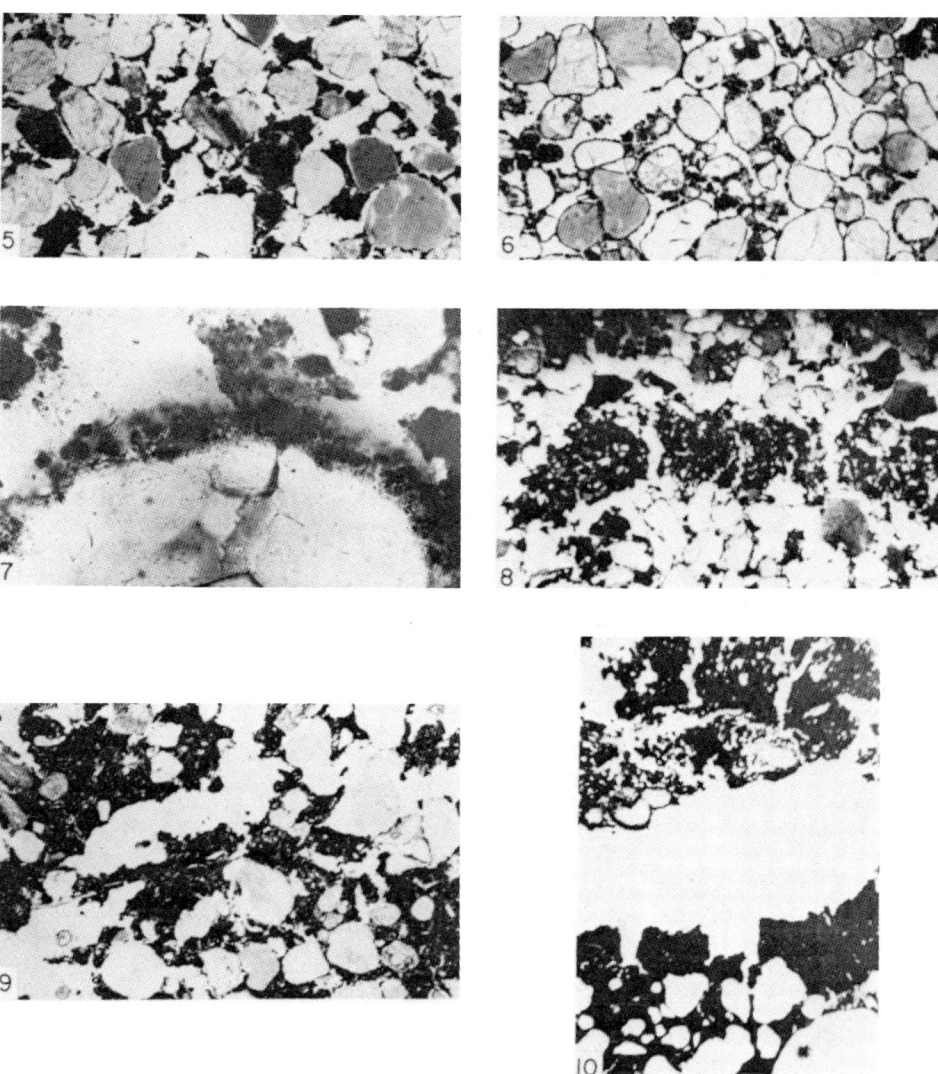

Figure 5. A_{11} horizon; x 30. Strongly coalesced pellets and polymorphic organic matter. (Linked related distribution).

Figure 6. $B_{22}h$ horizon; x 30. Monomorphic organic matter forming uniform monomorphic coatings and concentrations. (Coated related distribution).

Figure 7. B_2h horizon; x 200. Polymorphic organic matter, plant remains and fungal hyphae forming a compound coating.

Figure 8. B_2h horizon; x 40. Porous aggregates forming a cleavage striotubule. (Separated related distribution).

Figure 9. A_{11} horizon; x 30. Pellets and transformed plant remains. (Agglomerated related distribution).

Figure 10. $B_{21}h$ horizon and thin iron pan; x 40.
Upper part ($B_{21}h$) Transformed and non transformed plant remains, pellets and polymorphic organic matter. (Separated related distribution).
Lower part (thin iron pan) Monomorphic organic matter. (Filled related distribution).

(b) *Uniform polymorphic coatings* which are composed of polymorphic organic matter.

2. Compound coatings. These are composed of different kinds of organic matter (Fig. 7).

3. Complex coatings. These are composed of both organic and mineral units.

Concentrations (Figs. 3, 6).

Units composed of monomorphic organic matter, without distinct boundaries, but mostly fragmented by a network of angular or subangular cracks.

The third level is the related distribution between these units and the coarse mineral fraction, that in these soils always forms the most important part of the mineral material.

Separated (Figs. 8, 10). The organic units are present in distinct domains with little or no mixing with coarse mineral grains. This distribution is typically present in organic tubules.

Juxtaposed (Fig. 4). The organic units and the coarse mineral grains are randomly distributed, without any definite contact between each other, in a fabric that is always porous.

Agglomerated (Fig. 9). The organic units envelop the coarse skeleton grains, forming units whose external boundaries have no relation to those of the coarse grains.

Coated (Figs. 3, 6). The organic units are present in the form of coatings, covering all or part of the coarse mineral grains or other organic units.

Linked (Fig. 5). Concentrations or accumulations form bridges between the coarse skeleton grains.

Filled (Fig. 10). Monomorphic organic matter fills most of the space between coarse mineral grains or the other organic units.

Summary of above considerations

Kinds of organic matter. No subdivision has been made in the *plant remains* according to their nature or origin but only according their degree of transformation, because this criterion allows one to distinguish between either a rapid turnover into fecal pellets and amorphous organic matter by the soil fauna or a slow transformation into amorphous organic matter without intervention of the soil fauna. In the first case, most of the plant remains will be non-transformed, in the second case, many will be transformed.

The organic materials that have been transformed so strongly, that both their original cellular structure and their original shape have completely disappeared are called *amorphous organic matter*. In the soils we have studied, this amorphous organic matter can basically have two origins: (1) strong transformation of the remains of plants and microflora, but without any solubilization or translocation in a liquid phase, (polymorphic); (2) solubilization or dispersion and precipitation out of the liquid phase, (monomorphic). As a result of this precipitation, coatings or bridges are formed on coarse mineral or organic particles. The polymorphic form, due to the fact that it is composed of elements of different origin and composition, never has a uniform appearance, but shows differences in color and density in the same unit and in different units. Moreover, the shape of these units and of the elements in the units is always subangular or rounded.

The monomorphic form, as a result of its transition through a liquid phase, has a more uniform color and density and can show a fine layering. Since it never forms separate units but is always present in connection with coarse grains or other units, the drying out and subsequent shrinkage results in its contraction into smaller fragments with angular or subangular shape.

Units. The different units are formed under the influence of two processes:

(a) the fauna transforms part of the organic and possibly of the clay- and silt-size mineral material into excrements, which are the origin of fecal pellets, pellets, aggregates and accumulations.

(b) solubilization of organic matter and reprecipitation giving rise to coatings and concentrations.

Fecal pellets are these characteristic excrements that show a very regular shape with a size normally between 25μ and

150μ. They are composed of plant residues that have been consumed only once and that contain no mineral material. Since one consumption does not deeply transform the plant tissues, their structure and birefringence can still be recognized clearly, at least shortly after their formation (fresh). Biochemical activity can destroy these features but without altering the regular shape (transformed). They are always present in the interior of, or very close to plant remains.

Particles of about the same size as the fecal pellets, but with a more irregular shape and a more diffuse boundary and composed of polymorphic organic matter are very common. They have been called *pellets*. Micromorphological evidence clearly shows that their formation is biological.

(1) Fecal pellets are attacked by fungi: hyphae penetrate into the fecal pellets and cause both a partial disintegration partially destroying the regular shape, and a biochemical alteration of the tissues into polymorphic organic matter.

(2) They can be formed by a subsequent consumption of fecal pellets, transformed plant remains, and even monomorphic organic matter and in this way compose the building units of clusters, together with plant remains, and possibly fine mineral material. These clusters can be porous to very porous, with the pellets present as discrete particles. However, they normally show a tendency to coalesce so that the pellets lose their individual shapes. The pellets are very common in 0, A1, A2 and in many of the spodic B horizons.

In the clusters, mentioned earlier, *aggregates* and *accumulations* are distinguished, since these two units have different related distributions and different porosities, the aggregates being more porous than the accumulations.

Pellets, aggregates, and accumulations are doubtless of biological origin, i.e. excrements, originally forming granotubules or striotubules, the formation of which can be due only to the soil fauna. During the evolution, these tubules are gradually transformed into isolated aggregates and later into accumulations by compression and coalescence of the pellets. Both aggregates and accumulations are very common not only in the O but also in the A1 and A2 horizons and in most of the spodic B horizons, especially in the upper parts.

The presence of these clusters in the B horizons of the spodosols has been described extensively but mostly without a clear hypothesis about their formation: aggregation or coagulation of organic materials (Altemuller, 1962), microcrumbs or microaggregates (Kowalinski, 1969) coagulated plasma substances (Racz, 1968). Some authors, as Jongerius (1957), consider that these clusters have been mechanically translocated.

Besides the fact that the aggregates and accumulations are commonly present in tubules, which can have an horizontal orientation, more evidences can be given for the faunal origin of these units.

1. They are commonly in direct contact with decaying roots, and clearly recognizable fragments of these roots are present in the units. If they were formed by coagulation of humic substances, the presence of these fragments cannot be explained.

2. Admitting the coagulation hypothesis or the hypothesis of mechanical movement through the soil, the destruction of the plant roots can only be explained by a gradual biochemical or chemical transformation of these residues. Since spodosols are reputed to have a very low microbiological activity, this destruction would be very slow, and root remains that clearly show the original cellular structure, or at least the original shape, would be very common. But in reality, it is very exceptional to find such remnants in the A or even in the B horizons, although the rooting can be very intense. Moreover, these hypotheses cannot explain the presence of non-transformed fecal pellets at depths of 20 cm or even more; their very regular shape would surely not withstand this mechanical translocation without damaging their outer boundaries. However, some kind of mechanical transport is possible by the fauna through the formation of the numerous tubules.

3. These different points are very well illustrated by the micromorphological aspects of the spodosols that have a placic horizon or thin iron pan, between the B21h horizon and the B22 hir. Due to the impervious nature of this placic

horizon, a strong rootmat always develops above it, containing the whole range of organic materials; more or less strongly decayed roots, root remains, fresh and transformed fecal pellets, pellets and fungi, many times clustered together into aggregates and accumulations, or even forming tubules. This means that all the compounds of an O1 or an A1 horizon are present in this B horizon. Moreover the transformation of one kind of organic matter or one unit into another can be established very clearly, for example the transformation by fungi of fecal pellets into pellets.

4. Monomorphic organic matter, that is the other component of the spodic B, clearly shows the features of a material, translocated through a liquid phase: it forms layered coatings and bridges, which are very similar with clay illuviation cutans or argillans.

These different units of organic matter: at one side the polymorphic aggregates and accumulations, at the other side the monomorphic coatings and concentrations, can be present at the same place in the same horizon, closely adjacent to each other. For example, almost all the coarse grains of an horizon can be covered with a monomorphic coating but a few, forming a fine channel, lack it and have few aggregates. Admitting that these two forms of organic matter have been formed by illuviation, they must have been formed at the same time since they occupy the same level in relation with the coarse grains. This would mean that these two forms would be formed in the same conditions, which seems very improbable if not impossible.

The composition of the *coatings* can be widely divergent. *The uniform monomorphic coating* is composed of what has been called extensively "amorphous" or "dispersed" humus. It is given as the classical micromorphological characteristic of the spodic B horizons, although many of them lack it completely. When present, it is best developed in the lower part of the spodic B (B22h-B3) and in the lamellae which are very frequent underneath the B. However, we have found it in organic lamallae in the A2 of the spodosols, although rather exceptionally. As mentioned before, these coatings show the most striking features of illuviated material covering all the coarse elements with a layer of constant thickness, smoothing the irregularities of the surfaces of these elements and commonly showing a layering. Scanning microscope indicates that they are composed of irregular angular platelets, lying on the surface of the sandgrains. The surface of these platelets shows a certain similarity with the surface of an argillan.

The uniform polymorphic coating mostly has a composition of very dark brown to black, rounded to globular particles in a paler matrix. It can consist of more or less coagulated pellets sticking to the surfaces of coarse elements. They are present almost exclusively in the A horizons and the upper part of some spodic B horizons. Their formation is due either to the adhering of pellets to the surface of the skeleton grains and their subsequent transformation or to the evolution of monomorphic coatings. Indeed many times it seems that in the upper part of the profile, these coatings and pellets are losing part of their composing materials by leaching. The dark particles present in the polymorphic coatings are composed of the substances resisting the dissolution.

Compound coatings are formed by inclusion of plant remains or fragments of fungi in monomorphic or polymorphic coatings, whilst *complex coatings* are probably formed by compression of aggregates and accumulations, containing clay and silt-size particles, on the coarse mineral particles.

In many cases, the coarse skeleton grains are covered by a free grain argillan before the formation of the organic complex coating. Since these two coatings belong to a different pedogenetic process, such a feature cannot be considered as a complex coating, but has to be distinguished as a double coating. The concentrations are clusters of fragments formed by the cracking of bridges of monomorphic organic matter between coarse mineral grains or other organic units.

[*Editors' Note:* Material has been omitted at this point.]

REFERENCES

[*Editors' Note:* Only the references cited in the preceding excerpt are reproduced here.]

Altemuller, H. J., 1962. Beitrag zur mikromorphologischen Differenzierung von dulchsohlämmter Parabraunerde, Podsol-Braunarde und Humus-Podsol. Z. Pflanzenernahr., Düng., Bodenk., 98 (143): 247-258.

Jongerius, A., 1957. Morfologische onderzoekingen over de bodemstruktuur. Bodemkundige studies no. 2. Stichting voor Bodemkartering, Wageningen.

Kowalinski, St., 1969. Interdependence between micromorphological and chemical properties in some zonal soils of the Karkonosze Mountains (Poland). Geoderma, 3: 89-115.

Racz, Z., 1968. Podzols on the Territory of Croatia (Jugoslavia) and their micromorphological properties. Geoderma, 2: 41-55.

Part V
MICROSTRUCTURE

Editors' Comments
on Papers 14 and 15

14   BECKMANN and GEYGER
Entwurf einer Ordnung der natürlichen Hohlraum-, Aggregat- und Strukturformen im Boden

15   DUMANSKI and ST. ARNAUD
A Micropedological Study of Eluvial Soil Horizons

Although Kubiëna (1938) considered aggregates and cleavage blocks as fabric members of a higher order, the description of these structures was neglected for a long time in soil thin-section studies. In fact, the importance of investigating pore patterns and microstructures in thin sections was not emphasized until in the 1960s. One of the earliest and most valuable research works published in this field is that of Jongerius (1957), who also discussed the definitions of soil structure in use, and proposed definitions that could be used in soil micromorphology:

> There are 4 groups of definitions of the term soil structure, viz. 1. those which only emphasize the arrangement and aggregation of the solid constituents of the soil; 2. those in which the soil cavities are an important feature; 3. definitions which include the water in the soil; 4. the "soil fabric" concept.
>
> In our view none of these definitions is entirely satisfactory. In the first place, practically all definitions in groups 1-3 limit the term soil structure to the structural elements (i.e. the aggregates defined by natural planes). We would, however, regard all structures in the soil, with the exception of concretions and crystals, as belonging to the soil structure, including, for example, hole structures, single-grained structures, finely laminated sediments of varying grain size (chapter III) and the sand structures (chapter V). Moreover, the liquid phase is not an essential part of the soil structure. The term, "soil fabric," as employed by Kubiëna, is much too comprehensive, too great an emphasis being placed on genesis.
>
> In defining soil structure the following points should be taken into account:
>   1. The soil constituents are usually aggregated (into structural elements or otherwise) but sometimes this is not the case (single-grained structures).

2. The arrangement of the soil particles in relation to each other is essential to the structure (single-grained structures; elementary fabrics; the relative position of any structural elements).
3. The cavities in the soil (porosity of structural elements; hole structures).

From this it results that "soil structure is the spatial arrangement of the elementary constituents and any aggregates thereof, and of the cavities occurring in the soil."

Soil structures may be divided into macrostructures (any structure that can be discerned with the help of a magnifier not exceeding 4× lin.) and microstructures (structures only visible by means of more powerful magnification).

In Brewer's system (1964), voids are considered for the first time as full-bodied fabric elements, on the same level as plasma and skeleton grains. One of the difficulties encountered in describing voids is that they form a continuum. Brewer solved the problem by treating the voids, as seen in thin sections, as individuals. He proposed a subdivision, now widely used, based on their morphology, distinguishing packing voids, vughs, channels, chambers, vesicles, and planar voids. Planar voids are further subdivided according to their shape, distribution pattern, or both. Brewer's classification of ped-types was not much used, as it was not explicitly meant for thin-section work.

The micromorphometric research carried out by a group of Kubiëna's coworkers in Germany (Kubiëna, 1967) emphasized the need for a morphological classification of pores and structures. This research served as the impetus for Paper 14. In it, W. Beckmann and E. Geyger distinguish three morphological types: cracks, cavities, and intergrain pores. According to their smoothness and configuration, they can be further subdivided (Fig. 12-7). Aggregates are also subdivided into three groups: fragments, crumbs, and excrements, which are further categorized according to the roughness of their surface and their internal porosity (Fig. 12-16). The combination of pore types and aggregate types results in a set of six microstructure types: three microstructures determined by the presence of fissures (cracked, jointed, and fragmented microstructure), and three types characterized by the presence of cavities (porous, spongy, and crumbly) (Fig. 12-24). Many authors used this classification for their micromorphological descriptions in spite of a few imperfections: no distinction is made among vesicles, vughs, and channels, although this would be rather important both from a genetic and a practical point of view, and packing pores and single-grain structures are not considered, which excludes a range of materials.

Unfortunately, no one has created a more advanced descriptive system for pores and microstructures. The terminology proposed by

FitzPatrick (1980) for microstructures is an extension of the terminology used in the field for the description of soil structure.

Much attention has been given, however, to the analysis and identification of pore patterns by automatic image analyzers (e.g., Jongerius et al., 1972; Murphy, Bullock, and Turner, 1977), but no final results have been published that can be used for descriptive purposes. A discussion of this very important micromorphometric work is beyond the scope of this volume.

The classifications were developed for the general micromorphological description of any type of soil material. More limited schemes were developed by researchers studying specific aspects of soil components. For example, J. Dumanski and R. J. St. Arnaud (Paper 15) describe the microstructure and general fabric of some soils from northern Canada. Specifically, they described isoband and banded fabrics, characterized by a parallel arrangement of planar voids that separate, in the maximal stage of development, lens-shaped peds with a clay-rich upper part and a sand-rich lower part. Later it was found that these fabrics were characteristic of soils influenced by alternations of freezing and thawing (see, e.g., Romans, Stevens, and Robertsen, 1966; FitzPatrick, 1956; Van Vliet, 1975; 1976). As a result, Paper 15 has been frequently cited as one of the first micromorphological descriptions of this cryic structure. It also illustrates clearly how pore patterns and internal fabric are frequently interrelated.

REFERENCES

Brewer, R., 1964, *Fabric and Mineral Analysis of Soils,* J. Wiley & Sons, London, New York and Sydney, 470p.

FitzPatrick, E. A., 1956, An indurated soil horizon formed by permafrost, *Jour. Soil Sci.* **7:**248-254.

FitzPatrick, E. A., 1980, *The Micromorphology of Soils,* Department of Soil Science, University of Aberdeen, 186p.

Jongerius, A., 1957, *Morfologische onderzoekingen over de bodemstruktuur,* Meded. Stichting voor Bodemkartering, Bodemkundige Stud. 2, Wageningen.

Jongerius, A., D. Schoonderbeek, A. Jager, and St. Kowalinski, 1972, Electro-optical soil porosity investigation by means of Quantimet-B equipment, *Geoderma* **7:**177-198.

Kubiëna, W. L., 1938, *Micropedology,* Collegiate Press Inc., Ames, Iowa, 242p.

Kubiëna, W. L., 1967, *Die mikromorphometrische Bodenanalyse,* Ferdinand Enke Verlag, Stuttgart, 196p.

Murphy, C. P., P. Bullock, and R. H. Turner, 1977, The measurement and characterization of voids in soil thin sections by image analysis. Part I. Principles and techniques, *Jour. Soil Sci.* **28:**498-508.

Romans, J. C. C., J. H. Stevens, and B. Robertsen, 1966, Alpine soils of North-east Scotland, *Jour. Soil Sci.* **17:**184-199.

Van Vliet, B., 1975, Quelques observations à propos de structure "lamellaire" triée, *Pédologie* **3:**211.

Van Vliet, B., 1976, Traces de ségrégation de glace en lentilles associées aux sols et phénomènes périglaciaires fossiles, Biul. Peryglacjalny **26:**41-55.

14

Copyright © 1967 by Ferdinand Enke Verlag
Reprinted from pages 163–188 of *Die mikromorphometrische Bodenanalyse*,
W. L. Kubiëna, ed., Ferdinand Enke Verlag, Stuttgart, 1967, 196p.

Entwurf einer Ordnung der natürlichen Hohlraum-, Aggregat- und Strukturformen im Boden

Von Walter Beckmann und Erika Geyger

Einleitung

Klassifizierungen der Bodenstruktur können nach den verschiedensten Gesichtspunkten durchgeführt werden. Hier soll vor allem das morphologische Erscheinungsbild der Struktur und der Elemente, die sie aufbauen (Hohlräume und Aggregate) zur Unterscheidung und Klassifizierung benutzt werden.

Die Formen der natürlichen Strukturbildungen im Boden sind überaus mannigfaltig. Wir finden verschiedenste Aggregatformen, verschiedenste Hohlraumbildungen und in Kombination von diesen beiden äußerst vielfältige Strukturbildungen im Boden. Das gilt schon bei der Betrachtung der Bodenstruktur mit bloßem Auge. Die Mannigfaltigkeit nimmt noch erheblich zu, wenn Lupe und Mikroskop benutzt werden.

Eine Typisierung der natürlichen Hohlraum- und Aggregatformen ist recht schwierig. Wir kennen eine Anzahl von Versuchen, in denen unter verschiedensten Gesichtspunkten eine solche Typisierung durchgeführt wurde. Die Literatur darüber ist recht umfangreich. Hier kann daher keine Übersicht über die Literatur gegeben werden, auch soll nicht über die Richtigkeit des einen oder anderen Klassifizierungsschemas oder über Definitionsfragen diskutiert werden. Eine solche umfassendere Darstellung wird später an anderer Stelle veröffentlicht werden. Hier soll jetzt vielmehr ein erster Versuch gemacht werden, unter Auswertung der früheren Arbeiten und unseres in den letzten sieben Jahren zusammengetragenes Archives über die natürlichen Strukturbildungen im Boden eine möglichst umfassende Ordnung zu erreichen.

Eine Einteilung der Hohlraum-, Aggregat- und Strukturbildungen im Boden sollte sowohl in makroskopischen als auch in mikroskopischen Größenordnungen gültig sein. Der Name, mit dem wir ein bestimmtes Strukturelement oder einen Strukturtyp selbst charakterisieren wollen, sollte unabhängig von dessen Größenordnung sein. Eine detaillierte Charakterisierung der Bodenstruktur ist nur dann folgerichtig durchführbar, wenn — beginnend mit der makroskopischen Untersuchung, fortschreitend in immer feinere Details — gleiche Erscheinungsformen auch gleiche Namen bekommen. Eine Klassifizierung, in der auch die Größe des Objektes Unterscheidungsmerkmal sein soll, ist daher unzweckmäßig.

Unter diesen Voraussetzungen ergibt sich natürlich eine große Mannigfaltigkeit der zu ordnenden Formbildungen. Das zwingt dazu, zunächst sehr einfache und umfassende Oberbegriffe zu suchen und erst dann innerhalb dieser weiter zu differenzieren.

1. Die Hohlraumformen im Boden

Beginnen wir bei diesen Untersuchungen mit den natürlichen Hohlraumbildungen im Boden. Die Erfahrung hat gezeigt, daß es zweckmäßig ist, zunächst zu unterscheiden zwischen R i s s e n und H ö h l u n g e n.

1.1. Risse

R i s s e sind solche Hohlräume, die (trivial gesagt) durch Schrumpfung des Bodenmaterials entstanden sind. Sie sind mit bloßem Auge, mit der Lupe und dem Mikroskop leicht zu identifizieren. Ihre Formen können sehr mannigfaltig sein, aber immer kann man sich bei einem Riß vorstellen, daß man die feste Bodensubstanz wieder zusammenschieben könnte; dann würden die gegenüberliegenden Wandungen lückenlos ineinanderpassen.

Es erscheint zweckmäßig, diese rißförmigen Hohlräume nach der Art ihrer Wandausbildung weiter zu unterteilen: Wir kennen Risse, deren Wandungen auch bei höherer Vergrößerung noch vollkommen glatt sind; wir kennen aber auch solche Risse, deren Wandungen stets sehr rauh erscheinen, sie haben viele kleine Vorsprünge und Nischen, aber die gegenüberliegenden Seiten passen immer noch ineinander. Wir wollen solche nach ihrer Wandausbildung differenzierten Hohlräume als g l a t t w a n d i g e R i s s e und als r a u h w a n d i g e R i s s e bezeichnen.

G l a t t w a n d i g e R i s s e sind solche Hohlräume, deren gegenüberliegende Wände zusammenpassen; die Wandungen sind bis in hohe Vergrößerungen noch glatt und ungegliedert.

R a u h w a n d i g e R i s s e sind solche Hohlräume, deren gegenüberliegende Seiten im großen und ganzen zusammenpassen. Die Wandungen haben jedoch in kleinsten Abschnitten schon eine unregelmäßige Formbildung, also viele kleine Vorsprünge und Nischen; diese passen im einzelnen dann nicht mehr ineinander.

Eine weitere Unterteilung wollen wir nun nach der Art des Verlaufes der Risse innerhalb der festen Bodensubstanz vornehmen: Wir kennen Risse, die sehr geradlinig oder allenfalls im Zickzack durch den Boden führen. Risse können aber auch in Bögen oder sonst irgendwie ungerade verlaufen und dabei ihre Richtung ständig wechseln. Wir unterscheiden daher g e r a d e R i s s e und k r u m m e R i s s e. Natürlich können diese beiden Formen jeweils auch mit den verschiedenen Wandausbildungen vorkommen.

G e r a d e g l a t t w a n d i g e R i s s e sind also solche Hohlräume, deren gegenüberliegende Seiten zusammenpassen, deren Wände auch bei höherer Vergrößerung noch glatt erscheinen und deren Verlauf innerhalb des Bodens geradlinig ist.

Krumme glattwandige Risse sind solche Hohlräume, die zusammenpassende gegenüberliegende Wandungen haben; diese Wandungen erscheinen auch bei Betrachtung mit hoher Vergrößerung noch glatt, ihr Verlauf in der Bodenstruktur ist aber krummlinig.

Gerade rauhwandige Risse sind solche Hohlräume, die zusammenpassende gegenüberliegende Wandungen haben; ihre Wände sind aber rauh und ihr Verlauf innerhalb der Bodenstruktur ist geradlinig.

Krumme rauhwandige Risse sind solche Hohlräume, deren gegenüberliegende Seiten zusammenpassen; ihre Wände sind aber rauh und ihr Verlauf durch die Bodenstruktur krummlinig.

Diese Klassifizierung der Hohlraumformen geschieht vorerst rein qualitativ. Möglicherweise werden sich nach weiteren Untersuchungen auch quantitative Abgrenzungen festlegen lassen. Natürlich gibt es bei den Rissen eine Fülle von Übergangsbildungen, die eine Unterscheidung schwierig machen. Als Unterscheidungsmerkmal von glatt- und rauhwandigen Rissen kann z. B. oftmals folgendes benutzt werden: bei glattwandigen Rissen passen die gegenüberliegenden Seiten auch bei hoher Vergrößerung noch genau ineinander; bei rauhwandigen ist das nicht mehr der Fall.

Betrachten wir diesen Hohlraumtyp **Risse** jetzt zunächst in einigen Abbildungen: Die Abb. 12-1 zeigt in mehreren Vergrößerungen einen **geraden glattwandigen Riss**. Wir erkennen seine typischen Eigenschaften: Er durchzieht die feste Bodensubstanz ohne Richtungsänderung, seine Wände sind selbst in hoher Vergrößerung noch glatt und wenig gegliedert.

Die Abb. 12-2 bringt Bilder von einem **krummen glattwandigen Riss** in verschiedener Vergrößerung. Wir erkennen bei geringerer Vergrößerung den vielfachen Richtungswechsel des Hohlraumes; er ist auch bei mittlerer Vergrößerung noch deutlich; in allen drei Bildern sehen wir zudem, daß die Wandausbildung stets glatt und wenig gegliedert ist.

Abb. 12-3 zeigt nun in verschiedenen Vergrößerungen einen **geraden rauhwandigen Riss**. Das Bild mit der geringsten Vergrößerung beweist, daß dieser Hohlraum die feste Bodensubstanz in gerader Richtung durchzieht; das Bild mit mittlerer Vergrößerung zeigt aber, daß die Wände des Risses sehr stark gegliedert sind.

Die Abb. 12-4 gibt einen **krummen rauhwandigen Riss** wieder. Wir erkennen, daß der Hohlraum die Richtung oft wechselt, daß aber die gegenüberliegenden Seiten noch gut einander angepaßt werden könnten. Die Rauhwandigkeit des Hohlraumes ist in allen drei Vergrößerungen zu erkennen.

1.2. Zwischenbemerkung

Bei den übrigen Hohlraumformen, die wir nicht im obigen Sinne **Risse** nennen können, müssen wir nun eine grundsätzliche Trennung vornehmen: Hohlräume in Böden oder Bodenschichten, deren feste Teile nur lose und eigentlich zufällig gelagert sind, nehmen wir aus unserer morphologischen

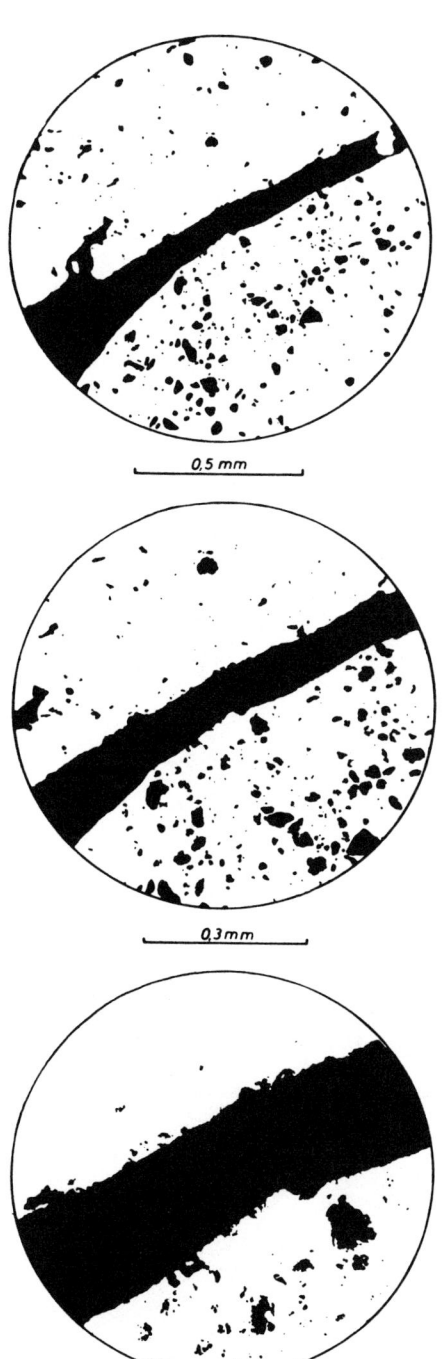

Abb. 12 - 1. Gerader glattwandiger Riß. Der gleiche Hohlraum ist in mehreren Vergrößerungen abgebildet.

Typische Eigenschaften sind:

1. Er durchzieht die feste Bodensubstanz ohne Richtungsänderung.

2. Seine Wände sind selbst in hoher Vergrößerung noch glatt und wenig gegliedert.

3. Die gegenüberliegenden Wände passen genau zusammen.

Ordnung der natürl. Hohlraum-, Aggregat- und Strukturformen 167

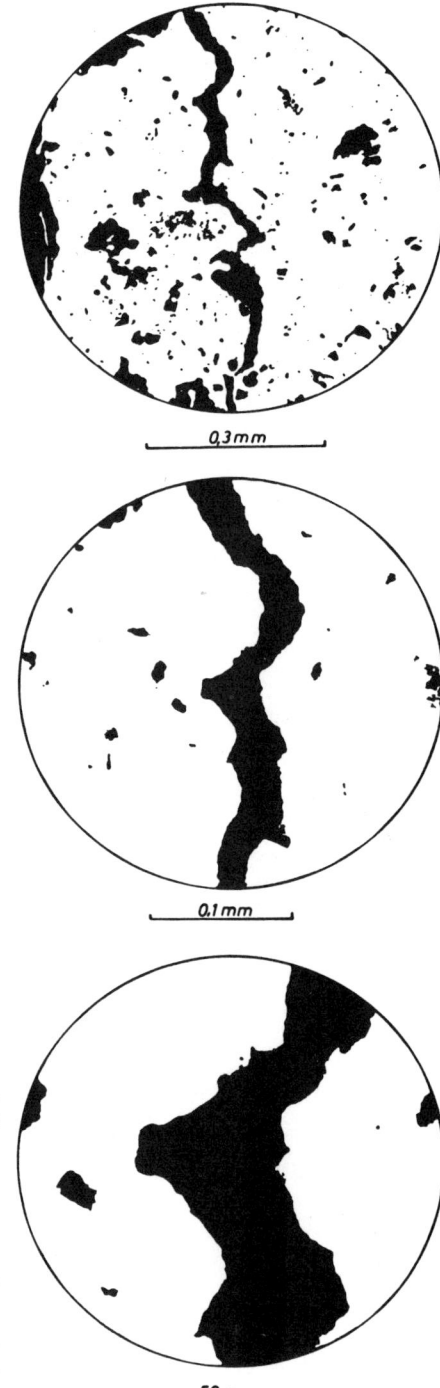

Abb. 12 - 2. Krummer glattwandiger Riß. Der gleiche Hohlraum ist in mehreren Vergrößerungen abgebildet.

Typische Eigenschaften sind:

1. Vielfacher Richtungswechsel.

2. Glatte und wenig gegliederte Wände auch in hoher Vergrößerung.

3. Die gegenüberliegenden Wände passen auch bei hoher Vergrößerung zusammen.

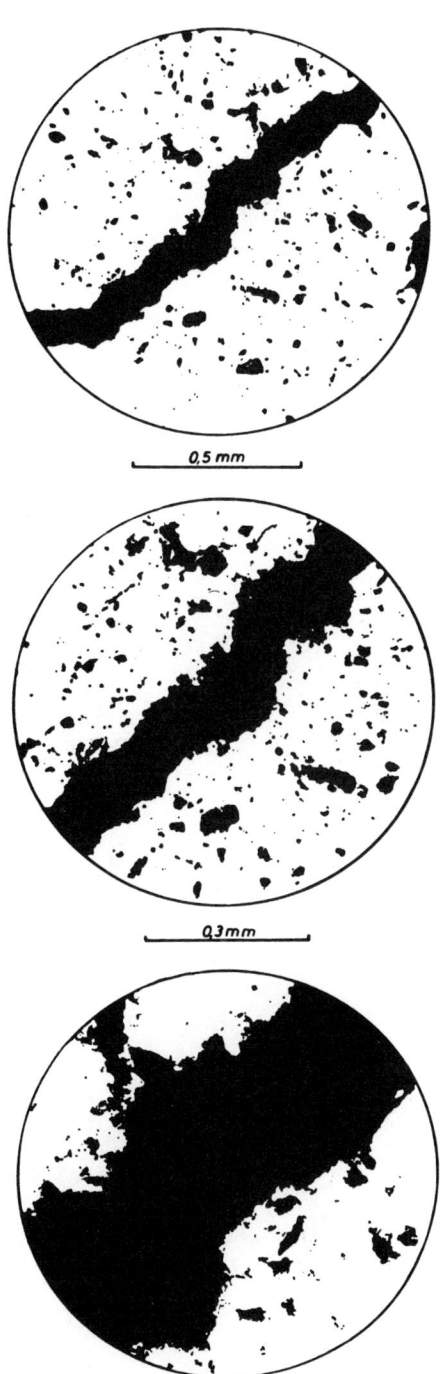

Abb. 12 - 3. Gerader rauhwandiger Riß. Der gleiche Hohlraum ist in mehreren Vergrößerungen abgebildet.

Typische Eigenschaften sind:

1. Der Hohlraum durchzieht die feste Bodensubstanz in gerader Richtung.

2. Die Wände sind rauh und stark gegliedert.

3. Die gegenüberliegenden Wände passen im großen noch zusammen.

Ordnung der natürl. Hohlraum-, Aggregat- und Strukturformen 169

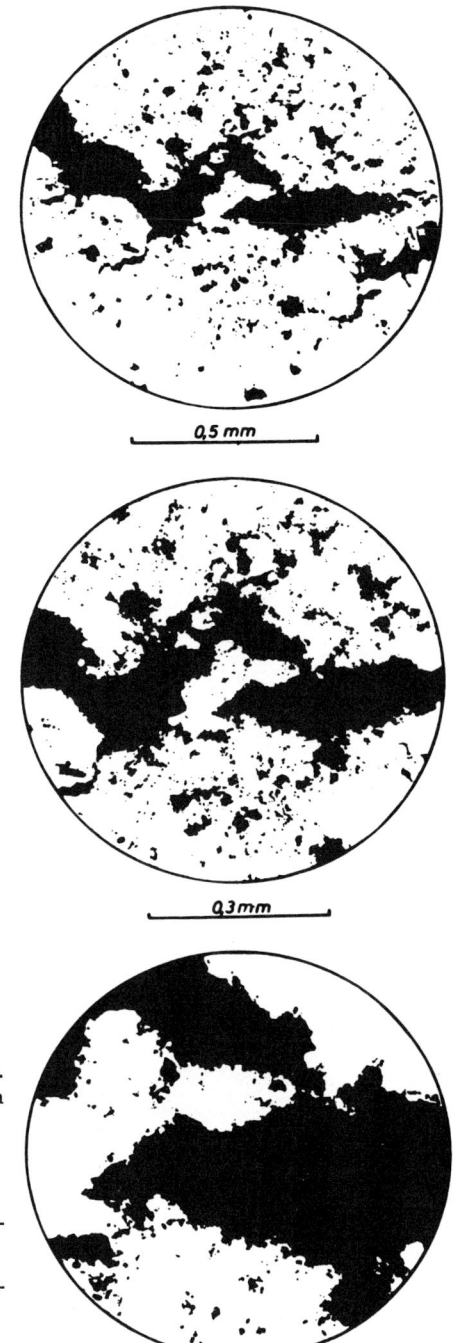

Abb. 12 - 4. Krummer rauhwandiger Riß. Der gleiche Hohlraum ist in mehreren Vergrößerungen abgebildet.

Typische Eigenschaften sind:

1. Der Hohlraum wechselt oft seine Richtung.
2. Die Wände sind rauh und stark gegliedert.
3. Die gegenüberliegenden Wände passen im großen noch zusammen.

Typisierung aus. Denn: Die Form, Größe oder Richtung dieser Hohlräume ist rein zufällig und durch die Umrißformen der festen Bestandteile und deren Lagerung bestimmt. Es sind dies z. B. Hohlräume zwischen losen Mineralkörnern, zwischen lose liegenden Pflanzenresten usw. Natürlich sind auch die Hohlräume in frischen oder halbzersetzten Pflanzen- und Wurzelresten nicht Gegenstand dieser Klassifizierung.

1.3. Höhlungen

Alle übrigen Hohlräume wollen wir Höhlungen nennen. Höhlungen sind also solche Hohlräume, deren gegenüberliegende Wandungen nicht ineinandergepaßt werden können. Ihre Formbildungen sind überaus mannigfach: Es gibt rundliche Höhlungen, allseits geschlossene Höhlungen, gangartige und vielfach verzweigte Höhlungen. Immer wechseln enge und breite Hohlraumstrecken oft und rasch ab.

Wir haben auch hier zu unterscheiden zwischen solchen Höhlungen, die eine glatte, und solchen, die eine rauhe Wandausbildung haben. Wir nennen sie glattwandige Höhlungen und rauhwandige Höhlungen:

Glattwandige Höhlungen sind solche Hohlräume im Boden, deren gegenüberliegende Wandungen nicht ineinanderpassen und deren Wände bis in höhere Vergrößerungen glatt sind.

Rauhwandige Höhlungen sind solche Hohlräume, deren gegenüberliegende Wände nicht ineinanderpassen, deren Wandungen aber schon bei geringer Vergrößerung rauh erscheinen.

In Anbetracht der stets wechselnden Breite dieser Höhlungen und wegen der sowieso schon oft wechselnden Richtung dieser Hohlräume wäre eine Unterteilung in gerade und krumme Höhlungen wenig sinnvoll. Ein anderes Merkmal zur Unterscheidung ist hier jedoch von großer Bedeutung: Für die Beurteilung einer Struktur ist es sehr wichtig, ob eine Höhlung im Boden allseits abgeschlossen ist oder ob diese Höhlung mit vielen anderen zu einem Höhlungssystem verbunden ist. Wir wollen daher unterscheiden zwischen abgeschlossenen Höhlungen und offenen Höhlungen.

Abgeschlossene Höhlungen sind demnach Hohlräume im Boden, deren Wände nicht ineinanderpassen, die innerhalb der Bodenstruktur allein und abgeschlossen sind und mit anderen Hohlräumen gar nicht oder nur durch — im Verhältnis zu ihrer eigenen Größe — wesentlich schmalere Hohlräume verbunden sind.

Offene Höhlungen sind demnach solche Bodenhohlräume, deren Wände nicht ineinanderpassen und die mit anderen Hohlräumen vielfach verbunden sind, und zwar durch Hohlräume gleicher oder ähnlicher Größenklassen wie sie selbst.

Die Unterscheidung von glattwandigen und rauhwandigen Höhlungen kann im Prinzip ähnlich geschehen wie bei den Rissen.

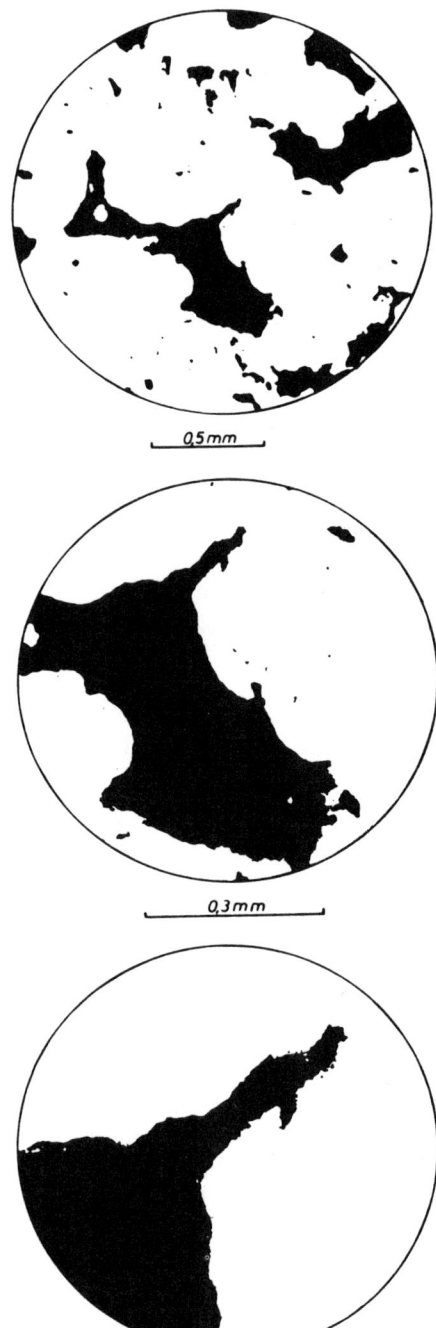

Abb. 12 - 5. Glattwandige Höhlung. Der gleiche Hohlraum ist in mehreren Vergrößerungen abgebildet.

Typische Eigenschaften sind:

1. Die Wände sind bis in hohe Vergrößerungen glatt.

2. Die gegenüberliegenden Wandungen passen nicht zusammen.

Bemerkung: Die Frage, ob es sich hier um eine abgeschlossene oder offene Höhlung handelt, kann am zweidimensionalen Schnittbild nicht sicher entschieden werden.

172   Walter Beckmann und Erika Geyger

Eine Trennung von offenen und abgeschlossenen Höhlungen kann aber nur dann sicher durchgeführt werden, wenn der tatsächliche dreidimensionale Aufbau der betreffenden Bodenstruktur bekannt ist; besonders bei der Interpretation von zweidimensionalen Schnittbildern durch die Bodenstruktur müssen wir uns davor hüten, Schnittflächen durch Höhlungen, die allseits von fester Bodensubstanz umgeben sind, schon für abgeschlossene Höhlungen zu halten. Sie können durchaus ober- oder unterhalb der Schnittebene mit anderen Höhlungen verbunden sein. Eine stereoskopische Betrachtung der Struktur und/oder die Auffindung sicherer Unterscheidungsmerkmale am Schnittbild sind zur Klärung dieser Frage unumgänglich (Beckmann und Geyger 1965).

Wiederum können wir nach der Art der Wandausbildung weiter differenzieren: Wir können sowohl bei den abgeschlossenen Höhlungen als auch

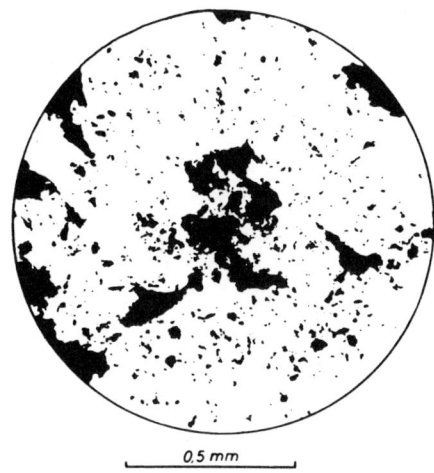

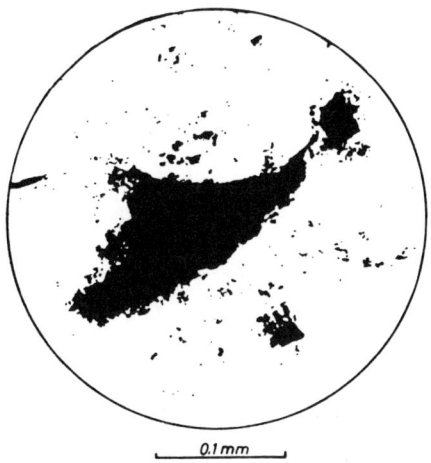

Abb. 12 - 6. Rauhwandige Höhlung(en). Die gleiche Stelle ist in zwei Vergrößerungen abgebildet.

Typische Eigenschaften sind:

1. Die Wände der Hohlräume sind sehr unregelmäßig und rauh.

2. Die gegenüberliegenden Wandungen passen nicht zusammen.

Bemerkung: Die Frage, ob es sich hier um eine abgeschlossene oder offene Höhlung handelt, kann am zweidimensionalen Schnittbild nicht sicher entschieden werden.

bei den offenen Höhlungen solche mit glatten oder mit rauhen Wänden erwarten. Wir haben also auch bei den Höhlungen vier verschiedene Formbildungen zu unterscheiden.

Betrachten wir nun den Hohlraumtyp H ö h l u n g e n in einigen Bildern: Die Abb. 12-5 zeigt — unter anderem — einige Höhlungen. Wenn wir voraussetzen, daß durch vorausgegangene vergleichende Untersuchungen des dreidimensionalen Aufbaues dieser Struktur sicher bekannt ist: die Höhlung in der Mitte des Bildes hat keine Verbindung zu den übrigen. Dann ist dieser Hohlraum eine a b g e s c h l o s s e n e g l a t t w a n d i g e H ö h l u n g. Denn wir erkennen: Die Wandungen haben keine geometrischen Beziehungen zueinander, sie passen in keiner Weise ineinander, auch sind die Wände bis in hohe Vergrößerungen glatt. Die Abb. 12-6 gibt — ebenfalls unter dem Vorbehalt dreidimensionaler Vergleichsuntersuchungen — ein Beispiel a b g e s c h l o s s e n e r r a u h w a n d i g e r H ö h l u n g e n. Die Wände der Hohlräume sind sehr unregelmäßig und rauh.

Nehmen wir hingegen an, daß Beobachtungen des dreidimensionalen Aufbaues der Bodenstruktur ergeben haben: Diese Hohlräume sind mit anderen verbunden. Dann können die beiden Bilder auch als Beispiele für o f f e n e g l a t t w a n d i g e H ö h l u n g e n (Abb. 12-5) und o f f e n e r a u h w a n d i g e H ö h l u n g e n (Abb. 12-6) gezeigt werden.

Die bei den vorherigen Bildbeispielen gemachten Vorbehalte weisen schon darauf hin, daß völlig abgeschlossene Höhlungen relativ selten sind. Wir finden weitaus mehr Höhlungen, die untereinander durch andere Hohlräume verbunden sind als solche, die keine oder nur sehr dünne Verbindungen mit anderen haben.

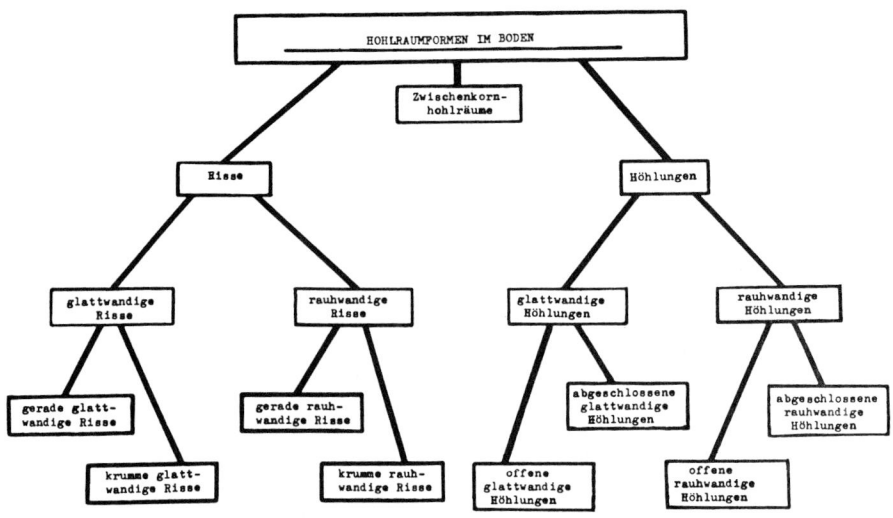

Abb. 12-7. Schema der natürlichen Hohlraumformen im Boden.

Die Formen der Hohlraumbildungen im Boden sind in einem Schema dargestellt und zusammengefaßt (Abb. 12-7).

2. Die Aggregatformen im Boden

Schließen wir jetzt die Typisierung der natürlichen Aggregatformen an. A g g r e g a t e sind Teilkomplexe der festen Bodensubstanz, die allseits abgelöst sind und nur an einer oder mehreren Stellen l o s e auf anderen Teilen der festen Substanz aufliegen.

Es ist klar, daß auch hier nur durch die Kenntnis des dreidimensionalen Aufbaues der Struktur zu entscheiden ist, ob es sich wirklich um Aggregate handelt oder nicht (B e c k m a n n und G e y g e r 1965).

Besonders bei der Interpretation von zweidimensionalen Schnittbildern durch die Bodenstruktur müssen wir uns davor hüten, Schnittflächen durch die feste Substanz, die allseits von Hohlraum umgeben zu sein scheinen, schon für Aggregate zu halten. Dies können durchaus Schnittflächen durch fest verbundene Strukturpartien sein: sie haben dann ober- und/oder unterhalb der Schnittebene Verbindungen miteinander. Eine stereoskopische Betrachtung der Struktur und/oder die Auffindung von sicheren Unterscheidungsmerkmalen sind zur Klärung dieser Frage unerläßlich (B e c k m a n n und G e y g e r 1965).

Die Typisierung der Aggregatformen muß sich in gewissem Sinne an die der Hohlraumformen anschließen, denn Aggregate und Hohlräume sind ja wie Positiv und Negativ zueinander. Wir unterscheiden also zunächst einmal Aggregate, die allseits von Rissen umgeben sind. Dies sind Absonderungsaggregate, wir nennen sie nach K u b i e n a (1953) B r ö c k e l. Ferner gibt es Aggregate, die nicht durch Absonderung entstanden sind, sie wollen wir K r ü m e l nennen (Ausnahmen sind in Abschnitt 2.2 genannt).

2.1. Bröckel

B r ö c k e l sind lose Teile fester Bodensubstanz, die durch Absonderung entstanden sind; ihre Form ist noch als die unregelmäßiger Vielecke erkennbar. Oftmals ist die Entscheidung darüber, ob es sich um Bröckel handelt, dadurch erleichtert, daß in ihrem Inneren als Hohlraumform Risse vorkommen.

Diese so charakterisierte Bröckelform kann selbstverständlich auch an Aggregaten auftreten, die innerhalb der Bodenstruktur jetzt nicht mehr von Rissen, also den ursprünglich ihre Form bedingenden Hohlräumen umgeben sind. Auch solche Aggregate werden Bröckel genannt.

Zur weiteren Unterteilung benutzen wir ähnliche Merkmale wie bei den Hohlraumformen. War vorher die W a n d ausbildung der Risse das Unterscheidungsmerkmal, so ist es jetzt die O b e r f l ä c h e n form der Aggregate. Wir unterscheiden also g l a t t f l ä c h i g e B r ö c k e l und r a u h f l ä c h i g e B r ö c k e l.

Glattflächige Bröckel sind solche Aggregate, die wie unregelmäßige Vielecke geformt sind; die Flächen stoßen mit geradlinigen Kanten zusammen; die Oberflächen sind bis in hohe Vergrößerung glatt.

Rauhflächige Bröckel sind solche Aggregate, die wie unregelmäßige Vielecke geformt sind; die Kanten zwischen den Flächen sind nicht mehr gerade; die einzelnen Oberflächenteile sind schon bei geringer Vergrößerung rauh.

Für die weitere Unterteilung der Aggregate muß jetzt als neues Merkmal die Innenstruktur herangezogen werden. Es gibt Teile der festen Bodensubstanz, die im Inneren völlig kompakt und dicht sind, aber auch solche, die im Inneren sehr hohlraumreich sind und daher aufgelöst erscheinen. Beide Fälle wurden sowohl bei glattflächigen als auch bei rauhflächigen Aggregaten beobachtet. Wir wollen daher allgemein zwischen geschlossenen und aufgelösten Aggregaten unterscheiden.

Geschlossene glattflächige Bröckel sind solche Aggregate, die wie unregelmäßige Vielecke geformt sind; ihre Oberflächen sind glatt; in ihrem Inneren sind sie hohlraumarm.

Aufgelöste glattflächige Bröckel sind solche Aggregate, die wie unregelmäßige Vielecke geformt sind; ihre Oberflächen sind glatt; in ihrem Inneren sind sie hohlraumreich.

Geschlossene rauhflächige Bröckel sind solche Aggregate, die wie unregelmäßige Vielecke geformt sind; ihre Oberflächen sind durch viele kleine Vorsprünge und Nischen rauh; in ihrem Inneren sind sie hohlraumarm.

Aufgelöste rauhflächige Bröckel sind solche Aggregate, die wie unregelmäßige Vielecke geformt sind; ihre Oberflächen sind durch viele kleine Vorsprünge und Nischen rauh; in ihrem Inneren sind sie hohlraumreich.

Betrachten wir nun diese verschiedenen Bröckelformen in Bildern:
Die Abb. 12-8 zeigt einige typische geschlossene glattflächige Bröckel. Die Oberflächen dieser Aggregate — es ist immer vorausgesetzt, daß solche Teile fester Bodensubstanz tatsächlich lose sind — sind glatt und wenig gegliedert; die allgemeine Umrißform kann man als unregelmäßig „vieleckig" bezeichnen; das Innere der Aggregate ist fast hohlraumfrei.

Die Abb. 12-9 bringt Beispiele von aufgelösten glattflächigen Bröckeln: Glatte Wände, mehr oder weniger „vieleckige" Form und sehr hoher Hohlraumgehalt im Inneren.

Die Abb. 12-10 zeigt nun geschlossene rauhflächige Bröckel. Wir erkennen, daß unter Beibehaltung der allgemein „vieleckigen" Form die Umrißlinien der Aggregate rauh erscheinen. Das Innere ist relativ dicht.

Es ist zu beachten, daß wir innerhalb von Bröckeln fast immer Risse als Hohlraumform finden, in rauhflächigen Bröckeln meist rauhwandige Risse und in glattflächigen Bröckeln meist glattwandige Risse. Es hat sich immer wieder gezeigt, daß darüber hinaus Bröckel mit sehr vielgestaltigem Umriß fast immer im Inneren krumme Risse aufweisen und daß Bröckel mit sehr einfachem Umriß im Inneren gerade Risse zeigen.

Die Abb. 12-11 zeigt nun einige **aufgelöste rauhflächige Bröckel**. Hier erkennen wir die geometrisch „vieleckige" Umrißform nur noch undeutlich. Der recht hohe Gehalt an Hohlräumen im Inneren — meist krumme rauhwandige Risse — führt auch zu einer unregelmäßigeren Umrißform der Bröckel. Wir haben hier bereits Übergangsbildungen zu anderen Aggregatformen.

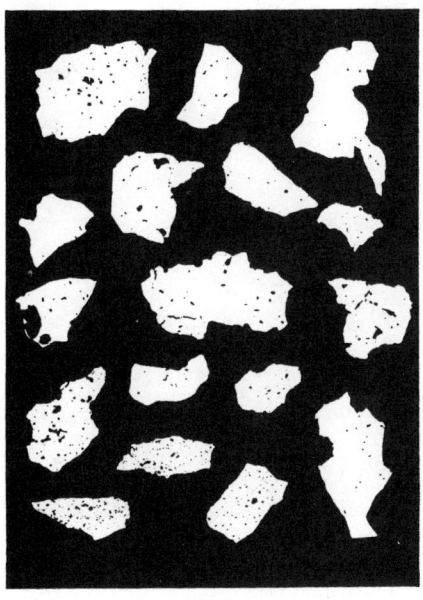 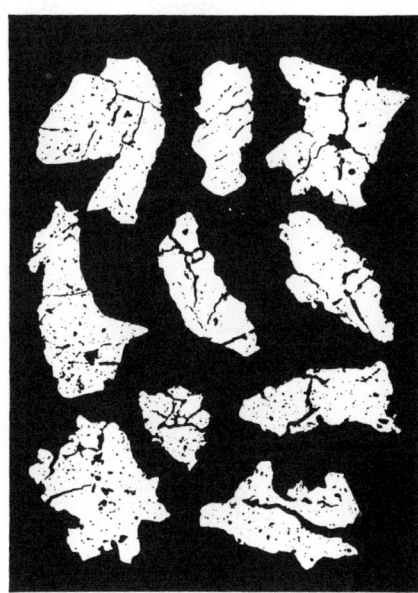

12 - 8   12 - 9

Abb. 12 - 8. Geschlossene glattflächige Bröckel.

Typische Eigenschaften sind:
1. Das Innere der Aggregate ist nahezu hohlraumfrei.
2. Die Oberflächen sind glatt.
3. Die allgemeine Umrißform ist unregelmäßig vieleckig.

Abb. 12 - 9. Aufgelöste glattflächige Bröckel.

Typische Eigenschaften sind:
1. Das Innere der Aggregate ist hohlraumreich.
2. Die Oberflächen sind glatt.
3. Die allgemeine Umrißform ist unregelmäßig vieleckig.

2.2. Zwischenbemerkung

Eine andere Gruppe von losen Teilchen innerhalb der Bodenstruktur sind die frischen, in ihrer Form noch unveränderten Losungen der Bodentiere. Sie werden hier nicht berücksichtigt, denn ihre Typisierung ist eine bodenzoologische Aufgabe. Wenn aber diese frischen Losungen durch Prozesse im Boden in ihrer Form verändert werden, wenn sie etwa aufgelöst oder mit anderen zu größeren Komplexen vereinigt werden, sind die neu entstehenden Aggregatformen wieder Gegenstand unserer Typisierung. (Die Grenzen sind also nur schwer zu ziehen.) Auch die Reste des Bestandesabfalles sowie die frischen oder halbzersetzten Wurzeln oder andere pflanzliche Produkte im Boden sind hier nicht zu berücksichtigen.

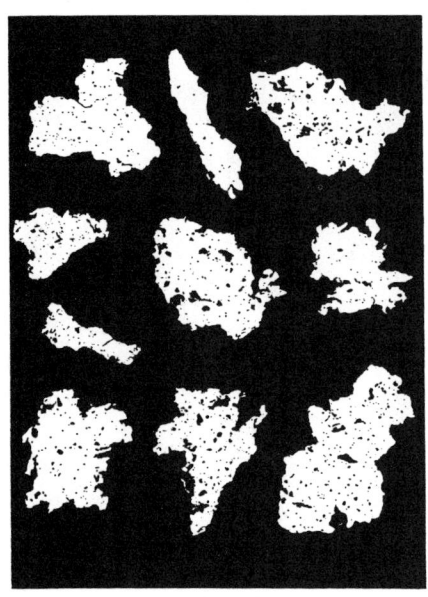
12 - 10

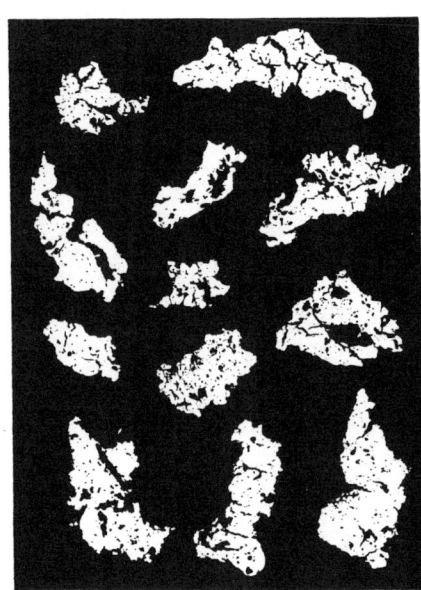
12 - 11

Abb. 12 - 10. Geschlossene rauhflächige Bröckel.

Typische Eigenschaften sind:
1. Das Innere der Aggregate ist nahezu hohlraumfrei.
2. Die Oberflächen sind rauh.
3. Die allgemeine Umrißform ist unregelmäßig vieleckig.

Abb. 12 - 11. Aufgelöste rauhflächige Bröckel.

Typische Eigenschaften sind:
1. Das Innere der Aggregate ist hohlraumreich.
2. Die Oberflächen sind rauh.
3. Die allgemeine Umrißform ist unregelmäßig vieleckig.

2.3. Krümel

Alle übrigen Teile fester Substanz im Boden nennen wir Krümel. Es sind dies ganz allgemein gesagt solche Aggregate, die innerhalb der oben genannten Höhlungssysteme liegen. Wir sagen also:

Krümel sind solche Aggregate, die sehr komplizierte Umrißformen haben und deren Seitenflächen keinerlei geometrische Beziehung zuein-

12 Kubiena, Die mikromorphometrische Bodenanalyse

178   Walter Beckmann und Erika Geyger

ander haben. Wieder können wir solche Aggregate auch losgelöst vom Strukturverband ihrer Entstehung finden, so z. B. also nicht nur in Höhlungen, sondern auch lose in Rissen liegend; auch dann sprechen wir von Krümeln.

Eine Unterteilung soll zunächst wieder nach der Oberflächenform erfolgen. Wir finden glattflächige Krümel und rauhflächige Krümel.

Glattflächige Krümel sind solche Bodenaggregate, die eine völlig ungeometrische Form haben und eine bis in hohe Vergrößerungen glatte Oberfläche besitzen.

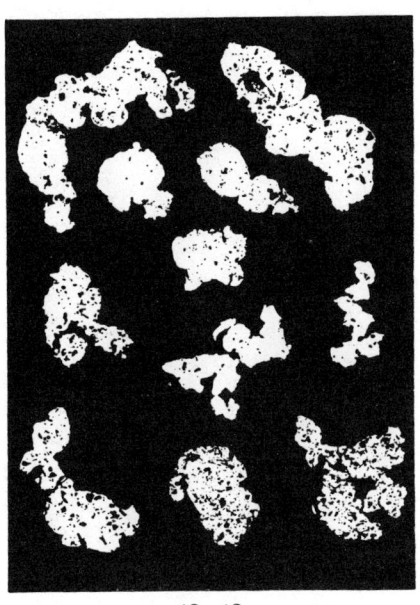
12 - 12

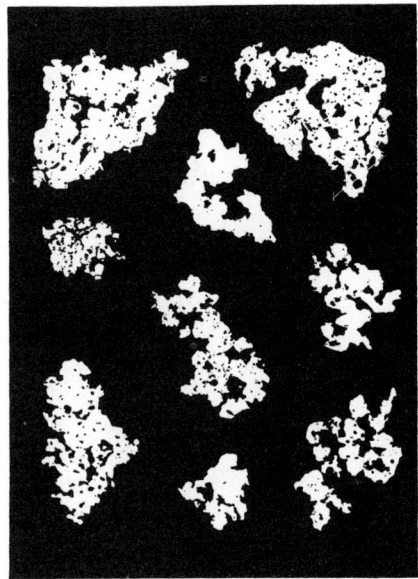
12 - 13

Abb. 12 - 12. Geschlossene glattflächige Krümel.
Typische Eigenschaften sind:
 1. Das Innere der Aggregate ist hohlraumarm.
 2. Die Oberflächen sind glatt.
 3. Die Umrißform ist ungeometrisch-vielgestaltig.

Abb. 12 - 13. Aufgelöste glattflächige Krümel.
Typische Eigenschaften sind:
 1. Das Innere der Aggregate ist hohlraumreich.
 2. Die Oberflächen sind glatt.
 3. Die Umrißform ist ungeometrisch-vielgestaltig.

Rauhflächige Krümel sind solche Aggregate, die kompliziert und völlig ungeometrisch ausgebildet sind und deren Oberflächen schon in geringer Vergrößerung rauh erscheinen.

Für die weitere Unterteilung soll — wie bei den Bröckeln — die Innenstruktur maßgebend sein. Es gibt Krümel, die nach außen eine recht geschlossene Umrißform haben und dann im Inneren sehr hohlraumarm sind und solche, die in ihrem Inneren sehr hohlraumreich sind und die dann gleichzeitig auch eine bedeutend kompliziertere Umrißform haben. Wir unterscheiden also **geschlossene Krümel** und **aufgelöste Krümel**.

Geschlossene Krümel sind also solche Bodenaggregate, die eine vielgestaltige ungeometrische Umrißform haben und im Inneren hohlraumarm sind.

Aufgelöste Krümel sind demnach solche Aggregate im Boden,

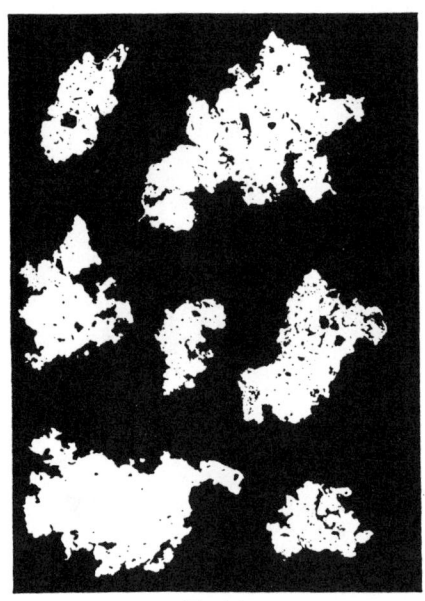 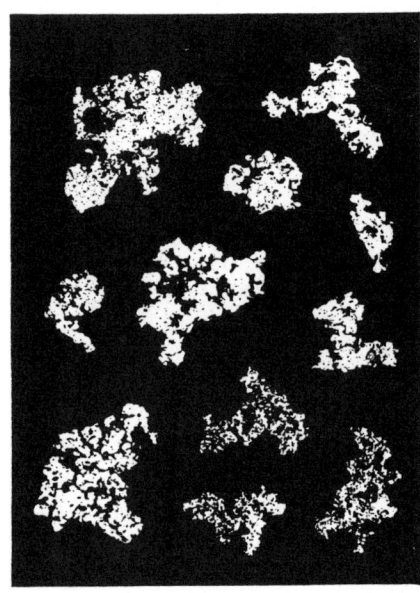

12 - 14   12 - 15

Abb. 12 - 14. Geschlossene rauhflächige Krümel.
Typische Eigenschaften sind:
1. Das Innere der Aggregate ist hohlraumarm.
2. Die Oberflächen sind rauh.
3. Die Umrißform ist ungeometrisch-vielgestaltig.

Abb. 12 - 15. Aufgelöste rauhflächige Krümel.
Typische Eigenschaften sind:
1. Das Innere der Aggregate ist hohlraumreich.
2. Die Oberflächen sind rauh.
3. Die Umrißform ist ungeometrisch-vielgestaltig.

die eine sehr vielgestaltige Umrißform haben und die im Inneren viele Hohlräume zeigen.

Es sind rauh- und glattflächige Krümel zu unterscheiden. Wir haben also — wie schon bei den Bröckeln — vier verschiedene Formen: **geschlossene glattflächige Krümel, geschlossene rauhflächige**

Krümel, aufgelöste glattflächige Krümel und aufgelöste rauhflächige Krümel.

Auch die verschiedenen Krümelformen sollen in einigen Bildern gezeigt werden.

Die Abb. 12-12 zeigt zunächst geschlossene glattflächige Krümel. Wir erkennen relativ komplizierte Umrißformen, aber nur sehr geringen Hohlraumgehalt im Inneren.

In Abb. 12-13 sehen wir jetzt einerseits einen bedeutend höheren Gehalt an Intra-Aggregat-Hohlräumen und andererseits auch kompliziertere Umrißformen der Krümel; dabei sind aber die Oberflächen der Krümel noch glatt. Dies sind aufgelöste glattflächige Krümel.

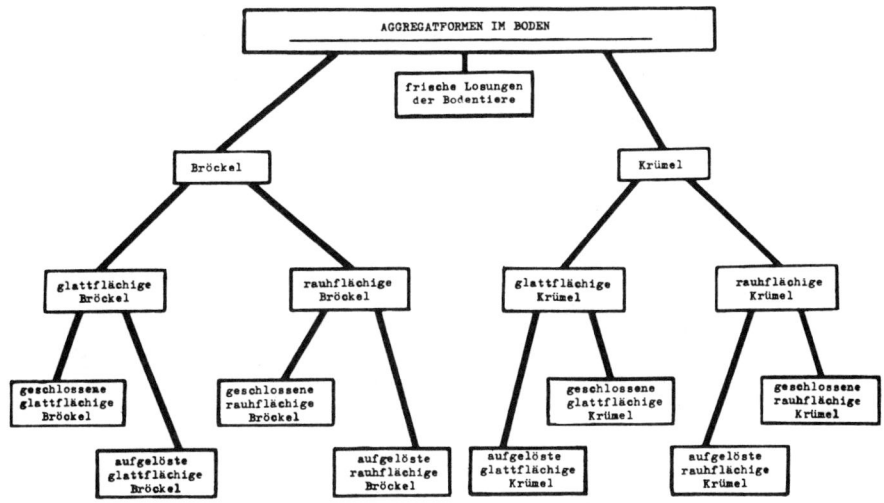

Abb. 12-16. Schema der natürlichen Aggregatformen im Boden.

In Abb. 12-14 beobachten wir jetzt Krümel, die eine sehr rauhe Oberfläche haben. Sie sind aber im Inneren dicht; es sind geschlossene rauhflächige Krümel.

In Abb. 12-15 sind die Aggregate jetzt aber im Inneren sehr stark durch Hohlräume aufgelockert; dies sind nun aufgelöste rauhflächige Krümel. Zwischen der mehr oder weniger stark gegliederten Umrißform der Krümel und dem Hohlraumgehalt im Inneren besteht oft eine sehr enge Beziehung.

Die natürlichen Aggregatformen sind in einem Schema zusammengestellt worden (Abb. 12-16).

3. Die Strukturformen im Boden

Die Kombination von diesen verschiedenen Hohlraum- und Aggregatformen führt nun im Boden zu mannigfaltigen Formbildungen der Struktur.

Ganz allgemein gilt: „Die Bodenstruktur ist das Ergebnis der Fähigkeit des Bodens, Hohlräume und/oder Aggregate zu bilden". Daher müssen wir dichte Bodenmassen, die weder Hohlräume noch Aggregate erkennen lassen, aus der Klassifizierung der Strukturformen ausnehmen: Solche Böden bzw. Bodenschichten sind strukturlos.

Das Auftreten von Aggregaten im Boden ist zwangsläufig mit dem Vorhandensein von Hohlräumen verbunden. Das Umgekehrte gilt jedoch nicht: In einer Bodenstruktur können durchaus Hohlräume vorhanden sein, ohne daß es zur Bildung von Aggregaten kommt. Aus diesem Grunde erscheint es sinnvoll, die Strukturformen des Bodens zunächst einmal nach der Art der Hohlraumbildung zu unterscheiden und erst dann, wenn tatsächlich Aggregate gebildet werden, diese Aggregate als weiteres Unterscheidungsmerkmal zu benutzen.

Enthält eine Bodenstruktur als Hohlraumform vorwiegend oder ausschließlich Risse, so sprechen wir ganz allgemein von einer Riss-Struktur. Sind dagegen die Hohlräume, die die feste Bodensubstanz ausschließlich oder in erster Linie durchziehen, Höhlungen, so sprechen wir von einer Höhlungs-Struktur.

Dabei sollen die vorher gegebenen Definitionen von Rissen und Höhlungen als Unterscheidungsmerkmale dienen.

In vielen Fällen mag die Unterscheidung Rißstruktur und Höhlungsstruktur schon ausreichen. Wir können aber weiter differenzieren; dabei soll wieder die Morphologie der Hohlräume und darüber hinaus auch der Verbund der Hohlräume herangezogen werden.

3.1. Durch Risse bestimmte Strukturformen

Beginnen wir bei der Klassifizierung zunächst mit Strukturbildungen, die durch Risse charakterisiert sind. Wir haben dabei folgende Fälle zu unterscheiden:

Sind die Risse in einer Struktur nicht miteinander verbunden, schneiden sie also nur in feste Bodensubstanz ein, ohne Systeme zu bilden, so sprechen wir von einer Spaltstruktur.

Verbinden sich die Risse dann zu Systemen, so wollen wir von einer Fugenstruktur sprechen.

Wird aber die feste Bodensubstanz von einem so engen Netz von Rissen durchzogen, daß sich Aggregate (Bröckel) bilden, so sprechen wir von einer Bröckelstruktur.

Eine Spaltstruktur ist also eine solche Bodenstruktur, in der die feste Bodensubstanz von Rissen durchzogen ist, in der die Risse aber voneinander getrennt sind und kaum Verbindung untereinander haben. Daher ist die feste Bodensubstanz nicht in Aggregate aufgeteilt.

Eine Fugenstruktur ist eine solche Bodenstruktur, in der sich die

Risse zu mehr oder weniger regelmäßigen Systemen vereinigen, ohne daß sich aber schon lose Aggregate bilden konnten.

Eine B r ö c k e l s t r u k t u r ist demnach eine solche Bodenstruktur, in der sich Risse zu einem so dichten Netz verbinden, daß zwischen den Hohlräumen lose Aggregate (Bröckel) entstanden sind.

Bei allen drei Strukturtypen können wir nun nach der Art der Wandausbildung und nach der Art des Verlaufes der Risse weiter unterscheiden. Gleichermaßen ist die Form der entstandenen Bröckel ein Unterscheidungsmerkmal.

Wir wollen uns nun diese Strukturtypen im Bild ansehen: Die Abb. 12 - 17 zeigt eine S p a l t s t r u k t u r. Hier erkennen wir Risse, die in die feste Bodensubstanz eingreifen, ohne sich zu vereinigen und ohne die feste

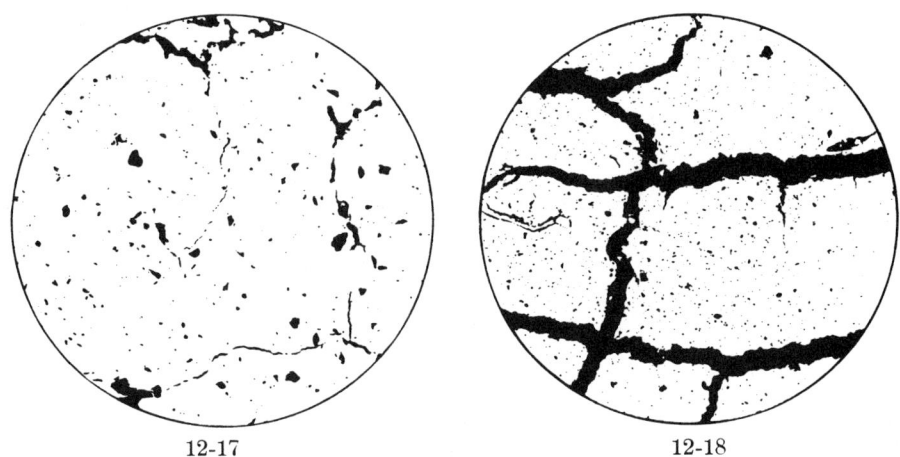

Abb. 12 - 17. Spaltstruktur.
Typische Eigenschaften sind:
1. Die zugehörige Hohlraumform sind R i s s e.
2. Die Risse sind voneinander getrennt und haben kaum Verbindung miteinander.
3. Lose Aggregate treten nicht auf.

Abb. 12 - 18. Regelmäßige Fugenstruktur.
Typische Eigenschaften sind:
1. Die zugehörige Hohlraumform sind R i s s e.
2. Die Risse sind zu regelmäßigen Systemen verbunden.
3. Lose Aggregate treten noch nicht auf.

Bodensubstanz in einzelne Teile (Bröckel) zu zerschneiden. In einer solchen Struktur können verschiedene Rißformen vorkommen. Wir können also eine Spaltstruktur noch genauer beschreiben, wenn wir gleichzeitig mit angeben, ob die Risse gerade oder krumm, rauh- oder glattwandig sind.

Vereinigen sich jetzt die Risse zu mehr oder weniger regelmäßigen Systemen, so erhalten wir folgendes Bild (Abb. 12 - 18). Die Risse sind hier geradlinig rauhwandig und sehr regelmäßig angeordnet, so daß der Fugencharakter dieser Struktur besonders gut sichtbar wird. Im allgemeinen sind die Rißsysteme weit unregelmäßiger. In Abb. 12 - 19 ist ein solcher Fall gezeigt. Die Risse sind verschieden geformt, wechseln oftmals ihre Breite und haben auch unterschiedliche Wandausbildungen. Auch hier ist keine Aggregatbildung vorhanden. Wir erkennen aber den Unterschied zur Spaltstruktur.

Bilden jetzt aber die Risse — welcher Art auch immer — ein so dichtes und engverzweigtes Netz, daß zwischen den Rissen Aggregate, d. h. Bröckel, gebildet werden, so haben wir eine B r ö c k e l s t r u k t u r (Abb. 12 - 20). Eine solche Struktur kann aus Bröckeln der verschiedensten Art aufgebaut sein.

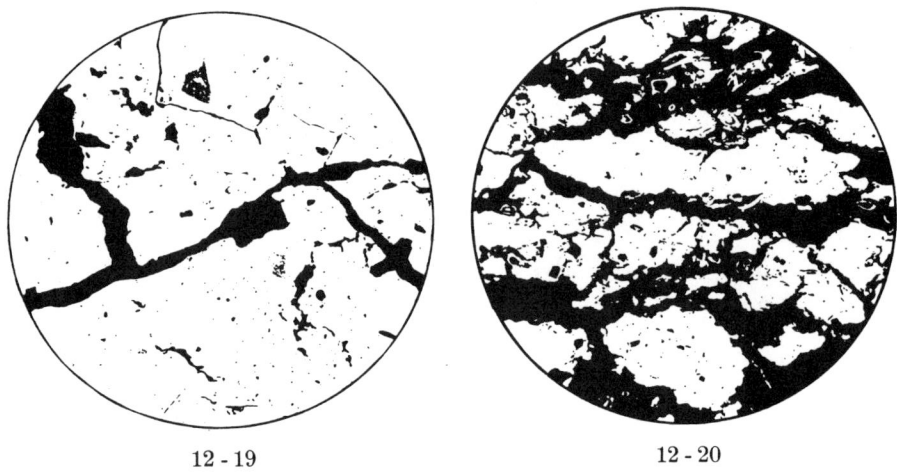

12 - 19   12 - 20

Abb. 12 - 19. Unregelmäßige Fugenstruktur.

Typische Eigenschaften sind:
1. Die zugehörige Hohlraumform sind R i s s e.
2. Die Risse sind zu unregelmäßigen Systemen verbunden, ihre Breite ist verschieden, ebenso die Wandausbildung.
3. Lose Aggregate treten noch nicht auf.

Abb. 12 - 20. Bröckelstruktur.

Typische Eigenschaften sind:
1. Die zugehörige Hohlraumform sind R i s s e.
2. Die Risse sind zu einem so dichten Netz verbunden, daß zwischen den Hohlräumen lose Aggregate entstanden sind.
3. Die gebildeten Aggregate sind B r ö c k e l.

3.2. Durch Höhlungen bestimmte Strukturformen

Wenn in einer Bodenstruktur Höhlungen die vorherrschende Hohlraumform sind, so sprechen wir von einer H ö h l u n g s s t r u k t u r. Das weitere Unterteilungsprinzip ist das gleiche wie bei den Rißstrukturen. Die Höhlungen können im obengenannten Sinne abgeschlossen sein, d. h. voneinander getrennt; sie können aber auch zu Höhlungssystemen verbunden sein;

schließlich kann das Netz der Höhlungen so eng werden, daß Aggregate (Krümel) abgetrennt werden. Wir wollen diese Strukturbildungen Porenstruktur, Schwammstruktur und Krümelstruktur nennen.

Eine Porenstruktur ist demnach eine Bodenstruktur, in der als Hohlraumform Höhlungen vorherrschen, in der aber diese Höhlungen voneinander getrennt sind. Aggregate fehlen.

Eine Schwammstruktur soll eine Bodenstruktur sein, in der die vorherrschende Hohlraumform Höhlungen sind, die aber zu einem schwammartigen Hohlraumsystem verbunden sind. Lose Aggregate gibt es nicht.

Eine Krümelstruktur ist eine Bodenstruktur, in der die vorherr-

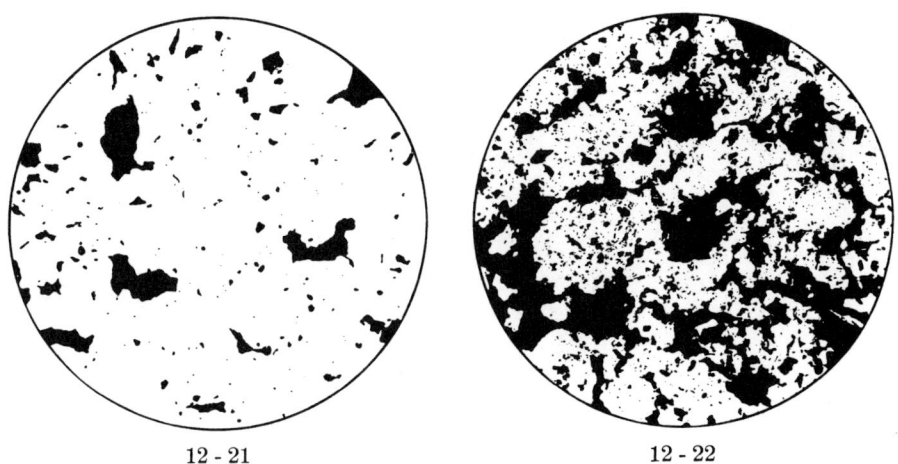

12 - 21   12 - 22

Abb. 12 - 21. Porenstruktur.

Typische Eigenschaften sind:
1. Die zugehörige Hohlraumform sind Höhlungen.
2. Die Höhlungen sind voneinander getrennt.
3. Lose Aggregate treten nicht auf.

Abb. 12 - 22. Schwammstruktur.

Typische Eigenschaften sind:
1. Die zugehörige Hohlraumform sind Höhlungen.
2. Die Höhlungen sind zu einem zusammenhängenden System verbunden.
3. Auch die feste Bodensubstanz hängt überall zusammen, lose Aggregate fehlen oder treten zurück.

schende Hohlraumform Höhlungen sind, in der aber das Höhlungssystem so eng verzweigt ist, daß sich Aggregate (Krümel) gebildet haben.

Auch diese Strukturtypen sollen in einigen Bildern gezeigt werden:

Die Abb. 12 - 21 gibt einen Ausschnitt aus einer Porenstruktur wieder. Wie schon bei der Beschreibung der abgeschlossenen Höhlungen betont wurde, ist die Bestimmung des wahren räumlichen Aufbaues der Bodenstruktur notwendig. Nur dann, wenn die im Bild sichtbaren Höhlun-

Ordnung der natürl. Hohlraum-, Aggregat- und Strukturformen 185

gen tatsächlich voneinander getrennt sind, dürfen wir von einer Porenstruktur sprechen. Sind sie untereinander verbunden, so zeigt dies Bild eine Vorstufe zur S c h w a m m s t r u k t u r.

Die Abb. 12 - 22 zeigt eine typische S c h w a m m s t r u k t u r. Wir erkennen ein vielgestaltiges Netz von verschieden breiten und sehr unregelmäßig geformten Höhlungen, die überall miteinander verbunden sind. (Wichtig ist, daß durch besondere räumliche Untersuchungen geklärt ist, daß keine losen Aggregate gebildet worden sind.) Die feste Bodensubstanz soll in einer Schwammstruktur überall durch viele Brücken und Stege zusammenhängen.

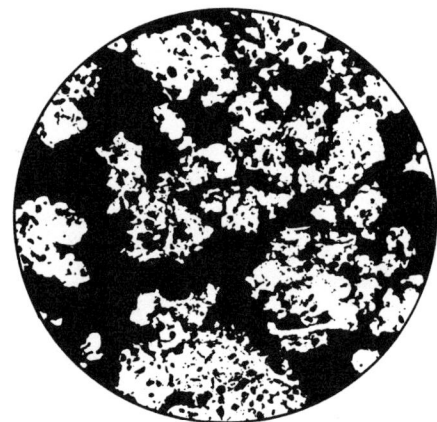

Abb. 12 - 23. Krümelstruktur.

Typische Eigenschaften sind:
1. Die zugehörige Hohlraumform sind H ö h l u n g e n.
2. Das Netz der Höhlungen ist so dicht, daß zwischen ihnen lose Aggregate zu finden sind.
3. Die vorkommenden Aggregate sind K r ü m e l.

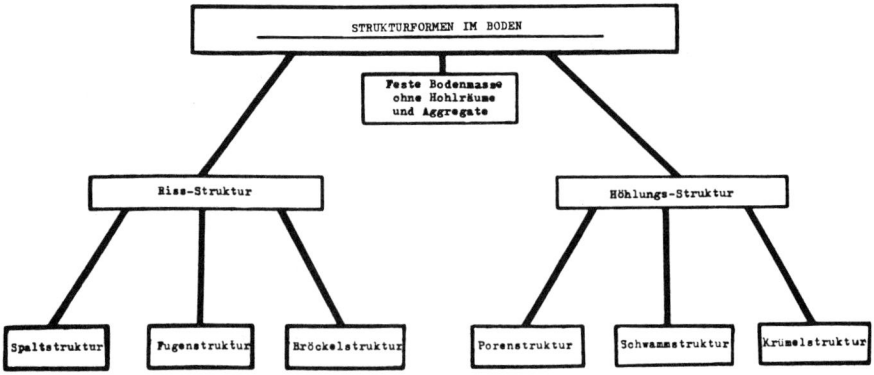

Abb. 12 - 24. Schema der natürlichen Strukturformen im Boden.

Wird dagegen der Gehalt an Höhlungen so hoch und das Netz der Hohlräume so dicht, daß lose Aggregate entstehen, so sprechen wir von einer K r ü m e l s t r u k t u r. Die Abb. 12 - 23 zeigt eine K r ü m e l s t r u k t u r.

Die natürlichen Strukturformen im Boden sind in einem Schema zusammengestellt (Abb. 12 - 24). Die Einteilung geht von der S p a l t -, F u g e n - und B r ö c k e l s t r u k t u r bis zur P o r e n -, S c h w a m m - und K r ü m e l s t r u k t u r. Eine noch weitergehende Unterteilung der Bodenstrukturformen kann durchaus noch unter Berücksichtigung der vorher beschriebenen verschiedenen Hohlraum- und Aggregatformen vorgenommen werden.

4. Bewertung der Strukturbildungen im Boden

Bei der Bewertung der natürlichen Hohlraum-, Aggregat- und Strukturformen im Boden wollen wir die langjährigen Vergleichsuntersuchungen heranziehen, bei denen immer die natürliche Formbildung zusammen mit dem Verhalten dieser Form betrachtet wurde.

Beginnen wir wieder mit den Hohlraumformen:

Es hat sich gezeigt, daß Risse wenig stabil sind; sie schließen sich mit zunehmender Befeuchtung. Bei ausreichender Feuchtigkeit kann ein Boden, in dem nur Risse vorkommen, völlig hohlraumfrei werden. Diese Erscheinung ist besonders deutlich, wenn gerade glattwandige Risse vorherrschen, ein wenig schwächer, wenn krumme glattwandige Risse überwiegen.

Wenn wir erkennen, daß die Wandausbildung der Risse rauh und gegliedert ist, dürfen wir mit einer etwas höheren Stabilität der Struktur gegen Wasser rechnen.

Wir können eine Reihe zunehmender Wasserstabilität von den geraden glattwandigen Rissen bis zu den krummen rauhwandigen Rissen aufstellen.

Eine entscheidende Verbesserung der Stabilität finden wir erst dann, wenn Höhlungen das Strukturbild bestimmen. Höhlungen sind im allgemeinen recht beständig. Ihre verschiedenartige Wandausbildung ist besonders wichtig für die Beurteilung des Bodenwasserhaushaltes.

Von ganz besonderer Bedeutung für den Luft- und Wasserhaushalt ist die Frage, ob es sich um abgeschlossene oder um offene Höhlungen handelt.

Bei den Aggregatformen gilt entsprechendes: Wir haben eine Reihe zunehmender Wasserstabilität von den geschlossenen glattflächigen zu den aufgelösten rauhflächigen Bröckeln. Die wirklich stabilen Aggregate sind aber erst die Krümel. Bei den Bröckeln müssen wir immer damit rechnen, daß sie sich bei Befeuchtung wieder zu großen Komplexen dichter Bodenmasse vereinigen; bei Austrocknung können dann wieder neue Bröckel gebildet werden, jedoch oft an anderer Stelle und von anderen Größenordnungen.

Anders verhält es sich bei den Krümeln, sie sind gegen Wasser sehr stabil. Jedoch muß bei ihnen berücksichtigt werden, daß die oft kompliziert geformten, über schmale Brücken und Stege sich fortsetzenden Krümel gegenüber mechanischer Beanspruchung sehr empfindlich sind, und zwar zunehmend von den geschlossenen glattflächigen bis zu den aufgelösten rauhflächigen Krümel. Diese Aggregate können z. B. durch die grabende und wühlende Tätigkeit der Bodentiere in ihrer Form stark verändert werden; das ist besonders in feuchtem und plastischem Bodenmaterial zu beobachten. Wo solcher Einfluß oft wiederholt wird, etwa in Tiergängen, führt er oft zu einer abgerundeten Form der Bodenaggregate.

Bei der Bewertung der Strukturformen werden die schon erkannten Beziehungen zwischen den Formen der Hohlräume und/oder Aggregate und ihren physikalischen Eigenschaften nun in der verschiedensten Weise kombiniert vorgefunden. Allgemein kann gesagt werden, daß eine Struktur, die überwiegend Risse als Hohlraumform und Bröckel als Aggregatform hat, instabil ist. Eine Struktur, in der Höhlungen und Krümel überwiegen, ist als relativ stabil anzusehen. Im einzelnen folgen die Stabilitätsreihen denen für Hohlraum- und Aggregatformen. Die Beurteilung einer Bodenstruktur muß also von den Mengenanteilen der verschiedenen Hohlraum- und Aggregatformen ausgehen.

Zusammenfassung

In dieser Arbeit wird versucht, die natürlichen Strukturformen im Boden nach ihrem morphologischen Erscheinungsbild zu klassifizieren und zu bewerten. Auch die Elemente, aus denen sich die Bodenstruktur aufbaut, Hohlräume und Aggregate, werden vor allem nach ihrer Gestalt unterschieden und typisiert. Zur Ordnung der Formenmannigfaltigkeit werden einige umfassende Oberbegriffe aufgestellt, die dann nach entsprechenden Gesichtspunkten weiter unterteilt werden; dabei gelten alle Begriffe unabhängig von der Größe der Einzelform.

Bei den Hohlräumen werden als Hauptformen R i s s e und H ö h l u n g e n unterschieden; die weitere Differenzierung erfolgt nach der Wandausbildung der Hohlräume und ihrem Verlauf innerhalb der Struktur. Bei den Aggregaten unterscheiden wir B r ö c k e l und K r ü m e l; beide werden nach der Oberflächenausbildung und nach der Innenstruktur weiter unterteilt. Die Strukturformen werden zunächst danach unterschieden, welche Hohlraumformen in ihnen auftreten; weiter, in welcher Weise diese Hohlräume in die Struktur eingreifen und ob sie die feste Bodensubstanz in lose Aggregate zerteilen.

Eine Bewertung der natürlichen Hohlraum-, Aggregat- und Strukturformen wurde durch Vergleiche ihrer Form und ihres physikalischen Verhaltens versucht. Es ergab sich bei den Hohlräumen, daß Risse bei Befeuchtung im allgemeinen nicht stabil sind, und zwar um so weniger, je glattwandiger und geradliniger sie sind; Höhlungen sind im allgemeinen wasserstabil; hier sind die Wandausbildung und der Zusammenhang der Hohlräume untereinander wichtig für den Luft- und Wasserhaushalt des Bodens. Entsprechendes gilt bei den Aggregaten bezüglich ihrer Oberflächenausbildung und ihrer Innenstruktur. Die Eigenschaften der natürlichen Bodenstrukturen ergeben sich aus den genannten Charakteristika der auftretenden Hohlräume und Aggregate und deren Mengenanteilen.

Résumé

Ce mémoire présente un essai de classification et d'évaluation des formes naturelles de structure dans le sol d'après leur apparence morphologique. Les éléments qui composent ces structures — vides et agrégats — sont aussi distingués et classés en types surtout selon leur forme.

Pour classer une grande variété de formes, certaines notions générales sont d'abord établies, qui sont ensuite subdivisées selon divers points de vue. Toutes les notions valent indépendamment de la grandeur prise par une forme donnée.

Quant aux vides, les principales formes distinguées sont les **fentes (fissures)** et les **cavités**. La différenciation se poursuit d'après la formation des parois qui entourent les vides et le cours que présentent ceux-ci dans la structure. Parmi les agrégats, on a classé par **fragments et grumeaux**, à subdiviser conformément à la constitution de leur surface et leur structure interne. Les formes de structure sont d'abord classées d'après les formes de vides qui s'y présentent, puis d'après la manière dont les vides interviennent dans la structure et démembrent — ou non — la substance solide du sol en agrégats détachés.

En comparant leur forme et leur comportement physique, l'auteurs tentent une évaluation des formes naturelles des vides, des agrégats et des structures. Chez les vides, il apparaît que les fentes ne sont en général pas stables lorsqu'elles sont humectées, et ceci d'autant moins que les parois sont plus lisses et droites. Les cavités accusent en général une bonne stabilité sous l'influence de l'eau; la formation des parois et leur enchaînement y sont importants pour le bilan de l'air et de l'eau dans le sol. Chez les agrégats, des déductions analogues peuvent être tirées de la constitution de la surface et de la structure interne. Les particularités des structures naturelles du sol sont données par les caractères énumérés des vides et des agrégats présents et par leur dosage.

Summary

This work is an attempt to classify and to evaluate the natural structural forms in the soil according to their morphological appearance. The elements which constitute a soil structure, such as pores and aggregates will be differentiated and classified according to their shape. Some general principles are proposed for the classification of the diversity of forms which occur. These principles or definitions are independant of the size of a single form.

Pores are divided into two major categories: **fissures** and **cavities**, which are further subdivided on the basis of the configuration of their walls and their deployment through the structure. Aggregates are divided into **fragments** and **crumbs**, these are further differentiated on surface outline and internal structure. Structure forms are divided into those with fissures and those with cavities and further subdivision depends on the way the pores traverse the structure and whether they separate the solid material into loose aggregates.

An evaluation of the natural forms of pores, aggregates and structures has been tried by a comparison of these forms with their physical behaviour. Investigations of pores revealed that fissures in general are not water stable and that the more smooth and straight they are the less stable they become. Cavities are, in general, water stable; here the forms of the walls and the interconnection of the cavities are important for the air-water regime of the soil. Similar results are found for aggregates depending on their surface and inner structure.

The properties of natural soil structures are due to the relationship between their different types of pores and aggregates.

Schrifttum

Beckmann, W., und Geyger, E., 1965: Möglichkeiten der stereoskopischen Untersuchung im Innern der natürlichen Bodenstruktur. Zeiss-Mitteilungen 3, 371—385.
— Kubiena, W. L., 1953: Bestimmungsbuch und Systematik der Böden Europas. Stuttgart, 392 S.

A MICROPEDOLOGICAL STUDY OF ELUVIAL SOIL HORIZONS[1]

J. Dumanski and R. J. St. Arnaud

Department of Soil Science, University of Saskatchewan, Saskatoon, Saskatchewan

Received April 12, 1966

ABSTRACT

Three distinctive types of banded fabric related to the degree of eluviation were observed in Ae horizons of Saskatchewan soils. In addition, a type of intertextic fabric and cleavage block fabric were found to be characteristic of the upper and lower portions respectively of Ae horizons. Although essentially similar fabrics can occur within Ae horizons of Chernozemic, Solonetzic, Podzolic, and Gleysolic soils, the occurrence of specific fabric types appears to be related to the degree of eluviation which the horizons have undergone and bears a general relationship to the macrostructure exhibited by the soil.

INTRODUCTION

Soil materials are composed of two broad groups of constituents that have very different properties, one which is relatively stable, and one which is capable of remarkable movement, concentration, and reorganization during soil formation (2). The first of these, called the 'skeleton' of the fabric, consists of mineral grains and resistant siliceous and organic bodies larger than colloidal size, whereas the second, referred to as 'plasma', includes all mineral and organic material of colloidal size. Pedogenic processes do not appreciably alter the size or shape of the stable constituents of the soil, but do affect the arrangement of these constituents. 'Fabric' specifically refers to the arrangement of the soil constituents and associated voids.

Osmond (7), after observing soils from many parts of the world, suggested that there may be only a few kinds of fabrics, and pointed out that attempts had been made to associate certain fabrics with specific soil types. Although a certain fabric may sometimes be associated with a particular soil, Kubiena (3) has given examples of the same fabric occurring in different soils. He also notes that there is a developmental sequence from one fabric type to another.

In Saskatchewan, eluviated horizons are characteristic of certain Podzolic, Gleysolic, Chernozemic, and Solonetzic soils (5). These horizons occur in the upper portion of the soil profile and are characterized by a bleached appearance, a friable consistence, and platy structure. They are generally neutral to slightly acid in reaction, low in soluble bases, and somewhat more siliceous and silty than accompanying horizons. Previous studies have established the presence of both banded fabric and isoband fabrics in Ae horizons, the former being associated with Podzolic soils (4, 10) and the latter with Solonetzic and Chernozemic soils (1, 4).

The purpose of this study was to characterize further the type and distribution of microfabrics in eluviated horizons and to relate these to macromorphological features.

MATERIALS AND METHODS

Eluviated (Ae) horizons from 57 soil profiles covering a range of Podzolic, Chernozemic, Gleysolic, and Solonetzic soils were sampled specifically for this study. Morphological descriptions of each were made following the terminology adopted by the National Soil Survey Committee (6). Generalized des-

---

[1]Contribution R6, Saskatchewan Institute of Pedology.

Table 1. Macromorphological characteristics of eluviated horizons studied

Soil Order	Av. thickness (in.)	Dry color	Typical structural features
Podzolic	5	10YR 6/2 to 10YR 7/2	Fine platy, easily crushed to fine granular or single grain; plates thickest in the middle, with lenticular ends; top of plate strongly coated, bottom porous and loose
Chernozemic	5	10YR 4/1 to 10YR 5/3	Coarse prismatic breaking to coarse platy, picked to medium platy, crushed to fine granular; plates of equal thickness throughout; peds generally uncoated
Solonetzic	6	10YR 4/1 to 10YR 7/1	Coarse columnar to coarse subangular blocky, breaking to coarse platy, picked to fine platy, crushed to fine granular; plates have both square and lenticular ends
Gleysolic	10	10YR 5/2 to 10YR 6/2	Massive to weak coarse prismatic, breaking to medium and fine platy, crushed to fine powder; plates have both square and lenticular ends; peds often coated, sometimes more on top than on bottom

criptions of macromorphological characteristics of Ae horizons of soils occurring within the various orders are given in Table 1. Air-dry soil colors were determined using a Munsell color chart.

Oriented peds or clods from each profile were selected from the exposed face of the sampling pit and, after being dried, were impregnated with Castolite X* under vacuum. Thin sections were prepared and examined under a light polarizing microscope. In addition, approximately 80 thin sections from Ae horizons already on file in this laboratory served as supplementary samples in the analysis of these soil fabrics.

Particle-size distribution as well as organic carbon analyses were conducted on selected horizons possessing specific fabrics in an attempt to characterize further the various types observed.

RESULTS AND DISCUSSION

Four major types of fabrics were found to occur within Ae horizons of the soils studied (Figs. 1 to 6); these included: (1) a type of intertextic (chernozemic) fabric, (2) isoband fabric, (3) banded fabric, and (4) cleavage block fabric. Characteristics of each fabric type are outlined below.

Modified Intertextic Fabric

This type of fabric (Fig. 1) which generally occurs in the upper portions of Ae horizons, or in Ahe horizons of eluviated Chernozemic soils in particular, is in many respects similar to chernozemic fabric (3, 10) which is characteristic of Ah horizons in Chernozemic soils. However, the well-flocculated appearance and the complete permeation of the matrix by dark-colored humic substances so typical of chernozemic fabric is lacking. Instead, highly humified aggregates of plasmic materials occur discontinuously throughout a dark grey to dark brownish grey matrix. Horizontal and vertical cleavages are few and

---

*Cold-setting plastic obtained from the Castolite Company, Woodstock, Ill.

very weakly expressed. It is postulated that this fabric represents a type transitory between chernozemic (3) and isoband (4) fabric and may indicate the zone of transformation from a humified to a leached horizon. It has previously been suggested (10) that such a fabric results from the leaching of material possessing a chernozemic fabric.

Isoband Fabric

Isoband fabric (Fig. 2) which was first described by McMillan and Mitchell (4) refers to a platy type of fabric in which the plates tend to be of equal thickness (1 to 2 mm) and throughout which the plasmic and skeletal materials are uniformly distributed. Such a fabric is invariably associated with macroprismatic structure in which the prisms can be separated into coarse to medium plates. When viewed in thin section, the finer plates can be observed more readily.

Banded Fabric

Banded fabric refers to a platy type of fabric in which the skeletal fraction is distributed relatively uniformly throughout each plate, but in which the plasma, composed primarily of clays, sesquioxides, and humic substances, is concentrated in bands at the top or within each plate (3, 4). Three distinct types of banded fabrics were observed in the eluvial horizons studied; for convenience, these are designated as types A, B, and C. All three types are characterized by textural gradients or plasmic accumulations within each band, but are differentiated on the basis of the distribution of the colloidal accumulations, as well as on the shape of the plates and degree of development of the cleavages.

1. *Type A banded fabric.*—This is very similar to isoband fabric, except that a thin, uniform band of colloidal material occurs along the upper surface of each plate (Fig. 3). The plates are generally of uniform thickness, although they may be slightly curvilinear or wavy in appearance. Type A banded fabric and isoband fabric generally occur in the same horizons, the two being associated with soils which possess a compound, weak prismatic and coarse platy macrostructure. Both types of fabric are very common in eluviated Chernozemic soils.

2. *Type B banded fabric.*—This fabric type is characterized by the presence of thicker and more diffuse bands of plasmic materials concentrated at the top, as well as within each plate (Fig. 4). It appears to be transitional to type C banded fabric, inasmuch as cleavages have not formed at the upper boundaries of accumulations within the plates.

3. *Type C banded fabric.*—This fabric differs from types A and B inasmuch as the plates are generally lenticular in shape and are usually less than 1 mm in thickness (Fig. 5). The plates often overlap, or if not, have a wavy appearance. There is a single pronounced band of plasmic material in the upper portion of each plate, which occupies up to one-half of the total volume of each plate. In Podzolic soils, where this fabric is often very well developed, there appears to be some segregation of skeletal material with the larger grains outlining the lower boundary of the plate. Type C banded fabric also occurs in Ae horizons of Gleysolic and Solonetzic soils, although to a much more limited extent.

Cleavage Block Fabric

This fabric is characterized by block-like peds (3) which, in eluvial horizons, form the mesostructure of coarse platy to prismatic macrostructures. It commonly occurs toward the bottom of an Ae horizon and in material containing more clay than that occurring in material having banded fabrics. To some extent, as indicated by the denser matrix, it appears to have formed as a result of leaching of the upper portion of a B horizon. The peds vary from 1 to 3 mm in cross-section, and are often not observed in macrostructure (Fig. 6). Although this fabric often occurs in horizons sampled as Ae horizons, the microfabric would indicate that it is more typical of AB horizons (1).

Although certain fabrics were predominant in Ae horizons of specific soils, essentially similar fabrics were exhibited by all the soils studied. On the basis of the profiles sampled, it was possible to determine the relative occurrence of the individual fabric types in soils separated at the order level (Table 2). For comparative purposes, the occurrence of each specific fabric within each group of profiles examined was reported as follows: low, less than 33% occurrence; medium, 33 to 66% occurrence; high, greater than 66% occurrence. It must be realized that the occurrence designations are based only on the presence of the fabric type in the various soils and give no indication as to their dominance in the horizons.

The results indicate that banded fabric occurs in Ae horizons of nearly all Podzolic soils and is common in Solonetzic and Gleysolic soils. In soils grading from Chernozemic to Podzolic, the following changes were observed with increasing degree of podzolic development: increasing occurrence of banded fabric, decreasing occurrence of isoband and modified chernozemic fabric, more pronounced and finer platy structure, and more pronounced gradation in particle sizes within the plates.

In general, a sequence of fabric types was observed within each Ae horizon. The uppermost portion of this horizon usually has a modified intertextic fabric. This is commonly followed by isoband and banded fabric in Podzolic soils, and by isoband, with or without weak banded fabric, in Chernozemic and Solonetzic

Table 2. Occurrence of various fabrics in Ae horizons of soils

Fabrics	Podzolic (17)	Chernozemic (18)	Solonetzic (20)	Gleysolic (4)
Banded (types A, B, and C)	High	Low	Medium	Medium
Isoband	Medium	High	High	High
Modified (Chernozemic)	Low	High	Medium	Low
Cleavage block	Low	Low	Low	Medium

*Numbers in parentheses refer to the number of profiles examined.

Fig. 1. Modified intertextic (chernozemic) fabric in Ae horizon of an Eluviated Dark Brown soil (Weyburn Association); plain light. The scale marker represents 1.0 mm, here and in subsequent illustrations.
Fig. 2. Isoband fabric in Ae horizon of a Solodized Solonetz soil (Waseca Association).
Fig. 3. Type A banded fabric in Ae horizon of a Podzol soil from Cypress Hills Park.
Fig. 4. Type B banded fabric in Ae horizon of an Orthic Grey Wooded soil (Waitville Association).

PLATE I

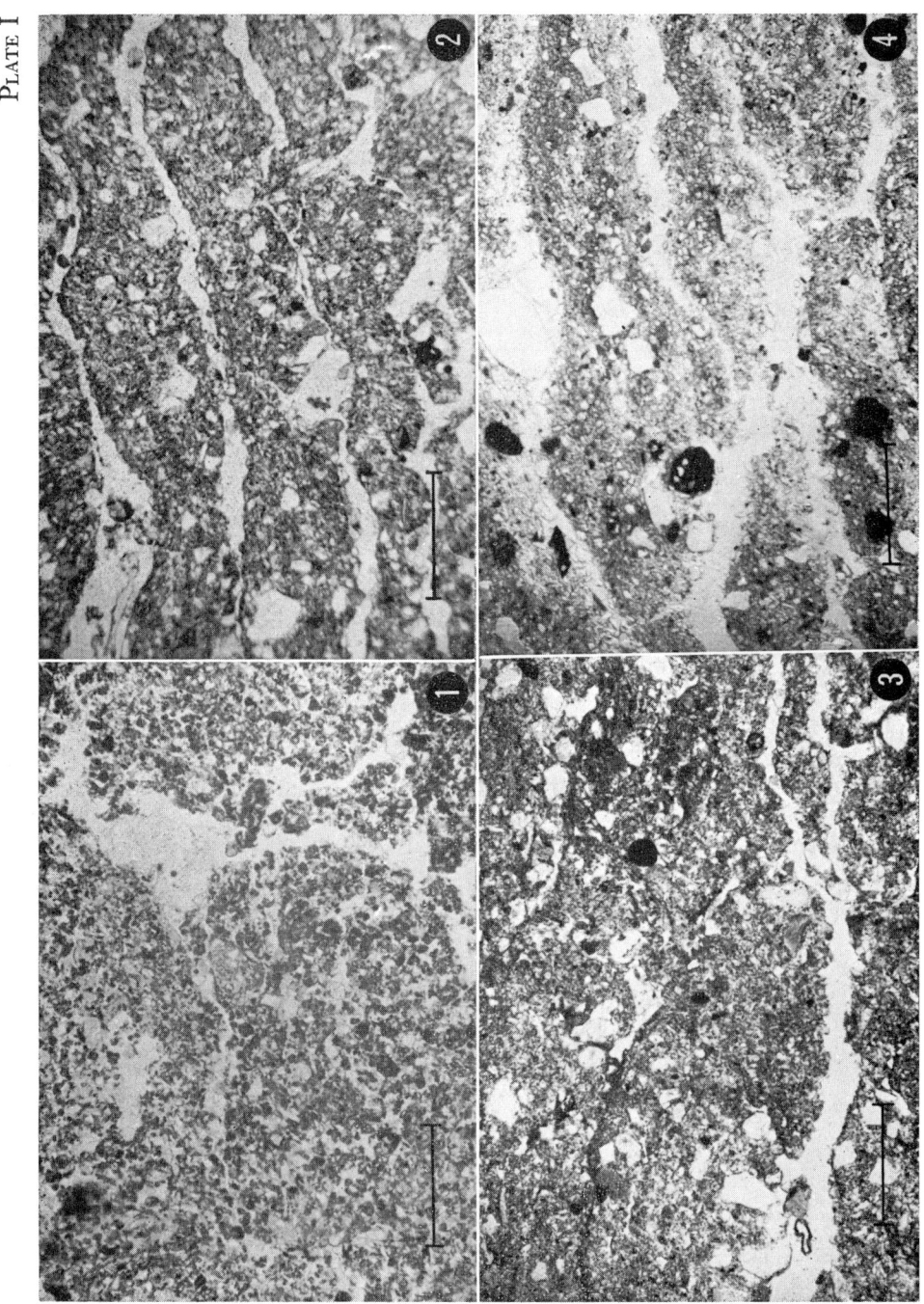

214

PLATE II

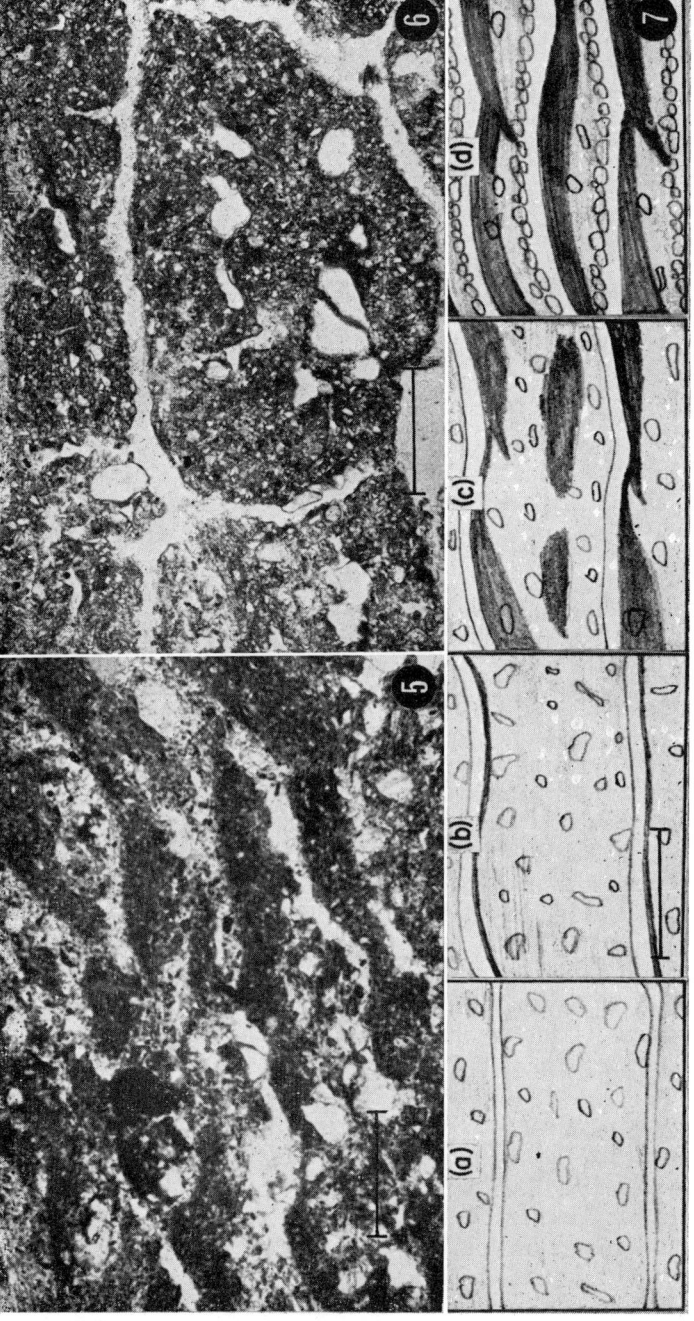

FIG. 5. Type C banded fabric in Ae horizon of an Orthic Grey Wooded soil (Loon River Association).
FIG. 6. Cleavage block fabric in Ae horizons of an Eluviated Dark Brown soil (Claybank Association).
FIG. 7. Schematic presentation of isoband and banded fabrics: (a) isoband, (b) type A banded, (c) type B banded, and (d) type C banded fabrics.

215

Table 3. Organic carbon contents and fine clay : total clay ratios associated with different fabric types

Fabric type	Organic carbon (%)		Fine clay : total clay ratio	
	Mean	Range	Mean	Range
Isoband and modified Intertextic (9)* (Chernozemic)	1.2	0.8–1.8	0.5	0.4–0.8
Banded type A (6)	0.9	0.4–1.5	0.4	0.2–0.6
Banded type B (5)	0.7	0.2–1.0	0.4	0.2–0.5
Banded type C (9)	0.6	0.4–0.6	0.2	0.1–0.3

*Numbers in parentheses refer to number of profiles analyzed.

soils. This zone of isoband, or banded fabrics, occupies three-quarters to two-thirds of the thickness of the Ae horizon and is the zone of maximal expression of structure development. The lower part of the Ae is characterized by cleavage block fabric. This sequence of fabrics describes a modal soil type, but does not necessarily occur in every case. However, it indicates inter-relationships and the possible developmental sequence of the various fabric types (Fig. 7).

Analytical Data

Analyses of selected horizons indicate a general relationship between fabric type and mean fine clay ($< 0.2\ \mu$) : total clay ($< 2\ \mu$) ratios, as well as between fabric type and mean organic carbon content (Table 3). Although such relationships are apparent from the mean values listed, it should be noted that the ranges of values for the different fabric types overlap to some extent. The reason for this no doubt lies in the fact that each horizon usually contains more than one fabric type. Although attempts were made to select horizons in which one fabric type was dominant, the proportions of other fabric types would no doubt affect the overall composition of the horizon. The data do indicate that, as the organic matter content of Ae horizons decreases, there is a greater tendency to have some type of banded fabric. The fine clay : total clay ratios serve as an indication of the degree of eluviation of clay which has occurred in the Ae horizon since, in general, the parent materials of Saskatchewan soils have a fine clay : total clay ratio of 0.5 or higher (8). The preferential movement of fine clays in contrast to coarse clays (9) makes it possible to use the ratio of fine clay : total clay as an 'eluviation index'; thus, a ratio $\ll 0.5$ would indicate a highly eluviated soil. The type of fabric occurring in Ae horizons is apparently associated with the degree of eluviation which has occurred.

The exact mechanisms involved in the development of the different soil structures are not as yet clearly understood. Platy peds have often been ascribed to the physical action of ice lenses which force the soil apart along horizontal planes. The effects of drying of the soil from the top down have been variously indicated as the means of development of banded fabrics in surface horizons (3, 4). Limited studies in this laboratory indicate that repeated freezing and thawing, and wetting and drying, may in fact be important processes in the development of platy structure and banded fabric.

CONCLUSIONS

The types of fabric occurring within Ae horizons appear to be related to the organic matter content, the degree of eluviation to which the horizon has been subjected, as well as to the macrostructure exhibited by the soil. Modified chernozemic and isoband fabrics are associated mainly with Ahe and Ae horizons of Chernozemic and Solonetzic soils and possibly result from the leaching of horizons formerly possessing chernozemic fabric. Isoband fabric is generally best expressed in eluvial horizons displaying a macroprismatic structure. Banded fabrics occur in Ae horizons that are low in organic matter and that have undergone considerable leaching as evidenced by the low proportion of fine clay. The three subtypes of banded fabric, A, B and C, are suggestive of a developmental sequence and reflect the degree of development of the profiles in which they occur. Ae horizons or portions of Ae horizons possessing cleavage block fabric generally have a remnant subangular blocky structure, indicating that they have formed within former B horizons.

ACKNOWLEDGMENTS

The authors express their appreciation to the National Research Council for the financial assistance which made this study possible.

REFERENCES

1. Acton, D. F. and St. Arnaud, R. J. 1963. Micropedology of the major profile types of the Weyburn catena. Can. J. Soil Sci. **43**, 377–387.
2. Brewer, Roy. 1964. Fabric and mineral analysis of soils. John Wiley and Sons Inc., New York.
3. Kubiena, W. L. 1938. Micropedology. Collegiate Press, Inc., Ames, Iowa.
4. McMillan, N. J. and Mitchell, J. 1953. A microscopic study of platy and concretionary structures in certain Saskatchewan soils. Can. J. Agr. Sci. **33**, 178–183.
5. Moss, H. C. 1965. A guide to understanding Saskatchewan soils. Sask. Inst. Pedology, Publ. M 1., Univ. of Saskatchewan.
6. National Soil Survey Committee of Canada. 1963. Rept. 5th Natl. Meet., Winnipeg.
7. Osmond, D. A. 1958. Micropedology. Soils and Fert. **XXI**, 1–6.
8. Saskatchewan Institute of Pedology. Unpublished data.
9. St. Arnaud, R. J. and Mortland, M. M. 1963. Characteristics of the clay fractions in a Chernozemic to Podzolic sequence of soil profiles in Saskatchewan. Can. J. Soil Sci. **43**, 336–349.
10. St. Arnaud, R. J. and Whiteside, E. P. 1964. Morphology and genesis of a Chernozemic to Podzolic sequence of soil profiles in Saskatchewan. Can. J. Soil Sci. **44**, 88–99.

Part VI

CLAY REARRANGEMENTS

Editors' Comments
on Papers 16 Through 20

16 MINASHINA
Optically Oriented Clays in Soils

17 STEPHEN
Clay Orientation in Soils

18 BUOL AND HOLE
Clay Skin Genesis in Wisconsin Soils

19 NETTLETON, FLACH, and BRASHER
Argillic Horizons without Clay Skins

20 FEDOROFF
Excerpt from *Classification of Accumulations of Translocated Particles*

The distribution and orientation patterns of the clay fraction can be considered one of the most important and characteristic microscopic features of most temperate and many tropical soils. This is so not only because the clay fraction is generally important—indeed, often dominant—in these soils, but mainly because the clay tends to be the component of these soils that is most easily affected by soil-forming processes and therefore that best reflects the influence of these processes.

Although individual clay particles are too small to be visible with the aid of an optic microscope, zones, or even small domains (a few micrometers in diameter), of parallelly oriented clays can be recognized with the petrographic microscope when observed between crossed polarizers because of their birefringence. The patterns of interference colors thus created in the clayey mass of soils were recognized early on by Kubiëna (1948), who described them as "birefringent streaks" or "fluidal structures." He saw them as proof of the mobility of fine material (plasma) when it is present in a peptized state. Initially, researchers made no distinction between different types of patterns; as a result, birefringent streaks were interpreted to be clay reorientations *in situ* (the plasma separations of Brewer

[1964b] as well as clay illuviations (part of the plasma concentrations *sensu* Brewer).

The first type of clay orientations that were positively recognized and whose genesis was understood were the clay coatings, which are found, for instance, in most argillic horizons. One could say that different authors more or less simultaneously discovered these features and recognized their significance. Two pedologists, N. G. Minashina (Paper 16) and I. Stephen (Paper 17), recognized the enormous importance of these microscopic features for soil genesis and classification, and reviewed and related most pertinent data that had been published.

Minashina offered the first detailed review of the occurrence and formation of optically oriented clay. One of her major contributions was the clear distinction between different "habits" of oriented clay: incrusted, or "secondary," clays (clay coatings) on pore walls; random-fiber clay and clay scales; films of clay on weathering primary minerals and rock fragments; and accumulations of clay around inclusions of silt and sand. Minashina tried to relate the occurrence of these habits to different soil-forming conditions. In this way she demonstrated that incrusted clay features (i.e., clay coatings) are generally found beneath eluvial horizons, and concluded that they are formed by illuvial clay, rather than be neoformation *in situ* of clay, an opinion then defended by several Soviet authors. Other habits she explained by processes such as sedimentation and weathering. Although Minashina mentioned the contribution of alternative wetting and drying to the orientation of the clay particles, no one had yet recognized the influence of mechanical pressure, nor made a clear distinction between clay accumulations and reorientations (separations *sensu* Brewer). She recognized also a negative correlation between the presence of oriented clay and calcite in the profile. Paper 16 is considered an important step toward a better understanding of the meaning of oriented clay in soil thin sections. It is also interesting to note Minishina's use of the terms *clay fibers* and *scales* to indicate more or less oriented clay domains. These terms, derived from old Russian sedimentologic petrographic nomenclature, have been used by Russian micromorphologists for many years (see also Paper 22). The clay fibers seem to correspond to Brewer's (1964a) sepic plasmic fabrics (e.g., random fibers correspond to omnisepic) and the scales partly to his asepic plasmic fabrics. The incrusted, or "secondary," clay corresponds to Brewer's (1964b) clay cutans, argillans, and ferri-argillans.

Paper 17, by I. Stephen, comprises three interrelated parts: a synthetic review of the early experimental work on the formation of clay-illuviation coatings in soil materials; in view of these results, a discussion on clay orientations as described by Kubiëna (1948) in different soil types; and a review of clay illuviation features observed in natural soils, and recognized as such. The paper is included here mainly because of the section on Kubiëna's concepts. It was indeed an interesting contribution to a better understanding of the purely micromorphologically defined fabric types, especially the Lehm-fabric.

In the United States in the 1960s, much attention was given to the presence of clay coatings in some soils, as an indication of active illuviation processes, and in view of their significance for soil fertility. Paper 18, by S. W. Buol and F. D. Hole, is one of the first modern studies on clay coatings, dealing with their micromorphological characteristics, some of their chemical aspects, and their formation in vitro. From the point of view of micromorphology, two new observations were very important for future work: (a) most recognizable clay skins are found in the C horizon, and (b) clay skins are best preserved in the deeper horizons (B_3 and C_1), whereas they tend to be destroyed by fracturation in the upper horizons, and incorporated into the ped interior. This latter observation is especially important, as it explains why some B horizons the clay increase can be much higher than could be inferred from the presence of only a limited amount of clay skins on ped surfaces. This paper is thus significant in that it introduced the important concept of "fragmented features."

From a historical point of view, the morphometric determination of the total volume of clay skins by Buol and Hole should be mentioned, as should the technique used (cutting and weighing). Both *in situ* and fragmented clay skins were counted without distinction. Miedema and Slager (1972) showed much later that a separate counting of both can yield important informations, and McKeague et al. (1978) clearly demonstrated the "relativity" of such "absolute" determinations. It is interesting to note that, even with a rough estimation, and without making a distinction between the different types of clay orientations, Reuter (1964) found a relationship between the amount of oriented clay in thin sections and the degree of profile development in the podzolic soils of western and central Europe.

Buol and Hole's conclusion that clay skins may become incorporated into the ped interior by means of soil expansion and contraction or other methods of turbation was further investigated by W. D. Nettleton, K. W. Flach, and B. R. Brasher (Paper 19) in soils from the arid and Mediterranean zones of the United States. These soils have horizons with a clay accumulation, but without distinct clay skins. After studying a relative large number of samples, these authors concluded that a clear relationship exists between the total amount of clay in the horizon and its linear extensibility on the hand and the absence of clay skins and the development of plasmic fabrics on the other. These data clearly indicate that under suitable climatic conditions clay skins are destroyed in horizons with a high shrink-swell potential, whereas stress-induced clay orientations may appear. Even more than Paper 18, this paper is the basis of the now generally accepted concept of mechanical disintegration of clay coatings and their transformation to striated birefringence patterns. Similar conclusions were reached later by several authors studying soils in the Mediterranean basin (e.g., Osman and Eswaran, 1974; Verheye and Stoops, 1974; Reynders, 1974). Fedoroff (1968) described the same type of phenomenon in the "dynamic B" horizons of relatively wet temperate soils.

Although considerable research has been done on the mircomorphological aspects of clay illuviation features in soil materials, until the mid-1970s little attention was given to their textural composition. True, some researchers noticed that coarser clay particles and even silt could be translocated, but few studied these features. Fedoroff (Paper 20) was the first to take a systematic approach to this phenomenon, in 1974. He tried to classify the different types of mechanical accumulations in soil based on their texture and layering and their distribution within the soil material. He also emphasized the relationship between different types of textural accumulations and the argillic horizons of different soil orders of *Soil Taxonomy*. Fedoroff's proposed classification, though incomplete, was a good start toward a more precise analysis of the typology of translocated particles in soils, but much research remains to be done. Much information can be gathered from detailed descriptions and comparisons of different types of horizons.

REFERENCES

Brewer, R., 1964a, Classification of plasmic fabrics of soil materials, in *Soil Micromorphology*, A. Jongerius, ed., Elsevier, Amsterdam, pp. 95–107.

Brewer, R., 1964b, *Fabric and Mineral Analysis of Soils*, J. Wiley and Sons, London, New York and Sydney, 470p.

Fedoroff, N., 1968, Génèse et morphologie de sols à horizon B textural en France atlantique, *Sci. Sol*, 1, 29–65.

Kubiëna, W. L., 1948, *Entwicklungslehre des Bodens*, Springer-Verlag, Vienna, 215p.

McKeague, J. A., R. K. Guertin, F. Page, and K. W. G. Valentine, 1978, Micromorphological evidence of illuvial clay in horizons designated Bt in the field. *Canadian Jour. Soil Sci.* **58**:179–186.

Miedema, R. and S. Slager, 1972, Micromorphological quantification of clay illuviation. *Jour. Soil Sci.* **23**:309–314.

Osman, A. and H. Eswaran, 1974. Clay translocation and vertic properties of some red mediterranean soils, in *Soil Microscopy,* G. K. Rutherford, ed., Limestone Press, Kingston, Ontario, pp. 846–857.

Reuter, G., 1964, Vergleichende Untersuchungen an lessivierten Böden in verschiedenen Klimagebieten, *8th Intern. Congr. Soil Sci., Trans.* (Bucharest) **5:**723–732.

Reynders, J. J., 1974, A study of a soil type with a textural B-horizon and expanding clays, *10th Intern. Congr. Soil Sci. Trans.* **7:**218–227.

Verheye, W., and Stoops, G., 1974, Micromorphological evidences for the identification of an argillic horizon in Terra Rossa soils, in *Soil Microscopy,* G. K. Rutherford, ed., Limestone Press, Kingston, Ontario, pp. 817–831.

OPTICALLY ORIENTED CLAYS IN SOILS

N. G. Minashina, *V. V. Dokuchayev Soil Institute, Academy of Sciences, USSR*

OPTICALLY oriented clays occur in different types of soils. They consist of aggregates of individual clay particles all possessing the same orientation. In polarized light such an aggregate behaves as one crystal, for which it is possible to determine the index of optical refraction, the birefringence, and other optical constants.

Optically oriented clays are easily detected in thin sections of soil samples with undisturbed structure under crossed Nicol prisms with simultaneous or periodic extinction. Optically unoriented clays are not visible when the Nicol prisms are crossed.

Optically oriented clays in soils were first discovered and described by Polynov (9) in thin sections of "ortstein-producing horizons of gray sandy soils." The method of soil study using thin sections came to be extensively used along with improvements of the methods applied to the preparation of the sections, after which optically oriented clays attracted the attention of many investigators: Frei and Cline (22), McCaleb (21), Brewer (19, 20), Altemüller (18), and others. Some research workers gave different names to these formations. In papers by Feofarova (17), Romashkevich (15), and Starykh (16), optically oriented clays are described as "secondary clays." Parfenova and Yarilova (7, 8) call them "mobile beidellite," "beidellite synthesized in the soil," and "iron beidellite." Foreign authors have used the designations "striated," "incrusted," "ribbon-shaped" and most often, "optically oriented clays."

The author (6) has detected optically oriented clays in the gray, cinnamon-brown carbonate-free soils of the Kirovabad massif and also when studying thin sections of the following soils: mountain-podzolic forest; mountain-forest brown; mountain-forest cinnamon-brown; shrub-dry steppe cinnamon-brown; sierozems and younger sierozem-meadow and meadow alluvial soils; and cultivated, watered soils on irrigation alluvium. It was shown that in soils in which the leaching is greatly hindered or completely lacking (such as cinnamon-brown soils and sierozems) there are no incrusted forms of "secondary clays." Instead there are forms of optically oriented clays which are quite specific in habitus. "Habitus" is used to designate the external form of accumulation of clays as distinguished from the "structure" associated with the crystal lattice of a mineral.

Accumulations of optically oriented clays in various soils differ not only quantitatively but also with respect to form, dimensions, and orientation with respect to other soil components. In the soils described below we have found the following types of oriented clays: 1) incrusted or "secondary" clays in the form of films on the walls of passages and cracks; 2) random-fiber clays in the form of fibers, extremely small scales and shells lying in different planes; 3) small accumulations as films on the surfaces of weathering primary minerals and rock fragments; and 4) accumulations of clay around inclusions of silt or sometimes of sandy particles or root fragments (in fresh deposits). The sizes of such accumulations may vary from microscopic to macroscopic layers consisting of poikilitic clays formed geologically. This is not a complete list of the abundant varieties of clays. Many other types are known, for example, those with a habitus of parallel or perpendicular fibers, and others.

Below are given descriptions of the microhabitus of clays and of some other specific characteristics important in clarifying the genesis of soils which developed both under moist climatic conditions in which the percolation of water was predominant and also under relatively dry climatic conditions in which leaching was greatly hindered.

Mountain-Forest Podzolic Soils (Altay, profile 48, A. N. Rozanov)

For all podzolized soils, including those of plains areas, the presence of optically oriented clays in the illuvial horizon in the form of incrustations along the cracks and pores ("secondary clays") is characteristic. Figure 1 shows a cross-section through one of these incrustations. By transmitted light the clay has a brownish-yellowish color. In some places it has a high concentration of iron and also of organic matter which, in transmitted light, have a dark-brownish and reddish color and include opaque limonitized areas. The incrustations are usually laminated. Sometimes the lamination is inconspicuous and the clay accumulations have a gelatinous habitus. Under crossed Nicols the extinction is periodic and the color reddish. Within the clays of this habitus there are usually extremely small darkly-stained particles of micron size.

It is important to note that the presence of "secondary clays" in illuvial horizons is always accompanied by eluvial horizons from which much clay has been removed. In the eluvial horizons mineral fragments usually do not have films visible under the microscope: as a result the podzolized eluvial horizon is light in color. In podzolic soils, in addition to clay in the form of incrustations, there are in the basic mass both of the illuvial and of the upper horizons very small quantities of optically oriented clays in the form of extremely fine aggregates, and also optically unoriented clays. (According to Kubiena [23] the basic mass is a very thin part of the soil consisting of humus and clay particles with an admixture of other extremely fine crystals. It is distinguished from the skeletal part of the soil which consists of fragments of primary minerals and rocks.)

The presence of round concretions with a concentric habitus is also characteristic of podzolic soils. These forms occur as particles of soil cemented together by brown iron oxides with manganese compounds. Iron accumulations always occur in soils in which there is the greatest accumulation of the incrusted forms of optically oriented clays, not only in podzolized soils but also in solodized.

There is no uniform opinion among investigators regarding the origin of the incrustations of optically oriented clays. Frei and Cline (22), who were the first to describe them in detail, accept as the most probable origin the eluviation of clays from the upper horizons as colloidal solutions under the protection of organic and other hydrophilic colloids (for example, SiO_2). Parfenova and Yarilova (8) consider that "the incrustational character of this mineral is evidence that it is synthesized from solutions percolating from higher levels of the soil." To this conclusion they have added the following consideration: "Mineral elements reach the soil surface in large quantities with the leaf litter. The easily mobile elements are leached from

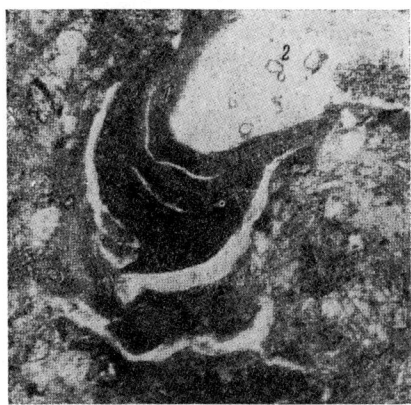

FIG. 1.—Incrustations of optically oriented clays along the walls of the pores of the illuvial horizon of a mountain-forest podzolic soil, profile 48, depth 85 to 95 cm. (magnification 5 × 20, Nicols parallel): 1—clay incrustations; 2—pores.

the litter and the more inert ones thus accumulate there: silica, iron and aluminum. Under the action of fulvic acids, which, in turn, are a product of the life processes of organisms, they are transferred along with the elements of the decomposing minerals into the depths of the profile, where their synthesis takes place" (7).

It is now generally known (1, 9, 10, 11) that the role of the higher and lower organisms in the soil clay formation is great, although the details of the process have not been fully clarified and we do not know the forms in which this occurs. Clay incrustations along cracks must not be considered only as a product of synthesis, and the incrustational nature of optically oriented clays cannot be considered as proof of their synthetic origin for the following reasons: 1. Optically oriented clays may be obtained from naturally-occurring optically unoriented clays by dispersing the particles and allowing them to settle undisturbed and dry out. Vikulova's method (2) for the optical determination of clays is based on this. 2. Incrustational, optically oriented clays are also detected in the illuvial horizons of solonetzes, solods and takyrs —that is, where there are suitable conditions for peptizing the clays and transporting them along the profile in colloidal solutions. Here, however, there is also peptization of the clay which has entered into the composition of the parent material. There is no basis for believing that clays are synthesized in takyrs where the

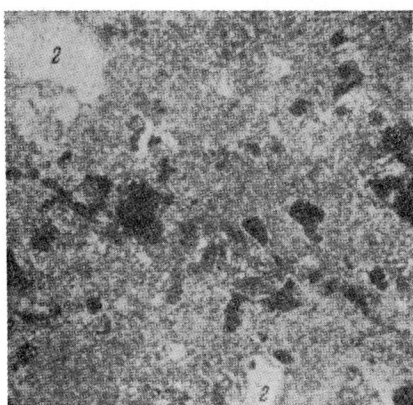

Fig. 2.—Optically oriented clay in brown mountain-forest soil of the Little Caucasus, profile 531, depth 65 to 83 cm. (magnification, 5 × 8, Nicols crossed): 1—pores; 2—residual minerals.

cycle of ash elements hardly exceeds a tenth of a kilogram per hectare. At the same time chernozems receive more than 1000 kg. per hectare per year of ash elements, but an accumulation of synthetic clay in the soil profile is not detected even though there is considerable growth of the soil. 3. Brewer (20) produced illuvial horizons with optically oriented clays artificially by putting suspensions of 60% illite and 40% kaolinite through columns of sand. Kubiena (23) in his experiments showed that clays from the illuvial horizon are easily transported further into the horizon by rainwater.

In consideration of all these facts, the author supports Frei and Cline's opinion that the presence of incrustational clays in the profile is an indication of illuvial processes in soils.

Brown Mountain-Forest Soils (Profile 531, E. M. Salayev)

From a study of brown mountain-forest soils of the Little Caucasus range which had developed on authigenous solid crystalline rock. It became clear that the occurrence of such optically oriented clays in almost all the horizons is characteristic of such soils, unlike the mountain-forest podzolic soils in which the optically oriented clays are concentrated mostly in the illuvial horizon. The habitus of most of the optically oriented clays of the brown mountain-forest soils is completely different. In addition to small accumulations of incrustational clays in the pores, they are distinguished by clays with a random-fiber habitus which make up the greater part of the soil. They are in the form of thin strips, fibers, granules and scales oriented in different directions, which gives the mass a net-like or random-fiber habitus which can be detected with crossed Nicols (Fig. 2).

The ratio of these two types of optically oriented clays changes, depending on the extent to which the soil has been leached: the more the soil has been leached, the more there are of the incrustational forms of clays in the pores. This is found not only in the soils of the Little Caucasus but elsewhere in the world—for example, in the soils of China. Thus, the small accumulations of incrustations of optically oriented clays in the pores of the brown mountain-forest soil being described is evidence of the slighter development of the leaching process than in mountain-forest podzolic soils. Further evidence for this is found in certain other peculiarities of the microhabitus of brown soils. For example, they have smaller accumulations of iron concretions although the soil is generally fairly rich in iron hydroxides which are deposited in the weathering minerals in the form of extremely fine granules of limonite, most often in the upper horizons. Evidently conditions are not suitable for the migration of iron from some places or its concentration in others. Further evidence of less leaching is given by the accumulations of fine needles of calcite in the lower part of the profile (at a depth of 90 cm.).

In general the brown soils of the Little Caucasus are characterized by a high clay content, the occurrence of small accumulations of clay incrustations in the middle and lower parts of the profile and a random-fiber habitus of most of the optically oriented clays.

Cinnamon-Brown Mountain Forest Soils (Profile 654, E. M. Salayev) and Cinnamon-Brown Shrub-Dry Steppe Soils (Profile 622, Described Jointly by the Author and A. N. Rozanov)

The cinnamon-brown soils investigated had developed on marly limestone (profile 654) and on porphyrites (profile 622). They are distinguished from mountain-forest brown and podzolic soils by the absence of illuvial clays in the form of incrustations on the pores and cracks (Fig. 3). They are characterized by an even higher clay content of the soil profile and by a negligible content of residual minerals and rock fragments. Most of the optically oriented clays have a random-fiber habitus. The scales and fibers of optically oriented clays are only microns thick. For cinnamon-brown soils, just as for the brown soils, the occurrence of a large number of branching pores with rounded edges is characteristic.

In the upper horizon most of the soil consists of aggregated humus-clay formations in which there are areas of optically oriented and unoriented clays mixed with humic matter and with undecomposed and half-decomposed plant residues. It is interesting to note that deposits of carbonate often occur in cinnamon-brown soils. They are sometimes present in the topmost horizon. When this occurs the clay usually has no optical orientation in places where the carbonates have accumulated. The absence of incrustational forms of optically active clays and the occurrence of carbonates in the profile is evidence that intensive leaching is not taking place. For the same reasons the iron hydroxides formed by weathering and soil formation remain in place. This was especially clear in the thin sections of profile 622, since the parent material—porphyrite—contained pyrite. In places where the pyrite had been oxidized, limonite remained and gypsum had crystallized out in the lower part of the soil profile.

Thus, surveying a series of soil from the mountain-forest podzolic soils through the brown soils to the cinnamon-brown mountain-forest and shrub-dry steppe soils, we can notice a gradual increase in the extent of clay formation and a decrease in the thickness of the incrustations of optically oriented clays down to the point of thin films and, in the cinnamon-brown soils, to the point of their complete disappearance. While there may be a considerable number of iron concretions in the podzolic soils, there are fewer in the brown leached soils, very few in the cinnamon-brown soils transitional to the brown soils, and none at all in the cinnamon-brown shrub-dry steppe soils. The carbonate content of the soils increases in the same direction. In the brown soils the carbonates are represented by accumulations of fine needles in the deepest horizons, in contact with the parent material. In cinnamon-brown soils they lie at a higher level in the form of accumulations of fine needles and microgranular aggregates, while in the cinnamon-brown shrub-dry steppe soils the micro-

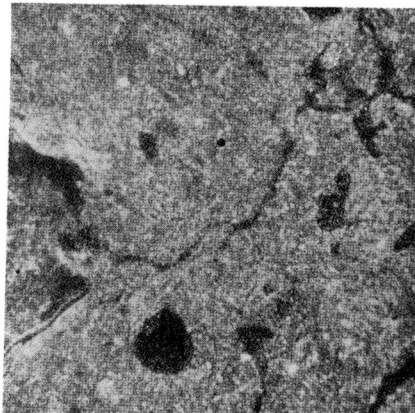

FIG. 3.—Random-fiber habitus of a high-clay horizon of cinnamon-brown mountain-forest soil of the Little Caucasus, profile 654 (magnification 8 × 20, Nicols crossed).

granular carbonate accumulations are detected at the surface in some places. All of this information confirms the fact, already known from soil-geographical observations, that there is a decrease in the intensity of the illuvial processes in the series extending from the mountain-forest podzolic soils to the cinnamon-brown soils.

In the sierozem soils on loess, where leaching is even more hindered and there are no conditions favoring dispersion of the soil mass, there are no incrustational forms of optically oriented clays. Clay formation is slight and optically oriented clays are detected in such soils in negligible quantities in the form of isolated films on the surfaces of primary minerals undergoing weathering. Most of the clays have no optical orientation and are assembled in microaggregates.

In sierozems on loess and in cinnamon-brown soils formed on authigenic rock, where the illuvial processes are hindered by specific hydrothermal conditions, there is no other way of forming optically oriented clays except by intra-soil weathering and clay formation in primary rock-forming minerals *in situ* under the influence of the soil-forming processes. The origin of the random-fiber habitus of optically oriented clays can be fully explained on this basis if it is considered that clay formation begins and continues along the finest cracks and pores which penetrate the rock in various directions. They proceed more easily along the surfaces of the soil-forming minerals which are more accessible to microorganisms, to the soil solution and to the action of the air. The random-fiber habitus of optically oriented clays and the occurrence of mutually isolated films on the surfaces of weathering minerals can be considered as signs of their formation *in situ*.

Conclusions regarding the possibility of formation of clays *in situ* by intrasoil weathering are not new. Intrasoil clay formation processes, which develop in occult form in sierozem formation, in more conspicuous form in the formation of gray cinnamon-brown soils (13, 14) and to a very emphatic extent in the formation of cinnamon-brown soils (3) were long ago described and proved by other methods. The micromorphological development of this process has been established by the use of the microscope.

In all the cases noted above the formation of optically oriented clays has been related to soil formation. Such clays may arise during the formation of parent material by sedimentation from water carrying alluvium or deluvium if the clays are peptized. Under natural conditions the clay particles are more often found in coagulated condition (that is, they are transferred by the water in the form of microaggregates), so that the alluvial and deluvial deposits consist more often of unoriented clays. Conditions favoring peptization also make possible the formation of optically oriented clays when the suspensions dry out. Such clays usually include particles of silt and fragments of plants which are also transported by water. Sometimes they form microfilms and microlamellae which are fairly pure. We were able to observe such formations in the sierozem-meadow soil of the Zeravshan valley. As seen from Figure 4 these formations have a completely different habitus and are distinct from all the forms listed previously.

Especially favorable conditions for optically oriented clays occur in the development of takyrs. The water stagnating on them has an alkaline reaction, which favors the peptization of the clays and crust formation (5). Thus it is no coincidence that in takyrs one may observe the two forms of optically oriented clays described by Feofarova (17)—"secondary clays" and "mixed clays." It may be assumed that the latter type was formed during sedimentation, so that it includes particles of silt and sometimes of sand. They are all subsequently buried by alluvial deposits.

Thus, optically oriented clays have different forms of habitus related to the different conditions under which they were formed. Brewer (20) proved the possibility of forming optically oriented clays not only by illuviation, but *in situ* as a result of the weathering of rock and also sedimentation from the water transporting alluvium. However, Brewer's article does not distinguish the specific characteristics of the habitus for optically oriented clays of various origins.

The mineralogical nature of the optically oriented clays of the soil has not yet been fully clarified. Thermal and x-ray analyses of soils containing an abundance of optically oriented clays show a preponderance of beidellite. Parfenova and Yarilova (7, 8) found this to be iron beidellite, while Brewer (20), Stremme (24) and others noted that the incrustations of optically oriented clays could be formed of illite, iron illite, or a mixture of kaolinite with illite. Popov (12) observed optically oriented clays in samples consisting of montmorillonite. He noted that kaolinite does not provide such formations. His investigations show that the character of the clay habitus is strongly affected by the adsorbed cations—specifically, that clays saturated with calcium form anisotropic aggregates of rectangular isometric shape. In water they neither expand nor lose their anisotropy. Sodium saturated clays form odd-shaped aggregates which are anisotropic when dry but which swell, disintegrate and lose their anisotropy when wet.

We may consider that optically oriented clays can be formed by montmorillonite, beidellite, illite, minerals of the polygorskite group and by other highly disperse minerals. It is generally known that aggregates of oriented particles can easily be formed from individual particles of clays. The orientation is caused by the flaky shape of these particles: on sedimentation they tend to occupy the most stable position possible so that they all lodge in one plane. When they dry out the aggregates of these oriented particles become hard. If we adhere to Rebinder's terminology we can call such a structure "crystallizational-condensational."

On the basis of his own investigations, Popov (12) concluded that clay particles may become oriented under the influence of surface tension. Other investigators have noted this factor. The surface tension of water may prove to be one of the important causes in the formation of the incrustational forms of optically oriented clays.

"Secondary clays" may evidently be formed not only from descending flow of colloidal solutions but also from rising and lateral capillary flow. Brewer considers the alternate wetting and drying to be the principal cause of the orientation of clay particles formed *in situ* by the weathering of rock. Here, too, the influence of surface tension is also evident.

Optically oriented clays are also formed under the influence of mechanical pressure. This phenomenon is known to petrographers from sedimentary rocks, and Popov (12) carried out appropriate experiments on it. Under the natural conditions of soil formation this process is evidently not important. The orientation of clay particles in soils takes place in general under the influence of surface tension and with periodic alternate wetting and drying. Mechanical pressure may be important as a factor in the formation of optically oriented clays when the soils are tilled, because of the effect of the tillage implements.

These differences in the microhabitus of soils are manifested also in their hydrophysical properties. It is not difficult to see that horizons with maximal accumulations of incrustational optically oriented clays (the illuvial horizons of podzolic soils, solonetzes, solods and also of takyrs) have hydrophysical properties which are extremely unfavorable for plants. They are very slightly permeable, of extremely hard consistence when dry and are difficult to till mechanically. Soils in which optically oriented clays are observed in minimal quantities, such as chernozems, sierozems and others, are distinguished by more favorable properties. The formation of very slightly permeable crusts during irrigation is obviously related to the formation within them of optically oriented clays as a result of intense dispersion of the tilled soils, alternating moistening and vigorous drying of the surface and the sedimentation of clays in the soils. We are far from knowing all the causes and conditions favoring the formation of different forms of optically oriented clays in soils. A study of them is of definite practical and theoretical interest.

Conclusions

1. Optically oriented clays are noted in many soil types. The habitus of their aggregates (form, size, mutual orientation, and other specific characteristics) varies and is determined by the conditions under which they were formed.

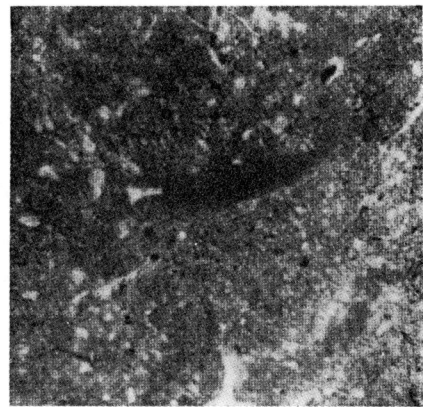

FIG. 4.—Optically oriented clay formed during sedimentation, profile 62, depth 0 to 50 cm. (magnification 5 × 8).

2. Optically oriented clays are formed by different processes: the migration of clay suspensions along the soil profile and the concentration of clays in the illuvial horizon in the form of crusts or incrustations on the walls of cracks and pores; they are formed *in situ* by weathering and soil formation; as a result of sedimentation from and the drying of suspensions carrying deluvium.

3. Aggregates of optically oriented clays may be formed by various highly disperse minerals. Their orientation is favored by the flaky habitus of the individual particles, their ability to cohere to each other when dry, the surface tension of water, mechanical pressure and other factors.

Received May 22, 1957

BIBLIOGRAPHY

1. AYDINYAN, R. KH. 1949. Obmen veshchestv i obrazovaniye mineral'nykh kolloidov v pervykh stadiyakh pochvoobrazovaniya na massivnokristallicheskikh porodakh (Metabolism and the formation of mineral colloids in the first stages of soil formation on solid crystalline rock). Doklady Akad. Nauk SSSR 67(4).
2. VIKULOVA, M. F. 1955. O noveyshikh metodakh issledovaniya glinistykh mineralov (The latest methods for investigating clay minerals). Vsesoyuz. Soveshchaniye Sotrudnikov Mineral.-Petrograf. Lab., Ministerstvo Geol. i Okhrana Nedr SSSR. Gosgeolizdat.
3. GERASIMOV, I. P. 1949. Korichnevye pochvy sukhikh lesov i kustarnikovykh lugostepey (Cinnamon-brown soils of dry forests and shrub meadow-steppes). Trudy Vsesoyuz.

Pochvennogo Inst. im. V. V. Dokuchayeva Akad. Nauk SSSR, Vol. 30. Izdatel'stvo Akad. Nauk SSSR, Moscow and Leningrad.
4. GORBUNOV, N. I. 1956. Zakonomernosti rasprostraneniya glinistykh mineralov v glavneyshikh tipakh pochv SSSR (Regular distribution patterns of clay minerals in the principal soil types of the USSR). Doklady VI Mezhdunarod. Kongressa Pochvovedov. II Komissiya po Khim. Pochv. Akad. Nauk SSSR, Moscow.
5. GORBUNOV, N. I., AND YE. M. LABENETS. 1956. Pochvennaya korka pri oroshenii takyrov v svyazi s ikh melioratsiyey. Sb.: Takyry Zapadnoy Turkmenii i puti ikh sel'skokhoz. Osvoyeniya (Soil crusts of irrigated takyrs as related to their improvement. In the symposium: Takyrs of Western Turkmenia and methods for their agricultural reclamation). Nauk SSSR, Moscow.
6. MINASHINA, N. G. 1955. Sero-korichnevye gazhevye (gipsonosnye) pochvy Kirovabadskogo massiva (Gray cinnamon-brown cement-forming [gypsiferous] soils of the Kirovabad massif). Dissertatsiya na Soiskaniye Stepeni Kandidata Nauk (Manuscript). Moscow.
7. PARFENOVA, YE. N. 1956. Issledovaniye mineralov podzolistykh pochv v svyazi s ikh genezisom. Kora vyvetrivaniya (An investigation of the minerals of podzolic soils in connection with their genesis. The crust of weathering). Vol. 2, Akad. Nauk SSSR, Moscow.
8. PARFENOVA, YE. I. AND YE. A. YARILOVA. 1956. Obrazovaniya vtorichnykh mineralov v pochvakh i rasteniyakh v svyazi s migratsiyey elementov (Formation of secondary minerals in soils and plants in relation to the migration of elements). Doklady VI Mezhdunarod. Kongressa Pochvovedov, II Komissiya. Akad. Nauk SSSR, Moscow.
9. POLYNOV, B. B. 1956. Vtorichnye mineraly ortshteynogennykh gorizontov pochv (Secondary minerals of soil horizons which produce ortstein). Izbrannye Trudy. Akad. Nauk SSSR, Moscow.
10. POLYNOV, B. B. 1948. Rukovodyashchiye idei sovremennogo ucheniya ob obrazovanii i razvitii pochv (Leading ideas of the contemporary doctrines of soil formation and development). Pochvovedeniye No. 1.
11. POLYNOV, B. B. 1956. O geologicheskoy roli organizmov (The geological role of organisms). Izbrannye Trudy. Akad. Nauk SSSR, Moscow.
12. POPOV, I. V. 1949. Mineralogicheskiye issledovaniya struktur glinnykh porod (Mineralogical investigations of the structure of argillaceous rocks). Problem. Sovremennogo Pochvovedeniya, Collection 15. Akad. Nauk SSSR, Moscow.
13. ROZANOV, A. N. 1951. Serozemy Sredney Azii (Sierozems of Central Asia). Akad. Nauk SSSR, Moscow.
14. ROZANOV, A. N. 1954. Zonal'nye pochvy ravnin i predgoriy Kura-Araksinskoy nizmennosti (Zonal soils of the plains and foothills of the Kura-Araks lowland). Trudy Vsesoyuz. Pochvennogo Inst. im. V. V. Dokuchayeva Akad. Nauk SSSR, Vol. 44. Izdatel'stvo Akad. Nauk SSSR, Moscow.
15. ROMASHKEVICH, A. I. 1956. Burye lesnye pochvy Krasnodarskogo kraya (Brown forest soils of Krasnodar region). Dissertatsiya na Soiskaniye Stepeni Kandidata Nauk (Manuscript). Moscow.
16. STARYKH, S. N. 1956. Mineral'nyi sostav i fiziko-khimicheskiye svoystva krasnozemov i zheltozemno-podzolistykh pochv Chernomorskogo poberezh'ya Kavkaza (Mineral composition and physicochemical properties of krasnozems and zheltozem-podzolic soils of the Black Sea shore of the Caucasus). Dissertatsiya na Soiskaniye Stepeni Kandidata Nauk (Manuscript), Moscow.
17. FEOFAROVA, I. I. 1956. Mikromorfologicheskaya kharacteristika takyrov. Sb.: Takyry Zapadnoy Turkmenii i puti ikh sel'skokhoz. osvoyeniya (A micromorphological characterization of takyrs. In the symposium: The takyrs of western Turkmenia and methods for their agricultural reclamation). Akad. Nauk SSSR, Moscow.
18. ALTEMÜLLER, H. S. 1956. Microskopische Untersuchungliniger Lössbodentypen mit Hilfe von Dünnschliffen (The microscopic investigation of loessial soil types by means of thin polished sections). Z. Pflanzenernähr. Düng. Bodenk. 72(2).
19. BREWER, R. A. 1956. A petrographic study of two soils in relation to their origin and classification. J. Soil Sci. 7(2).
20. BREWER, R. 1956. Optically oriented clay in thin sections of soils. VI Congr. intern. sci. sols (Comm. I–II). Paris.
21. MCCALEB, S. B. 1953. Profile studies of normal soils of New York. IV. Mineralogical properties of the grey-brown podzolic sequence. Soil Sci. 77: 319.
22. FREI, E. AND M. G. CLINE. 1949. Profile studies of normal soils of New York. II. Micromorphological studies of the grey-brown podzolic–brown podzolic soil sequence. Soil Sci. 68: 333.
23. KUBIENA, W. 1938. Micropedology.
24. STREMME, H. 1951. Quantitative Untersuchungen über Zersetzung und Bildung von Mineralen im Braunen Walboden (Quantitative investigations of the decomposition and formation of minerals in brown forest soils). Z. Pflanzenernähr. Düng. Bodenk. 53/4(3).

CLAY ORIENTATION IN SOILS

By I. STEPHEN, B.Sc., Ph.D., Rothamsted Experimental
Station, Harpenden.

THE lower limit for the study of the shape and optical properties of mineral particles with the petrological microscope is of the order of 5 μ, but the clay material of soils and other sediments may show optical properties such as birefringence and extinction phenomena similar to those of larger crystalline individuals. This was recognised by the early workers in soil microscopy, and Fry [11] stated that " between crossed nicols, the aggregates of soil colloids may or may not be doubly refracting. Those showing some degree of double refraction are perhaps more common than those showing none. Absence of double refraction, isotropy, seems usually associated with high iron content." Attention was drawn to differences in the optical properties of the aggregates: degree of completeness of the extinction directions, character of the interference figure and refractive indices.

The clay fraction of the majority of soils consists mainly of the so-called clay minerals (hydrous alumino-silicates) with subsidiary amounts of finely divided oxides and organic matter. Clay particles differ in shape, but many of the common clay minerals are micaceous in habit and form thin hexagonal flakes or shreds with well defined basal cleavage. Advantage has been taken of their anisodimensional shape to prepare aggregates for X-ray analyses by allowing clay suspensions to sediment with the consequent orientation of the (001) planes parallel to the surface of deposition. The optics of such aggregates have been studied by Williamson [34] using a clay consisting of 60 per cent. kaolinite, 30 per cent. illite and 10 per cent. quartz. Sections normal to the (001) planes showed a high degree of extinction parallel to these faces, the extinction direction being one of positive elongation. The same orientational relationships would hold for most clays, as the majority of the clay minerals such as micas (illites), kaolinite, vermiculites and montmorillonites have similar properties to muscovite in that they are optically negative with X about normal to (001); their extinction is essentially parallel to the cleavage traces and sections normal to the cleavage will therefore show sensibly straight extinction and positive elongation (Fig. 1). It can therefore be assumed that the optical anisotropy of clay aggregates in sediments results from the orientation of the individual platelets and, further, that the degree of extinction gives information about the perfection of the orientation, while the character of the elongation reveals the spatial relationships of the aggregates.

The degree and nature of the orientation of the clay which could be expected in soils has been investigated by several workers. Peterson [27] mixed montmorillonite and kaolinite with varying proportions of acid-washed fine sand, and the preparations were alternately wetted and dried. Thin sections showed areas of birefringent clay, but the patterns were different according to the type of clay. Montmorillonite formed a "sponge-like" structure with connecting wedges or bands of oriented clay between the sand grains, whereas with kaolinite vertical sections showed a laminated structure consisting dominantly of "islands" of oriented clay with

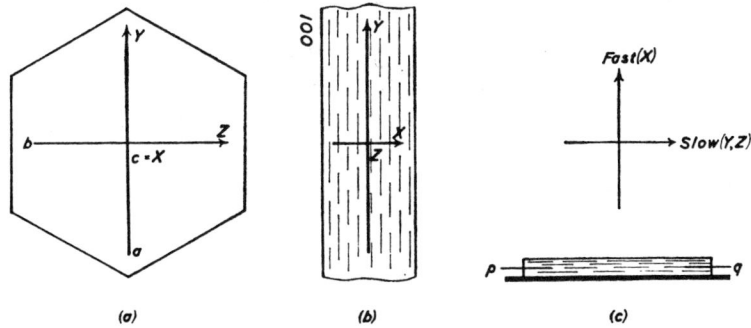

FIG. 1.—Orientation of muscovite. Sections (a) parallel to (001); (b) parallel to (010); (c) fast and slow vibration directions in sedimented clay; $p-q$ is trace of (001) planes.

threads of birefringent material penetrating an apparently unoriented matrix. These and other experiments together with field observations suggested that kaolinite is associated with the formation of platy structures in soils, but this has been questioned by McMillan and Mitchell [21] who described the occurrence of well developed platy A_2-horizons in podzolised and solonised soils in Saskatchewan, which contain very little kaolinite. Clearly insufficient is known at present to generalise about the effects of specific minerals in structure formation.

Brewer and Haldane [5] introduced clay-sized material, consisting of 60 per cent. illite, 30 per cent. kaolinite and 5 per cent. quartz, saturated with different cations, into quartz sand by percolation or capillarity. The clay formed strongly oriented coatings around the sand grains, and the orientation was unaffected by the method of introduction or the nature of the saturating cation. Another set of experiments concerned the drying of the clay after mixing with different amounts of coarser material of various size distributions.

When highly flocculated clay was dried in a mass, little or no orientation occurred, but bands and coatings of strongly oriented clay were produced in clay-sand mixtures when the amount of clay was insufficient to fill the spaces between the sand grains. A gradual decrease in the degree of orientation was observed with either increasing proportions of clay or the introduction of silt into the mixtures. The authors considered that, when soluble salts are not present in excessive amounts in a soil profile, clay migration would result in strongly oriented clay bands and coatings developing in the illuvial horizons, and evidence from natural profiles supported this conclusion. If, however, owing to a high percentage of soluble salts, eluviated clay is rapidly flocculated and deposited in large pore spaces in illuvial horizons it will be only weakly oriented; this type of orientation was observed in illuviated clay in some solodised solonetz soils of Southern Australia. The heavy-textured grey and brown soils of the Riverine plain containing appreciable amounts of silt show no textural differentiation and have a degree of clay orientation comparable with that produced in experiments with mixtures of similar mechanical composition, and the orientation appears to be due primarily to the conditions of deposition of the soil parent material.

In a comprehensive treatise on soil formation Kubiëna [14] included detailed descriptions of thin section analyses of a variety of soil groups. One of the most widespread microfabrics described is that occurring in the terra fusca (limestone braunlehm) associated with limestone and dolomite in Central Europe, and in the braunlehm formed by the weathering of silicate and other rocks under subtropical and tropical conditions. These soils have ochre-yellow, ochre-brown to reddish coloured (B) or B-horizons, which are compact and plastic with a pronounced waxy appearance. In section, the uniform fabric is dense, with the very finely divided iron (probably mainly in the form of goethite) diffusely dispersed, giving the clay a light yellowish brown colour. The clay is capable of becoming mobile and may be strongly oriented when deposited in situations where it can dry, such as in cracks, around grains or on walls of cavities. The general groundmass is anisotropic and mainly weakly oriented; small areas, however, are moderately well or strongly oriented, but are randomly arranged in relation to surrounding units. This distribution somewhat resembles the decussate fabric of crystalline rocks and probably results mainly from small scale plastic deformations caused by the alternate wetting and drying of the clay *in situ* effecting the orientation of the particles with respect to one another and to the coarser-grained minerals in the soil.

Kaolinite and halloysite (and occasionally mica and montmorillonite) appear to be the common clay mineral constituents of the braunlehms [1, 14, 29], and kaolinite is dominant in the rotlehms, the corresponding red-coloured tropical soil formations [14, 15]. The physical behaviour of the soils is explained by Kubiëna [14] on the basis of the presence of colloidal silica as a protective colloid, which acts as an efficient peptising agent and " confers on the kaolinitic and halloysitic clays their high swelling capacity and plasticity as well as their extraordinary hardness when dry."

Two distinct processes of transformation of braunlehm material have been described: rubefaction (formation of rotlehm) and laterisation [16]. Laterisation is regarded as a kind of regional diagenesis, generally under stable moisture conditions, leading to a slow separation of iron and its precipitation in fissures and other pore spaces producing a wide variety of structural forms. In contrast the process of rubefaction results in a comparatively uniform morphology and takes place in a strongly contrasting wet-dry climate, especially in well drained sites. Under these conditions the iron separates more quickly (probably as microcrystals of hæmatite) and becomes distributed in the groundmass of the fabric as localised deposits and flecks with diffuse edges (Iwatoka-precipitates). The fabrics of the rotlehms, therefore, are considered to consist of two phases: (1) the yellowish brown, anisotropic, mobile, silica-rich clay characteristic of the braunlehm and (2) the secondary, immobile, iron-enriched Iwatoka-precipitates which are brilliant red in reflected light and show little double refraction between crossed nicols. The rubefaction first results in the formation of single flecks and then extends progressively, so that it is possible to recognise a complete sequence from the non-rubefied braunlehms to rotlehms of the Iwatoka variety (earthy rotlehm), as described [13] from the Sudan, in which the fabric consists mainly of very fine deep red thrombus-like deposits, the yellowish brown mobile clay occurring only in small amounts, mainly in cavities protected from leaching. With progressive rubefaction the physical properties of the clay are greatly altered so that the Iwatoka rotlehm is not dense and waxy, but loose and crumbly with a uniform spongy fabric. The importance of micromorphological studies in investigating the products of laterisation and rubefaction is stressed, as chemical methods do not enable a distinction to be made between the products of these two different processes.

One of the most important uses of thin section analyses is in deciding whether clay migration has been an operative process in the development of textural profiles, as it is becoming abundantly clear that clay content increasing with depth is not proof of illuviation, as seems often to have been assumed in earlier pedological investigations. Several different reasons have been suggested for the presence of finer-textured subsoil horizons; these include deflocculation and downward movement of clay-sized material; the net difference between the amount of clay formed *in situ* and that destroyed being greater in the B-horizon than in the surface layers; and non-uniformity of parent material. Uniformity (or otherwise) of the parent material within the solum can usually be ascertained by standard petrological techniques, such as a comparison of the resistant heavy mineral suites of successive layers, but conclusive evidence about clay migration is less easy to obtain, although in some soils the surfaces of the peds have a visible coating of clay ("clay skins"), often with small ripple-like markings suggestive of its movement under the influence of percolating waters. In thin section, however, the microfabrics of many subsoil horizons show distinctive features which are difficult to reconcile with any process other than the deposition in them of illuviated clay. Such features are the presence of more or less homogeneous clay deposits, non-uniformly distributed and concentrated on ped surfaces, lining channels and cracks and partially or completely filling pore spaces (Fig. 2, A). This type of distribution was noted by Frei and

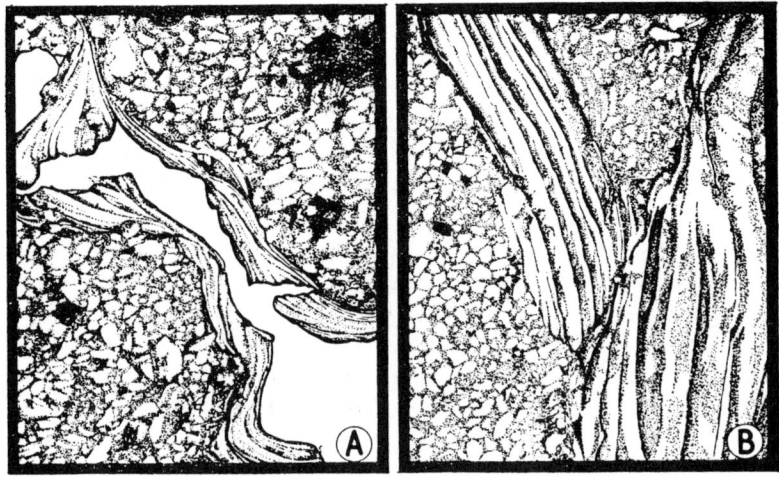

Fig. 2.

(A) Strongly oriented clay coating channel walls in lenticles of fine sandy clay-overlying "Clay-with-flints," near Chesham Bois, Bucks. (Grid Ref.: SU/972996.) Magnification: × 35.

(B) Laminated clay-layers in "brickearth," Prestwood brick-pit, Bucks. (SP/859014). Magnification: × 35.

(Drawings by Mr. D. V. JONES, Soil Survey of England and Wales.

Cline [9] in grey-brown podzolic soils developed from calcareous till in New York State, and it was argued that " if the clays were the weathering products of primary minerals originally in this part of the profile or were residual from clays originally present in the till, one would expect to find them more uniformly distributed throughout the horizon. . . . In addition, the layering and optical continuity of the clayey bodies would be unlikely if the clays were residual from either clays or other minerals originally present in this part of the profile. The mode of occurrence and optical continuity both suggest that the clays have been deposited on the walls of channels through which percolating waters pass."

Recent Russian investigations [25, 35] show that, in a wide variety of soils with illuvial horizons (podzolic soils, grey forest soils, degraded chernozems), there is a strong development of clay films that show orientation (colloform clay). Chemical, X-ray and d.t.a. examinations of the fine clay films led to the conclusion that the clay was in part a new formation, which was named polynite (a 2 : 1 layer lattice silicate), but another Russian worker [22] believes that the clay has been illuviated, the view that is generally accepted.

These illuviated deposits may show a fairly wide range in properties, but a discussion of their characteristics from a brick-pit at Prestwood, Bucks., where clay skins are well developed, will illustrate some of their micromorphological features [30]. At this locality a superficial deposit of flinty clay overlies weakly laminated " brickearth." The latter is seamed with narrow, roughly vertical, fissures several feet long but often less than $\frac{1}{10}$ in. thick practically filled with red-brown clay that contrasts strongly with the yellow-brown " brickearth." With increasing depth several fissures unite to form clay-joints up to $\frac{3}{4}$ in. thick. The deposits are often layered, the layering being parallel to the walls of the fissures (Fig. 2, B). The structure probably arises through clay, from the overlying drift, being carried in thin films of water and deposited on the mineral skeleton of the " brickearth "; once any crevices and irregularities have been filled deposition becomes regular, and the layered nature of the clay suggests a process of repeated (cyclic) sedimentation. Examination of thin sections cut normal to the layering reveals a very high degree of optical continuity, a field showing positive elongation changing to one showing negative elongation on rotation through 90°, the slow vibration direction of the clay aggregates being parallel to the channel walls; this is consistent with the drying out and consequent orientation of the (001) planes parallel to the surface of deposition as demonstrated in laboratory experiments (*see*

Fig. 1). In circular cavities the clay exhibits convex layering, and shows multiple dark interference bands between crossed nicols. The clay shows pronounced pleochroism from deep chocolate brown (Y, Z) to pale yellowish brown (X). The reason for the pleochroism is not clear as the iron, which is responsible for the colour, is probably not an integral part of the clay mineral structure, but is in the form of free iron oxide coating the particles; epitaxic growth of crystalline iron oxides in a definite orientation on the clay particles, as occurs in the system kaolinite-goethite [10], is probably responsible.

The favourable physical conditions at Prestwood, with the existence of well fissured strata, appears to have provided suitable conditions for the maximal development of clay skins showing a very high degree of preferred orientation and associated properties, but comparable phenomena, including layering and pleochroism, have been noted by several workers in the B-horizons of the soils of a number of great soil groups. Brewer [4] has pointed out that, although the term clay skins may be adequate for a macromorphological description, micromorphologically they may show considerable variation and " may be oriented to any degree from perfectly to practically unoriented. They may be composed of innumerable thin lamellæ or a few thick layers. The layers may be uniform in colour or vary from one edge to the other, and so on. Each of these different characteristics is due to some difference in genesis or conditions of formation." Further observations along these lines would be expected to yield additional information about the mechanisms of clay illuviation in soil horizons.

Microfabrics indicative of clay migration in colloidal suspension and its accumulation to form textural B-horizons have been described in a number of grey-brown podzolic soils developed from calcareous parent materials of glacial, fluvioglacial and loessial origin of the Great Plains and north-eastern United States [9, 12, 19, 26, 33], and in the red-yellow podzolic and related reddish brown lateritic soils in the Piedmont area of the south-eastern United States [6, 20, 24]. The red-yellow podzolic soils are intensely weathered with reddish, reddish yellow and yellow subsoil horizons; the yellow colours are associated with siliceous and feldspathic parent materials, whereas the redder members of the group occur on parent materials which contain larger amounts of ferromagnesian minerals and intergrade to the reddish brown lateritic soils on basic rocks. Whereas in the C_1 and B_3-horizons the clay accumulations are mainly *in situ* alterations of primary minerals, the B_2-horizons have prominent continuous and usually thick clay skins with illuviated clay in channels, pores and on ped surfaces. McCaleb [20] considers that "the genesis of red-yellow and grey-brown podzolic soils seems to be alike in kind, but differs considerably in the degree and intensity of expression of similar horizon sequences formed under quite different environments. The red-yellow soils of the United States are older genetically than the grey-brown podzolic soils. This maturity is expressed in terms of degree of primary mineral alteration, dominant clay mineral suites present, amount of clay and its distribution, profile development, and the extreme acid conditions resulting from base depletion."

Descriptions of fabrics apparently closely akin to those of the grey-brown podzolic soils have been given by Altemüller [2] and Kubiëna [17] for soils developed in loess in Western Europe. On the weathering of the loess a sequence of soils is formed from AC-soils (black earths, pararendzinas) to soils with an ABC-horizonation. In the early stages of development of the latter soils field evidence of clay migration is lacking, but thin sections of the subsoil horizons show the presence of yellowish brown oriented clay mainly on the walls of channels and in pore spaces. Further evolution leads to the formation of distinct A and A_e-horizons depleted of clay, and the whole fabric of the illuvial B-horizon is permeated by oriented clay deposits, which also fill numerous and well developed conducting channels. Mückenhausen [23] introduced the term parabraunerde for such soils, and they have been equated by Kubiëna [17] with the *sols lessivés* of the French pedologists and with a segment of the grey-brown podzolic soils of the United States.

From the few micromorphological studies that have been made on British soils some soils on limestone, *e.g.* on Devonian limestone in Devon [7], and on Oolitic limestone in the Cotswolds [31], have patterns of clay orientation comparable with that of the terra fuscas, and soils developed in loessial deposits and in parent materials containing loessial additions have features in common with the parabraunerde and the grey-brown podzolic soils of the United States, involving differential accumulation of clay by translocation within the profile. These include soils developed from the valley brickearths in Kent [8], deep silty deposits overlying the Carboniferous limestone in the Mendips [31], and calcareous drifts (Coombe deposits) which incorporate substantial amounts of material of loessial origin in the Chilterns [3]. Most of these soils may be correlated with the grey-brown podzolic soils as defined by Tavernier and Smith [32], but the soils of the Chiltern plateau developed on " Clay-with-flints " and allied deposits have thick intensely weathered subsoils that are generally low in bases throughout and conform in most respects to the red-yellow podzolic soils [3]. The soils have a superficial mantle of loessial material incorporated with the underlying "Clay-with-flints" which itself consists largely of the weathered remains of Chalk and Eocene sediments and has a locally rubefied braunlehm fabric probably formed under the hotter, wetter tropical or subtropical conditions of the early Pleistocene or Tertiary. Thin sections provide evidence of clay migration in these soils, which have developed by the superimposition of pedogenic processes on composite parent materials further accentuating the original textural differences.

From this brief review it is evident that several distinctive patterns of clay morphology exist in undisturbed soils, and detailed investigation of the micromorphology yields valuable information on the conditions of deposition of the parent material and on the genesis of the soils, and assists in their classification. The processes by which soils are formed are probably relatively few, but the resulting fabrics are influenced by their interaction on a diversity of parent materials under different climatic conditions. Features indicative of clay migration may be found in a variety of great soil groups, e.g. in the grey-brown podzolic soils of temperate regions, the red-yellow podzolic soils of the subtropics and the latosolic soils of the tropics [18, 28], but other fabrics such as those of the rubefied soils appear to be characteristic of particular climatic régimes. Should this relationship between specific fabrics and climate be firmly established, micromorphological studies should also throw much light on the problem of the conditions of formation of buried and relic soils.

References

1. ALBAREDA, J. M., ALEIXANDRE, V., and SÁNCHEZ CALVO, M. del C. (1955), *An. Edafol. Fisiol. veg.*, **14**, 543–63.
2. ALTEMÜLLER, H. J. (1956), *Z. PflErnähr. Düng.*, **72**, 152–67.
3. AVERY, B. W., STEPHEN, I., BROWN, G., and YAALON, D. H. (1959), *J. Soil Sci.*, **10**, 177–95.
4. BREWER, R. (1957), *Report of overseas visit.* C.S.I.R.O. (Australia).
5. —— and HALDANE, A. D. (1957), *Soil Sci.*, **84**, 301–9.
6. CADY, J. G. (1950), *Proc. Soil Sci. Soc. Amer.*, **15**, 337–42.
7. DALRYMPLE, J. B. (1957), *J. Soil Sci.*, **8**, 161–5.
8. —— (1958), *ibid.*, **9**, 199–209.
9. FREI, E., and CLINE, M. G. (1949), *Soil Sci.*, **68**, 333–44.
10. FRIPIAT, J. J., and GASTUCHE, M. C. (1952), *Publ. Inst. nat. agron. Congo belge, Ser. Sci.* No. 54.
11. FRY, W. H. (1933), *U.S.D.A. Tech. Bull.*, **344**.
12. JOHNSTON, J. R., and PETERSON, J. B. (1941), *Proc. Soil Sci. Soc. Amer.*, **6**, 360–7.

13. KUBIËNA, W. L. (1943), *Beitr. z. Kolonialforschung*, **3**, 48–58.
14. —— (1948), *Entwicklungslehre des Bodens*. Springer-Verlag, Wien.
15. —— (1953), *The Soils of Europe*. Murby & Co., London.
16. —— (1956), *Rep. 6th int. Congr. Soil Sci.*, E, 247–9.
17. —— (1956), *Eiszeitalter und Gegenwart*, **7**, 102–12.
18. LARUELLE, J. (1956), *Pédologie*, **6**, 38–58.
19. MCCALEB, S. B. (1954), *Soil Sci.*, **77**, 319–33.
20. —— (1959), *Proc. Soil Sci. Soc. Amer.*, **23**, 164–8.
21. MCMILLAN, N. J., and MITCHELL, J. (1953), *Canad. J. agric. Sci.*, **33**, 178–83.
22. MINASHINA, N. G. (1958), *Pochvovedenie* No. 4, 90–6.
23. MÜCKENHAUSEN, E. (1955), *Entwurf einer Systematik der Böden Deutschlands*.
24. NYUN, M. A., and MCCALEB, S. B. (1955), *Soil Sci.*, **80**, 27–41.
25. PARFENOVA, E. I., and YARILOVA, E. A. (1958), *Pochvovedenie* No. 12, 28–35.
26. PETERSON, J. B. (1937), *Proc. Soil Sci. Soc. Amer.*, **2**, 9–13.
27. —— (1944), *ibid.*, **9**, 37–48.
28. RUHE, R. V., and CADY, J. G. (1954), *Trans. 5th int. Congr. Soil Sci.*, **4**, 401–7.
29. SÁNCHEZ CALVO, M. del C. (1956), *Rep. 6th int. Congr. Soil Sci.*, E, 433–7.
30. STEPHEN, I., and OSMOND, D. A. (1957), *Rep. Rothamst. exp. Sta. for 1956*, 65.
31. —— —— (1959), *ibid., for 1958*, 62.
32. TAVERNIER, R., and SMITH, G. D. (1957), *Advanc. Agron.*, **9**, 217–89.
33. THORP, J., CADY, J. G., and GAMBLE, E. E. (1959), *Proc. Soil Sci. Soc. Amer.*, **23**, 156–61.
34. WILLIAMSON, W. O. (1947), *Amer. J. Sci.*, **245**, 645–62.
35. YARILOVA, E. A., and PARFENOVA, E. I. (1957), *Pochvovedenie* No. 9, 37–48.

Copyright © 1961 by the Soil Science Society of America
Reprinted from *Soil Sci. Soc. America Proc.* 25:377-379 (1961)

Clay Skin Genesis in Wisconsin Soils[1]

S. W. BUOL AND F. D. HOLE[2]

ABSTRACT

Clay skins separated from the B_3 horizon of the "Ockley-like" silt loam, a Gray-Brown Podzolic soil, were further analyzed and found to contain 186% as much total phosphorus and 177% as much total manganese as the bulk of the same horizon. More than 200 thin sections were made of samples from a variety of Wisconsin soils, including soils of the Podzol, Gray-Brown Podzolic, Brunizem and Humic-Gley great soil groups. A study of volume of clay skins, as determined from microscopic views of thin sections, and soluble salt concentration revealed that in the "Ockley-like" profile maxima of both occurred in the C_1 horizon. By alternately leaching with percolate from a leaching column and drying with a water aspirator, artificial clay skins were produced in "unweathered" loess material. From observations made during this study a definition of the term "clay skin" is proposed and it is concluded that clay skins in Wisconsin soils are formed by the percolation of dilute clay suspension, from which clay is deposited at or below the bottom of the solum as percolation ceases and the larger pores are emptied of water.

COATINGS on the surface of soil aggregates and coarse particles are referred to in field descriptions of soils representing many great soil groups, including Brunizems (9), Solonetz (12), Brown Podzolic (5, 6), Gray-Brown Podzolic (5), Humic-Gley (13), Planosol (11), Latosol (8), Red-Yellow Podzolic (10), Low Humic Latosol (14), and several other great soils groups (11). Other reference to clay skins have been listed previously (3). Reference to clay skins and clay skinlike structures is made in reports on thin-section studies of various parts of the profiles of the Miami soil (15), some Yellow Podzolic soils (1), two soils from the Piedmont region of the eastern United States (4), and in a Gray-Brown Podzolic–Brown Podzolic sequence (6). Artificial clay skins have been formed on sand particles in leaching columns and by blowing hot air over sand-filled columns wetted from the bottom by dilute suspensions of clay (2).

The present paper reports data on total phosphorus and manganese contents in clay skins previously separated from the B horizon of a Gray-Brown Podzolic soil (3), reproduces tracings of photomicrographs of thin sections made from the major horizons of a loess-derived Gray-Brown Podzolic soil, presents conclusions from field observations of clay skins in 58 Wisconsin soil profiles, proposes a definition for the term "clay skin," and describes a method of artificially producing optically oriented deposits of clay in soil material.

FIELD AND LABORATORY METHODS

Field observations and sampling.—Field evidence of clay skin development was studied in and samples were taken from 58 soil profiles during a 2-week trip over the state of Wisconsin in August, 1958. All profile sites were selected with the aid of the U. S. Soil Conservation Service personnel. Soluble salt content determinations were made on soil samples taken June 12 and September 24, 1959, from a freshly exposed profile of "Ockley-like"[3] silt loam in Dodge County, Wisconsin, and sealed in plastic bags immediately upon sampling. An attempt to form artificial clay skins in the field was made by injecting a suspension of reddish-brown clay from a Kewaunee soil through a large needle into B and C horizons of a Gray-Brown Podzolic and a Humic-Gley soil developed from loess.

Clay skin separation.—Methods of clay skin separation are the same as those reported by Buol and Hole (3). In addition, material was separated from the B_3 horizon of the "Ockley-like" silt loam with a Franz isodynamic separator, a device that envelopes an inclined vibrating table with a magnetic field. This separator was found useful in separating clay skins from air-dried, sieved samples. However, clay skin material separated by this separator was largely contaminated with iron concretions.

Methods of soluble salt content analysis.— A 30- to 40-mg. subsample was taken from each field sample and oven-dried to determine the moisture content. Another subsample was wetted to saturation and allowed to stand for at least 30 minutes. A subsample was then taken from each saturated sample and oven-dried to determine the water content. The remainder of the saturated sample was placed in a Buchner funnel, using a Whatman No. 2 filter paper, and the vacuum applied with a water aspirator. The electrical resistance of the extracted solution was determined with a Bouyoucos bridge. The amount of salt in each sample was calculated by the formula: % salt = 0.064 L mmhos per cm. × % $H_2O/100$, where L = specific conductivity of the solution and % H_2O is the percent of water in the saturated paste. % salt 10^4 = ppm. (7).

Method for the measurement of volume of clay skins.—Undisturbed cores 2 inches in diameter by 1.75 inches in depth were taken from the major horizons of the "Ockley-like" silt loam profile. From each core two thin sections were cut at right angles to each other, and ground to a thickness of about 20μ. Fields of view were selected at random on each thin section and projected through a petrographic microscope onto a ground-glass viewer, using Bausch and Lomb photomicrographic equipment. The areas of clay skins in each view were traced on translucent paper, cut out and weighed. Ten such views were found to cover about 1% of the area of a single thin section.[4]

Phosphorus and manganese analyses.—Total phosphorus content was determined by using the chlorostannous-reduced molybdophosphoric blue color method, after sodium carbonate fusion of the sample. Total manganese content was colorimetrically determined after hydrogen fluoride fusion of the sample (7).

---

[1] Contribution from the Soil Survey Division, Wisconsin Geological and Natural History Survey and the Soils Department, University of Wisconsin, Madison. Published with the permission of the Director of the Wisconsin Agr. Exp. Sta. This work was supported in part by the Research Committee of the Graduate School with funds from the Wisconsin Alumni Research Foundation. Received Oct. 17, 1960. Approved Apr. 5, 1961.

[2] Former Research Asst., Soil Survey Div., Wis. Geol. and Nat. Hist. Survey and Soils Dept., University of Wisconsin; now Asst. Professor of Agricultural Chemistry and Soils, University of Arizona, Tucson; and Associate Professor of Soils, in charge, Soil Survey Division, Wisconsin Geological and Natural History Survey, University of Wisconsin, Madison.

[3] Recent correlation has determined that the name Ockley cannot be used for this soil in Wisconsin. No new series name was available at the time this manuscript was prepared.

[4] Thin sections were made in thin-section laboratory of the Department of Geology, University of Wisconsin, by the senior author and Mr. D. M. Fadness, Laboratory Technician.

Artificial production of clay skins.—Loess containing 62% silt and 24% clay, sampled at a depth of 25 to 30 feet near the Mississippi River in Grant County, Wisconsin, was air-dried and passed through a 2-mm. sieve. The loess was then placed in leaching columns through which distilled water was passed. At intervals, 250-ml. portions of the percolate from the columns, which was found to contain suspended clay, was allowed to drop onto samples of sieved loess which had just been dried in a Buchner funnel for about 6 hours under suction from a water aspirator. By means of 10 such drying and wetting cycles, clay skins were formed around large pores which had developed in the sample of loess in the funnel.

RESULTS AND DISCUSSION

Definition of the term "clay skin."—It is proposed that the term "clay skin" be defined as *the assemblage of optically oriented clay (0.002 mm), with included coarser particles, formed on the walls of interstices in the soil and exhibiting abrupt internal and external boundaries.* Clay skins in well-drained soils also contain larger proportions of iron and/or organic matter than the surrounding matrix.

Phosphorus and manganese contents of clay skins.—Samples of clay skin material manually separated from the B_3 horizon of the "Ockley-like" silt loam, and samples of the entire B_3 horizon material were analyzed for total P and Mn content. P content was found to be 149 ppm. in the clay skin and only 80 ppm. in the entire B_3 horizon sample. Mn content increased from 98 ppm. in the entire B_3 horizon sample to 174 ppm. in the clay skins separated from the same horizon. It is of interest to note that clay skin samples separated with the Franz separator contained approximately the same amount of P as the manually separated clay skin material. No difference was found in pH between clay skins and the bulk sample of the "Ockley-like" B_3 horizon.

Distribution of clay skins in the "Ockley-like" soil profile.—Petrographic microscopic study of a series of thin sections of undisturbed samples taken from the major horizons of an "Ockley-like" silt loam, a Gray-Brown Podzolic soil developed from 69 inches of loess over outwash in Dodge County, Wisconsin, revealed that the maximum accumulation of clay skins, by volume, is in the C_1 horizon, as reported in the right hand portion of figure 1. It is important to note that the tracings in figure 1 do not show average conditions for the horizons in question, but represent only the conditions adjacent to interstices, which are spaced farther apart with increasing depth in the soil profile. The boundary between the B_3 and C_1 horizons in the "Ockley-like" silt loam soil is gradual. The C_1 horizon was found to be a silt loam texture and the B_3 horizon a silty clay loam texture. The ped size increases from coarse blocky in the B_3 horizon to very coarse blocky in the C_1 horizon. Much thicker clay skins are developed in the C_1 horizon than in the B_3 horizon. The thickness of the clay skins in the C_1 horizon accounts for the greater volume of clay skin material. The tracings illustrate that clay skins in the C horizon occur chiefly on the surfaces of peds and root channels, and that in overlying horizons clay skins are increasingly fragmented and remote from ped surfaces.

Distribution of soluble salts in the "Ockley-like" soil profile.—The soluble salt content was determined by 6-inch increments in the "Ockley-like" silt loam profile on two different dates, in June and September, 1959. The highest total content of soluble salt, 369 ppm. in June, was found in the C_1 horizon at a depth of 54 to 60 inches. In September, the C_1 horizon was found to contain 650 ppm. salts at a depth of 46 to 54 inches, and 423 ppm. salts at 54 to 60 inches. These figures are to be compared with from 31 to 102 ppm. for the A and B horizons, respectively, and 132 to 212 ppm. for the C_2 horizon. The highest concentration of soluble salts in the "Ockley-like" profile is found to be at the same depth as the maximum clay skin development.

Observations concerning some factors and processes of formation of clay skins.—The following observations are based on field investigations, on studies of representative

Figure 1—Tracings of views of microscopic thin sections of soil from horizons of "Ockley-like" silt loam, Dodge County, Wisconsin. Distinctly yellowish and reddish-brown portions of the views, showing parallel orientation of clay, have been reproduced in black in this figure, regardless of variation in density of color apparent in thin section. The "nearly continuous clay skins" reported in currently standard soil profile descriptions of Gray-Brown Podzolic soils in Wisconsin, appear in thin section as indistinct and very faint yellowish stains on the walls of voids, without the oriented structure of "clay skins," as defined in this paper.

thin sections of undisturbed samples from 58 soil profiles in Wisconsin, and on laboratory experiments with leaching columns:

1. Clay skins form chiefly in soils with relatively stable aggregates and root channels, as occur in soils of finer texture than sands, loamy sands, and coarse sandy loams in Wisconsin. Where clay skins do form in soils of coarse texture, the skins coat individual coarse mineral particles rather than ped surfaces.

2. Clay skins form chiefly in soils which contain an adequate supply of clay, as is the case in soils of finer texture than sands, loamy sands, and coarse sandy loams in Wisconsin. Coarser textured soils simply don't supply enough clay in the A horizon to produce appreciable quantities of clay skins.

3. Time, as a factor in genesis of clay skins, allows for their construction, disruption, and removal.

4. Percolating water moves the clay skin material in suspension and deposits it wherever the percolation is arrested. As seen in figure 1 the oriented clay deposits found in the upper B horizon are found in the interiors of the peds. The theory of clay skin formation indicates that these formations were originated on the walls of large pores. It is therefore postulated that the areas of oriented clay found in the upper B horizon were originally on the walls of larger pores but due to turbation processes in the soil they were incorporated into the ped interiors. The absence of clay skin material around the larger pores now found in the upper B horizon indicates that clay skins are not actively forming in these horizons and in fact clay is at the present time being eluviated from these horizons and deposited at a greater depth in the profile.

5. Natural clay skins are formed in very small increments of minute lamellae. Injections of reddish-brown clay suspensions into subsoil horizons in the field produced thick clay coatings on peds and around large pores, but these coatings lacked oriented, lamellar structure. Oriented clay appeared in thin sections of clay skins formed artificially in the laboratory by passing clay suspensions through sieved loess.

6. Soil expansion and contraction and soil turbation or mixing by freezing and thawing, by dessication and wetting, by slow mass movement, and by activities of organisms cause the fragmentation of clay skins in upper horizons, and the incorporation of the fragments into the interiors of peds. The severity of natural soil disturbances decreases with depth in soil profiles in Wisconsin. This permits clay skins to endure relatively undisturbed in the C horizon.

7. The C_1 horizon of the "Ockley-like" soil in Wisconsin is the zone of maximum salt accumulation and of maximum clay skin accumultion. It is proposed on the basis of this evidence that most of the effective percolating water ceases to move through the larger interstices in the soil upon reaching the C_1 horizon.

8. Conditions of clay skin formation in Wisconsin soils are such as to yield skins of uniform color (5YR 3/3-3/2, moist) and structure throughout the state. Percolating waters apparently make similar combinations of clay, organic matter and iron compounds from one soil to another.

CONCLUSIONS

Clay skins are thought to be formed, in Wisconsin soils, from clay carried in suspension by percolating water from the upper part of the solum and deposited throughout the lower B and upper C horizons in an oriented fashion as the water leaves the large pores.

LITERATURE CITED

1. Brewer, R. Mineralogical examination of a Yellow Podzolic soil formed on granodiorite. Com. Sci. Ind. Res. Org., Australia, Soil Publ. 5. 1955.
2. ———, and Haldane, A. D. Preliminary experiments in the development of clay orientation in soils. Soil Sci. 84:301-309. 1957.
3. Buol, S. W. and Hole, F. D. Some characteristics of clay skins on peds in the B horizon of a Gray-Brown Podzolic soil. Soil Sci. Soc. Am. Proc. 23:239-241. 1959.
4. Cady, J. G. Rock weathering and soil formation in the North Carolina piedmont region. Soil Sci. Soc. Am. Proc. (1950) 15:337-342. 1951.
5. Cline, M. G. Profile studies of normal soils in New York: I. Soil profile sequences involving Brown Forest, Gray-Brown Podzolic and Brown Podzolic soils. Soil Sci. 68:259-272. 1949.
6. Frei, E. and Cline, M. G. Profile studies of normal soils of New York: II. Micromorphological studies of the Gray-Brown Podzolic–Brown Podzolic sequence. Soil Sci. 68:333-334. 1949.
7. Jackson, M. L. Soil Chemical Analysis. Prentice-Hall, Inc., Englewood Cliffs, N. J. 1958.
8. Kellogg, C. E. and Davol, F. D. An exploratory study of soil groups in the Belgian Congo. Inst. Nat. Etude Agron. Congo Belge. Serie Sci. 46. 1949.
9. Matelski, R. P. Great soil groups of Nebraska. Soil Sci. 88:218-239. 1959.
10. McCaleb, S. B. The genesis of the Red-Yellow Podzolic soils. Soil Sci. Soc. Am. Proc. 23:164-168. 1959.
11. Minashina, N. G. Optically oriented clays in soils. Soviet Soil Sci. 4, 1958. Translated Dec., 1959.
12. Retzer, J. L. and Simonson, R. W. Distribution of carbon in morphological units from the B horizon of Solonetz-like soils. Agron. J. 33:1009-1013. 1941.
13. Schaefer, G. M. and Holowaychuk, N. Characteristics of medium and fine-textured Humic-Gley soils of Ohio. Soil Sci. Soc. Am. Proc. 22:262-267. 1958.
14. Sherman, D. G. and Alexander, L. T. Characteristics and genesis of Low Humic Latosols. Soil Sci. Soc. Am. Proc. 23:168-170. 1959.
15. Thorp, J., Cady, J. G. and Gamble, E. Genesis of Miami silt loam. Soil Sci. Soc. Am. Proc. 23:156-161. 1959.

Argillic Horizons Without Clay Skins[1]

W. D. NETTLETON, K. W. FLACH, AND B. R. BRASHER[2]

ABSTRACT

Although the clay in some moderately fine and fine-textured Bt horizons of soils of arid and mediterranean climates of the southwestern United States is highly oriented, no distinct illuvial clay skins can be recognized. The distribution of clay skins is related to shrink-swell potentials. Clay skins are absent in horizons having a shrink-swell potential of more than 4% or a masepic or omnisepic plasmic fabric; they are present in equivalent horizons having low shrink-swell potentials and an insepic or mosepic plasmic fabric. The clay content, mineralogy, and moisture regime of a Bt horizon in turn largely determine its potential to shrink and swell and hence determine its plasmic fabric.

Evidence that clay illuviation has indeed taken place in these finer textured Bt horizons is based on four pairs of geographically associated soils with horizons of clay accumulation. Bt horizons of the coarser textured members of pairs have clay skins and the finer textured members do not. The distribution of biotite pseudomorphs in some of these pairs parallels the distribution of clay skins, suggesting that oriented bodies of clay can be destroyed. Clay orientation in one of the horizons was reformed experimentally to show that the highly oriented soil fabrics do not acquire their orientation by illuviation of clay.

The studies further indicate that bodies of oriented clay in medium and fine-textured B horizons have been erroneously described as clay skins.

Additional Key Words for Indexing: clay movement, micromorphology, shrinking and swelling soils, soil fabrics, thin sections.

SOME SOILS of arid and mediterranean climates of the southwestern United States have horizons of clay accumulation but do not have distinct illuvial clay skins in the part of the profile having the highest clay content. Coarser textured, lower parts of these horizons may have clay skins but if the Bt horizon overlies a lithic contact or a petrocalcic horizon clay skins are not found anywhere in the profile. Absence of clay skins has been reported for the B horizon of a Mohave soil (Typic Haplargid) by Buol and Yesiloy (1964) and has been attributed to natural soil turbation. Preliminary studies by one of us showed that it was not possible to distinguish between coatings of illuviated clay and stress-oriented clay formed in place in moderately fine and fine-textured Bt horizons of soils of the southwestern United States. The definition of the argillic horizon in the 1967 supplement to the 7th Approximation therefore allows for argillic horizons lacking clay skins if there is evidence of stress in the soil matrix and if there is other evidence of clay illuviation. [Soil Survey Staff, Supplement to soil classification system (7th Approximation), March 1967.]

In this paper we show (i) that highly oriented bodies of clay that had been mistaken for clay skins in the field could have formed without illuviation, (ii) that the absence of clay skins in Bt horizons is consistently related to micromorphological evidence of stress in the fabric and to the shrink-swell potential of the horizon, and (iii) that clay illuviation has probably taken place in these horizons.

MATERIALS AND METHODS

We studied soils from Arizona, southern and central California, and Oregon (Table 1). All but one have a clay increase from the A to the B horizon that is sufficient to meet the clay requirement of argillic horizons. The exception, a Vertisol, was included as an extreme example of a stress-oriented soil fabric. All the soils are in arid climates or in mediterranean climates having a pronounced dry season.

The following methods were used: particle size distribution analysis—pipette method (Kilmer and Alexander, 1949); 15 bar water retention—pressure membrane apparatus (Richards, 1954); linear extensibility (LE)—from the bulk density of plastic coated soil fragments equilibrated at ⅓-bar tension and oven-dryness expressed as percent change in one dimension, $LE = [(Db_d/Db_m)^{1/3} - 1] \times 100$, where Db_d is bulk density of the < 2 mm fine earth fraction at oven dryness and Db_m is bulk density of the < 2 mm fine earth fraction at ⅓ bar water content (Brasher, et al., 1966, Grossman et al., 1968); thin sections—from undisturbed soil fragments and artificial briquettes. The briquettes were made by mixing a 1:1 (by volume) slurry of soil and water in a blender and allowing it to dry in an aluminum dish. Some samples were taken through several wetting and drying cycles.

The microfabric of all of the horizons had some kind of striated extinction pattern indicating some organization of the plasma by soil-forming processes. [Plasma is that part of a soil material which is capable of being or has been moved, reorganized, and/or concentrated by the processes of soil formation. It includes all the material, mineral or organic, of colloidal size and relatively soluble material which is free to move in the solution or suspension phase. (Brewer, 1964, p. 12)]. We used Brewer's terminology to classify these plasmic fabrics (Brewer, 1964, p. 308). With increasing orientation the following kinds of plasmic fabrics are recognized; asepic, insepic, mosepic, and masepic, lattisepic, and omnisepic. The last three are similar in degree of orientation but differ in the kind of orientation pattern.

The term "clay skins" in this paper is restricted to coatings of laminated and presumably illuviated clay on void peripheries. Only those coatings that differ distinctly from the underlying matrix in the apparent particle size distribution, or color, are designated as clay skins (illuviation argillans, Brewer, 1964).

RESULTS

The B horizon of Mohave (Fig. 1), a Typic Haplargid, fine loamy, mixed, thermic family, illustrates the difficulties in recognizing illuviated clay in these soils. Clay skins had been described in the field but none that were distinctly different from clay in the matrix of the ped interiors could be identified in thin sections. [Mimeographed material for Tour III, Int. Congr. Soil Sci., Trans. 7th (Madison, Wis.)]. Rather the whole matrix is highly oriented. Any of three conclusions seems possible: (i) all the clay in the B horizon is illuvial; (ii) all the clay in the B horizon formed

---

[1] Presented before Div. S-5, Soil Science Society of America, Stillwater, Okla. August 1966. Received May 21, 1968. Approved July 5, 1968.
[2] Research Soil Scientists, Soil Survey Lab., SCS, USDA, P.O. Box 672, Riverside, Calif.

Table 1—Location and classification of the soils

Soil series	Location	Subgroup	Family
Bonsall	San Diego County, California S64Calif-37-3	Haplic Natrixeralf	Fine, montmorillonitic, thermic
Bosanko	San Diego County, California S64Calif-37-8	Chromic Pelloxerert	Fine, montmorillonitic, thermic
Chualar	Monterey County, California S65Calif-27-13	Typic Argixeroll	Fine-loamy, mixed, thermic
Dayton*	Linn County, Oregon S62Oreg-22-1	Typic Albaqualf	Fine, montmorillonitic, mesic
Fallbrook	San Diego County, California S64Calif-37-2	Typic Haploxeralf	Fine - loamy, mixed, thermic
Frye	Cochise County, Arizona S64Ariz-2-18	Typic Durargid	Fine, mixed, thermic
Tehama	Yolo County, California S64Calif-57-19	Typic Haploxeralf	Fine-loamy, mixed, thermic
Mohave*	Maricopa County, Arizona S59Ariz-7-2	Typic Haplargid	Fine-loamy, mixed, thermic
Placentia	Monterey County, California S65Calif-27-14	Haplic Natrixeralf	Fine, montmorillonitic, thermic
Ramona*	Riverside County, California S63Calif-33-1	Typic Haploxeralf	Fine-loamy, mixed, thermic
San Ysidro	Solano County, California S64Calif-48-8	Typic Palexeralf	Fine, montmorillonitic, thermic
Solano	Solano County, California S64Calif-48-4	Typic Natrixeralf	Fine-loamy, mixed, thermic
White House	Cochise County, Arizona S64Ariz-2-25	Ustollic Haplargid	Fine, mixed, thermic
Woodburn*	Marion County, Oregon S62Oreg-24-4	Aquultic Argixeroll	Fine-silty, mixed, mesic

* Not included in the fabric study.

in place and acquired its orientation through stresses in the soil similar to the orientation produced in clays by Tressler and Williamson (1966. p. 339–410) using externally applied stresses; or (iii) some of the clay is illuvial but has been incorporated in the stress-oriented matrix.

As a first step we tried to eliminate the possibility that all oriented clay in this horizon is solely due to illuviation. We made briquettes by mixing a 1:1 (by volume) soil-water slurry in a blender and allowing it to dry in aluminum dishes. A single drying of the briquettes at 105C. produced clay orientation around grains and on peripheries of drying cracks (Fig. 2) similar to that of the undisturbed B21t horizon. Additional wetting and drying cycles produced little change in clay orientation. From this we concluded that the fabric of the B horizon of Mohave is primarily the result of orientation of clay in place.

We then tried to find whether it was likely that clay illuviation takes place in Mohave and similar soils. For this we studied the upper B horizons of pairs of geographically associated soils. Both members of each pair have clay skins in the B3 or C horizon, indicating that there is potential for clay movement, but only one member of each pair has clay skins in the horizon of maximum clay accumulation (Table 2). The soils were picked from characterization samples, the sampling sites of members of a pair may be as far apart as 50 km (the Ramona-Mohave or the Woodburn-Dayton pairs, for example), but the soil series of each member of the pairs may be found in close associa-

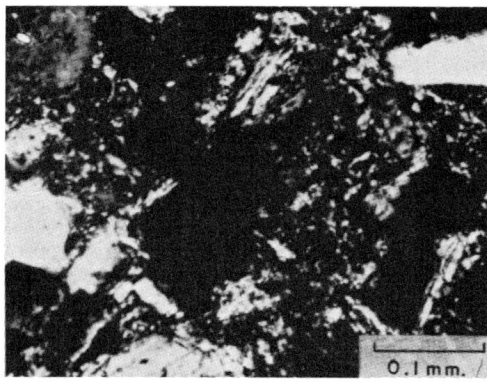

Fig. 1—Photomicrograph of the B21t horizon of the Mohave soil under crossed polarizers. The longitudinal void extending from the top right side to the bottom of the photomicrograph is surrounded by highly oriented plasma. The plasma has a skel-mosepic fabric. There are no clay skins in the void.

Fig. 2—Photomicrograph of a briquette formed of ground <2 mm soil from the Mohave B21t horizon shown under crossed polarizers. The oval void in the center of the photomicrograph is surrounded by highly oriented plasma (skelmosepic fabric). There are no clay skins in the void.

Table 2—Some B2t horizon properties of pairs of geographically closely associated soils

Soil series	Clay mineralogy	A1 & A2	B21t	Linear extensibility	Clay skins
		% clay		%	number
Ramona	vermiculite	8.9	18.1	-	common
Mohave	montmorillonite	3.1	24.2	-	none
Fallbrook	vermiculite	8.0	26.1	2.5	common
Bonsall	montmorillonite	10.2	38.3	4.5	none
Chualar	montmorillonite	10.8	17.8	1.1	common
Placentia	montmorillonite	13.5	47.6	7.3	none
Tehama	montmorillonite	21.2	31.4	4.3	few
San Ysidro	montmorillonite	16.3	40.3	6.2	none
Woodburn	montmorillonite	14.4	18.3	2.7	common
Dayton	montmorillonite	12.6	48.3	12.5	none

tion and under similar rainfall. Sites of one of the pairs, Chualar and Placentia, are within 6 km and sites of another, Bonsall and Fallbrook, are within a few meters. Hence, if clay moved into the B2 horizon of one member of each pair it probably also moved into the B2 horizon of the other member. If the B2 horizon of one member of each pair of soils does not have clay skins, they were either not formed or were formed and later destroyed. The horizons without clay skins consistently contain more clay and more expandable clay minerals and have a higher shrink-swell potential (LE) than the ones with clay skins.

The interrelations between presence of clay skins, clay content, and shrink-swell potential in Bt horizons were further tested on 28 subhorizons of 10 soils (Table 1) having a large range in clay content, and shrink-swell capacity. All the soils meet the clay increase requirement for argillic horizons and were considered to have argillic horizons in the field. They occur in desert or mediterranean climates and can be expected to reach field capacity and wilting point under field conditions. Hence, linear extensibility should give a good estimate of soil movement due to shrinking and swelling in the field. Clay skins are present (Fig. 3) predominantly in soils containing less than 40% clay, 20% 15 bar water retention, and 4% linear extensibility and are absent in horizons exceeding these values. As in the B horizon of Mohave, the distribution of clay skins is also related to the kind of plasmic fabric. Horizons with insepic and mosepic fabrics commonly have clay skins, while those with omnisepic and masepic fabrics do not. The kinds of plasmic fabric, in turn, are related to the potential of the soils to shrink and swell (Fig. 4). The shrink-swell potential (LE) of horizons with insepic fabric is lowest followed by horizons with mosepic fabric and those with lattisepic, omnisepic, and masepic fabrics. Differences in linear extensibility among the lattisepic, omnisepic, and masepic fabrics are not significant by Duncan's Multiple Range test (Duncan, 1955) but these three as a group differ significantly from the mosepic and insepic fabrics. Shrink-swell potential is, in turn, related to clay content (Fig. 4).

Evidence that clay skins may be destroyed can also be deduced from the distribution of biotite pseudomorphs of vermiculite, kaolinite, or montmorillonite. These pseudomorphs retain their size and shape in horizons having insepic fabric, but they are smaller and deformed in horizons having mosepic fabric and they are absent in horizons with masepic, lattisepic, and omnisepic fabrics.

DISCUSSION

These studies show that presence of clay skins, the degree of clay orientation in the soil plasma, and the potential of the soil to swell on wetting are interrelated. There are no clay skins in Bt horizons with high shrink-swell potentials where the plasmic fabrics show that the horizons have been under stress. Shrinking and swelling may prevent

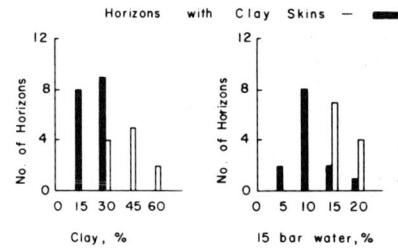
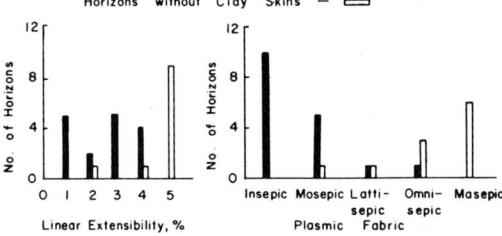

Fig. 3—Distribution of clay skins relative to some other soil properties. The soils are listed in Table 1. Only argillic horizons or horizons at equivalent depths were used.

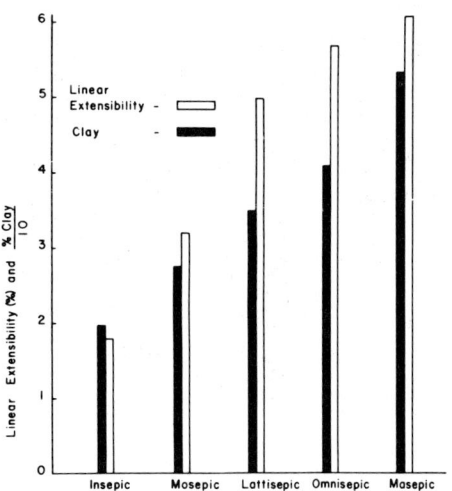

Fig. 4—Relation of plasmic fabrics to mean linear extensibility. The soils are listed in Table 1. Only argillic horizons or horizons at equivalent depths were used.

clay skins from being formed and it may destroy preexisting clay skins.

Stresses between peds may incorporate ped peripheries into a massive soil matrix. When the soil dries again new voids may form at different places. Accepting Ruhe's estimate (1967) that a well-developed Argid is older than 9,550 years, the clay accretion in the Bt horizon and the formation of clay skins must be exceedingly slow. If ped surfaces are not extremely stable and do not persist through many wetting and drying cycles, clay skins that are thick enough to be visible in thin sections cannot be formed. Hence clay may be illuviated but clay skins may not be formed.

Stresses in the soil fabric may also destroy preexisting clay skins. Some of the fine-textured members of the pairs of soils in Table 2, for example, may have gone through a stage of development like that of the associated coarser textured soils having clay skins. If this is so, the clay skins present in the youthful soils must have been destroyed later. Destruction of clay skins should start in the upper B horizon, or that part that contains the most clay and is exposed to the most frequent wetting and drying, and then proceed downwards with time. Figure 5 shows a clay skin from the B31 horizon of Bonsall that appears to be partly incorporated into the soil matrix. This horizon has 3.5% linear extensibility and is near the limit of extensibility for horizons having clay skins (Fig. 3).

The finer texture of Bt horizons also reduces permeability causing less leaching and higher base saturation and in turn favors the formation of swelling clays. When cracks become large enough to extend to the surface, the soil may become self-swallowing and a Vertisol may be formed.

The limits for the occurrence of clay skins (40% clay or 4% linear extensibility) established in this paper for Argids, Argixerolls, and Xeralfs of the southwestern United States depend on the presence of the pronounced dry seasons of desert or mediterranean climates. They should not be expected to apply to poorly drained soils or to soils in more humid environments where the lower part of argillic horizons may never, or seldom, dry to the wilting point. For example, the B3 horizon of Dayton, an Albaqualf from the Willamette Valley, Oregon with 5.3% linear extensibility, has well-expressed clay skins. (Presence of clay skins in this horizon does not contradict the findings of Parsons and Balster (1967) that the clay of the B horizon of Dayton is largely inherited from the parent sediment.)

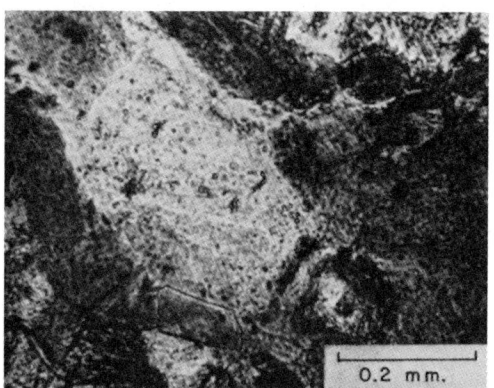
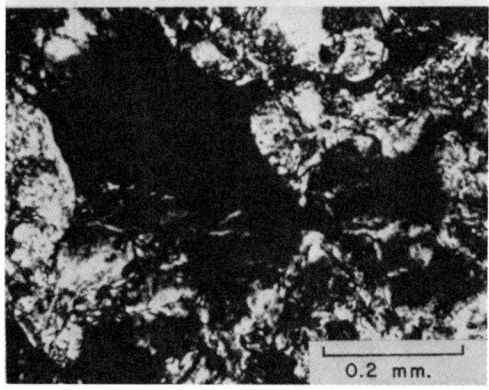

Fig. 5a and 5b—Photomicrographs of (*a*, plain light; *b*, crossed polarizers) a clay skin that is partly engulfed in the soil matrix (Bonsall soil, B3t horizon). The west wall of the large void is covered by a well oriented clay skin, with extinction brushes visible under crossed polarizers. A morphologically similar body of clay on the east side of the micrograph is now part of the soil plasma with skeleton grains embedded in it.

CONCLUSIONS

Moderately fine and fine-textured subsoil horizons in areas of desert or mediterranean climates of the southwestern United States often do not have distinct clay skins. Clay skins are absent in horizons of high shrink-swell potential; it seems reasonable to assume that stresses produced by shrinking and swelling prevent formation of ped faces that are sufficiently permanent for accumulation of clay skins and destroy any preexisting clay skins. Nevertheless there is evidence that these horizons were formed through illuviation of clay and they should continue to be recognized as argillic horizons. As a corollary, one should be conservative in identifying clay skins in these fine-textured horizons.

LITERATURE CITED

1. Brasher, B. R., D. P. Franzmeier, V. Valassis, and S. E. Davidson. 1966. Use of Saran resin to coat natural clods for bulk density and water retention measurements. Soil Sci. 101:108.
2. Brewer, R. 1964. Fabric and mineral analysis of soils. John Wiley and Sons, New York. 470 p.
3. Buol, S. W., and F. D. Hole. 1961. Clay skin genesis in Wisconsin soils. Soil Sci. Soc. Amer. Proc. 25:377–379.
4. Buol, S. W., and M. S. Yesiloy. 1964. A genesis study of a Mohave sandy loam profile. Soil Sci. Soc. Amer. Proc. 28:254–256.
5. Duncan, D. R. 1955. Multiple range and multiple F-tests. Biometrics 11:1–42.
6. Grossman, R. B., B. R. Brasher, D. P. Franzmeier, and J. L. Walker. 1968. Linear extensibility as calculated from natural-clod bulk density measurements. Soil Sci. Soc. Amer. Proc. 32:570–573.

7. Kilmer, V. J., and L. T. Alexander. 1949. Methods of making mechanical analysis of soils. Soil Sci. 68:15–24.
8. Parsons, R. B., and C. A. Balster. 1967. A depositional planosol, Willamette Valley, Oregon. Soil Sci. Soc. Amer. Proc. 31:255–258.
9. Richards, L. A. (ed.) 1954. Diagnosis and improvement of saline and alkali soils. U.S. Dept. Agr. Handbk. 60.
10. Ruhe, R. V. 1967. Geomorphic surfaces and surficial deposits in southern New Mexico. State Bureau of Mines and Mineral Resources, New Mexico Institute of Mining and Technology, Memoir 18.
11. Tressler, R. E., and W. O. Williamson. 1966. Particle arrangements and differential imbibitional swelling in deformed or deposited kaolinite-illite clay. Clay and Clay Minerals 13:399–410. (Pergamon Press, New York).
12. USDA Soil Survey Staff. 1960. Soil classification, a comprehensive system, 7th approximation.

Copyright © 1974 by G. K. Rutherford and The Limestone Press
Reprinted from pages 695-700, 711, and 712 of Soil Microscopy,
G. K. Rutherford, ed., Limestone Press, Kingston, Ontario, 1974, 857p.

CLASSIFICATION OF ACCUMULATIONS OF TRANSLOCATED PARTICLES

N. FEDOROFF

INTRODUCTION

From recent studies (Jongerius, 1970; Roose, 1968, 1970a, 1970b; Targulian, 1971) we know that particles of any size can be translocated in suspension through the soil and can be deposited in the soil after being sorted or unsorted.

In the field, as well as in a thin section, we see only the result of translocations; residuum in areas of dispersion (called also area of leaching) and deposits and coatings in areas of accumulation. We call all these residuum, deposits and coatings, mechanical accumulations.

We identify a mechanical accumulation by a granulometrical differentiation with a better sorting in comparison with the adjoining s-matrix and/or a microstratification. A continuous birefringence is not sufficient. Such a birefringence can be given by clays oriented by drying in contact with air (Dalrymple, 1969) or oriented by pressure.

A clayey mechanical accumulation is differentiated from neogenetic clays by (1) its hyalinity and in many cases by its microstratification and its colour, (2) a continuous or at least rather continuous birefringence or by, (3) a distribution related to voids (total or partial).

A clayey mechanical accumulation is also differentiated from a new formation of goethite either by its microstratification or by an absence of crystals (in general, visible at highest magnification) perpendicular to the border of voids.

EXISTING CLASSIFICATIONS OF MECHANICAL ACCUMULATIONS AND OF HORIZONS IN WHICH THEY ARE PRESENT

In most cases, mechanical accumulations can be studied accurately only under the microscope. But it is not sufficient to describe them under the microscope; it is necessary to relate the microscopic data with field observations.

CLASSIFICATION OF TRANSLOCATED PARTICLES

Brewer (1964) proposed for the first time a complete descriptive scheme and classification of mechanical accumulations at a microscopic level. He describes mechanical accumulations and their related distribution only in the lower level. More complex units are absent in his system. They would help to correlate his microscopial classification with the one of horizons, for instance with the 7th Approximation.

In Brewer's classification, the usual behaviour of mechanical accumulations (deposition in a void, integration to the s-matrix and finally dispersion in the s-matrix) is not taken into account. From this point of view of the size of particles of the mechanical accumulations, Brewer's classification is also inadequate.

The 7th Approximation is the only general soil classification in which horizons with mechanical accumulations are described with a relative accuracy. This classification system does not propose a classification of argillic horizons. But in fact, changeable characters of argillic horizons: morphology of the contact with the overlying horizon, degree of self-mixing, compaction, degree of hydromorphy are taken into account in the general classification.

However, we must make some criticisms about the argillic horizon. The 7th Approximation gives an exaggerated importance to coatings seen in the field. From micromorphological studies, we now know that what was described as clay skins in the B horizon of red Mediterranean soils are in fact only shiny faces. Under the microscope, some oxisols appear as having numerous clay skins, missed by field pedologists; they have also missed coatings in cracks and planes of weathered granites. In fact, the 7th Approximation is ignorant of minor types of horizons with mechanical accumulations. But we must recognize that when the first draft of the 7th Approximation was issued, our knowledge about translocations of solid particles in suspension through the soil was weak.

The 7th Approximation and Brewer's classification was not correlated. So we now present:

(1) a synthetical micromorphological classification of mechanical accumulations and a descriptive scheme for horizons

having these mechanical accumulations.

(2) a draft of a classification of horizons with mechanical accumulations, using field observations, but based on micromorphological background.

BASIC CLASSIFICATION OF MECHANICAL ACCUMULATIONS

First, simple mechanical accumulations must be distinguished from complex mechanical accumulations.

Simple mechanical accumulations

Residual areas. They are only or mostly constituted of grains, from silts to sands, rather well sorted. They are always defined in relation to an adjoining s-matrix. They fill totally or partially major voids.

Coarse deposits. They are constituted of grains, from silts to sands, in general well sorted. They can cover the upper side of gravels and stones or they fill totally or partially major voids.

Bedded deposits. They are constituted of alternate beds of coarse grains and fine particles, both are well sorted. They fill partially, sometimes totally, major voids. Laterally they can merge into layered coatings.

Poorly sorted deposits. These are poorly sorted, in general silty clay mechanical accumulations fill partially, sometimes totally major voids. Laterally they can merge into layered coatings.

Layered coatings. These are clayey layered mechanical accumulations. The layered morphology is given by alternate layers of quasi pure clays and clays with coarse grains imbedded in them (organic and ferruginous). They coat natural surfaces independently of their position, but they can fill partially, or even totally, a void.

Homogeneous coatings or deposits. These are clayey mechanical accumulations coating natural surfaces independently of their position, but they can fill partially, or even totally, a void.

CLASSIFICATION OF TRANSLOCATED PARTICLES

Complex mechanical accumulations

Complex mechanical accumulations are a combination of simple mechanical accumulations more or less regularly superposed.

Importance and distribution of mechanical accumulations in a horizon

Optical electron methods are not yet adapted for counting (perimeter and surface) mechanical accumulations. The most sophisticated opto-electronic microscopy, even complimented by additional optical methods are only able to count mechanical accumulations well differentiated from the s-matrix and having an internal homogeneous fabrics. In all other cases, we still must estimate roughly the importance of mechanical accumulations.

Even if an opto-electronic method will be elaborated for counting mechanical accumulations, we only could obtain for most horizons an indication of the importance of accumulations since translocations have started. We must remember that mechanical accumulations are integrated to the s-matrix with variable speed due to biological activity or to pressure. The result is that we can identify nearly all of them in horizons with slow or no integration, but their identification becomes impossible just after their deposition in horizons with strong integration. In fact, it is only possible to determine exactly the importance of accumulations in stable horizons without fine mass.

PROPOSED CLASSIFICATION OF MECHANICAL ACCUMULATIONS

We propose to classify horizons with mechanical accumulations into:

Horizons without mechanical accumulations. Such horizons have no identifiable mechanical accumulations and any character in the fine mass which would suggest that mechanical accumulations could have been integrated.

Horizons with some mechanical accumulations. In such horizons, identifiable mechanical accumulations, bound to surfaces and those already integrated to the s-matrix occupy a surface of less than 5 percent.

Horizons with mechanical accumulations. In such horizons, identifiable mechanical accumulations, bound to surfaces and those already integrated to the s-matrix occupy a surface between 5 and 20 percent.

Horizons with numerous mechanical accumulations. In such horizons, identifiable mechanical accumulations, bound to surfaces and those already integrated to the s-matrix occupy a surface greater than 20 percent.

The above classes apply only to horizons with no or slow integration. Horizons with strong integration can be classified only into two classes:

Horizons without distinct mechanical accumulations. Such horizons do not show distinct mechanical accumulations but some characters of the fine mass suggest that translocations do happen or have had happened, for instance, an orientation of micro-birefringent zones parallel to voids or to some peculiar directions.

Horizons with mechanical accumulation. This has mechanical accumulations, but they integrate too rapidly into the s-matrix that their importance cannot be estimated.

Distribution of mechanical accumulation in a horizon

The distribution of mechanical accumulations in a horizon depends on the porosity, the stability of its pores, the pressures which can occur and also on the importance of accumulations. On this basis, it is possible to distinguish three main types of distribution; generalized, diffuse and localized.

In a *generalized distribution*, mechanical accumulations can be present on the surfaces of pores, but must be present in the s-matrix. A generalized distribution can be split according to the degree of integration of mechanical accumulations, to the nature of the s-matrix or to their degree of deformation and fragmentation (karstification or cryoturbation).

CLASSIFICATION OF TRANSLOCATED PARTICLES

In a *diffuse distribution*, mechanical accumulations are only present on surfaces of pores (on all or only on some of them). There is no or very little integration of mechanical accumulations into the s-matrix (which is often absent).

In a *localized distribution*, thick mechanical accumulations either fill partially, or totally, a few major voids, or thin mechanical accumulations are localized in only some zones of the horizon.

[*Editors' Note:* Material has been omitted at this point.]

REFERENCES

[*Editors' Note:* Only the references cited in the preceding excerpt are reproduced here.]

Brewer, R., 1964. Fabric and mineral analysis of soils. J. Wiley Inc., 470 p.

Dalrymple, J. B., 1969. Experimental micromorphological investigation of iron oxide-clay complexes and their interpretation with respect to the soil fabrics of paleosols. Third International Working-Meeting on Soil Micromorphology, Wroclaw. Polish Academy of Sciences, pp. 583-594.

Jongerius, A., 1970. Some morphological aspects of regrouping phenomena in Dutch soils. Geoderma, 1970-4, pp. 311-331.

Roose, E. J., 1968. Un dispositif de mesure du lessivage oblique dans les sols en place. Cah. Orstom. Sec. Pedol., Vol. VI, No. 2, pp. 235-249.

Roose, E. J., 1970a. Erosion, ruissellement et lessivage oblique sous une plantation d'hévéa en basse Côte d'Ivoire. Mémoire polycopié Orstom et IRCA, 115 p.

Targulian, V. O., 1971. Soil formation and weathering in cold humid regions (on massive crystallina and sandy polymictic rocks), (in Russian). Publishing house "Nanka", Moscow, 268 p.

Part VII

AMORPHOUS AND CRYSTALLINE NEOFORMATIONS

Editors' Comments
on Papers 21 Through 25

21 KUBIËNA
Die taxonomische Bedeutung der Art und Ausbildung von Eisenoxyhydratmineralien in Tropenböden

22 PARFENOVA and YARILOVA
Characteristic Features of Certain USSR Soils in Thin Sections

23 FLACH, CADY, and NETTLETON
Pedogenic Alteration of Highly Weathered Parent Materials

24 TURSINA, YAMNOVA, and SHOBA
Combined Stage-by-Stage Morphological, Mineralogical and Chemical Study of the Composition and Organization of Saline Soils

25 MIEDEMA, JONGMAN, and SLAGER
Micromorphological Observations on Pyrite and Its Oxidation Products in Four Holocene Alluvial Soils in the Netherlands

The placement of a soil material within Kubiëna's morphoanalytical classification system is generally determined by the amount and type of iron oxyhydrates present and the state (peptized or flocculated) of the fine material. Whereas Paper 17 dealt mainly with the fabric of the fine material (plasma), Paper 21 treats the evolution of the iron compounds. Together these two papers give a fairly good impression of the interrelationship between soils as proposed by Kubiëna (1948, 1953). The central focus is the Braunlehm material, formed by weathering of limestone in temperate climate or of any rock type in the tropics, and characterized by a high mobility of both the iron compounds and the clay fraction. Depending on drainage and climate conditions, this Braunlehm may be transformed into a Pseudogley, Rotlehm, Roterde, or Braunerde. Rubefaction takes place under warm, alternatively wet and dry conditions, and results first in the formation of a Rotlehm (fabric), ultimately in that of a nonlateritic Roterde, due to the crystallization of iron compounds and flocculation of the plasma. Flocculation in a more temperate climate, without formation of crystalline iron minerals, would lead to the formation of a Braunerde (fabric). Kubiëna considered latosols and laterites to be the end products of hematitization of a Braunlehm under hot, humid conditions. These theories, however, were not supported by published chemical or mineralogical data, and they are no longer accepted. The formation of laterites, for instance, has proved to be a much more complicated polygenetic process.

One of the main difficulties and drawbacks of Kubiëna's system is the fact that it gives soil types, materials, and fabric types the same name and relates them genetically; e.g., the term Braunlehm may refer to a soil type (namely a terra fusca) or to a fabric (always found in a Braunlehm soil, but as a relict also in a Braunerde or Rotlehm), or to a material (Braunlehm-Teilplasma, which corresponds to clay illuviation coatings).

Whereas until the end of the 1960s many European micromorphologists were using the concepts of Kubiëna's morphogenetic system, soil scientists in the Soviet Union were taking a more objective, sediment-petrographical approach to describing soil thin sections. An excellent review of their methods and terms was given in the book *Mineral Investigations in Soil Science* by E. I. Parfenova and E. A. Yarilova, an excerpt from which is Paper 22. Because these authors were better acquainted with semiarid and arid soils, they attributed more importance to the description of different crystalline materials, such as calcite and gypsum, whereas they somewhat neglected tropical soils. The short discussion (p. 281) with regard to the origin of carbonates in the soil, and their subdivision into "primary" and "secondary" carbonates, quoted later by many authors, can be considered as a start of more detailed work in this field (e.g., Seghal and Stoops, 1972; Bal, 1975a, 1975b; Wieder and Yaalon, 1974). Another interesting point is the description of the formation of clay pseudomorphs after vegetable tissue (see, for instance, Parfenova, Mochalova, and Titova, 1964).

An important field of investigations in soil science is that of the formation of the soil material. Both geologists and soil scientists have studied rock weathering, but mainly mineralogical and geochemical aspects of it (e.g., Carrol, 1970; Loughnan, 1969); even when thin sections are studied, most attention is given to the mineralogical changes (e.g., Delvigne, 1965). Only very few studies have analyzed the fabric transition from weathered rock (saprolite) to soil (Romashkevitch, 1965), even though in many cases this transition

Editors' Comments on Papers 21 Through 25

partially determines soil characteristics. K. W. Flach, J. G. Cady, and W. D. Nettleton (Paper 23) made a significant contribution in this field, proving that soil characteristics depend not only on the chemical and mineralogical composition of the material, but also on its fabric. The process responsible for the transformation of the rock fabric to a soil fabric is called "pedoplasmation," referring to the formation of the "plasma" (i.e., the randomized fine material). Since then little work has been done in this field (e.g., Beaudou and Chatelin, 1979). One reason may be that the transition is either very abrupt, so that it is practically impossible to follow the process, or so gradual and heterogenous that a comparison between fabric and physicochemical aspects becomes difficult.

Since the Second World War, micropedological studies have been merely restricted to morphological observations on thin sections, and although several authors emphasized in their papers the need to keep in close contact with the macromorphology and the field, this sensible idea was mostly neglected. The application of submicroscopic techniques had the seemingly contradictory result that more and more micromorphologists rediscovered the old binocular microscope to study their samples prior to preparation. Others, however, immediately switched to submicroscopic techniques. Paper 24 (by T. V. Tursina, I. A. Yamnova, and S. A. Shoba) has been included mainly because of its excellent plea for a step-by-step investigation of the soil with increasing magnification, and its clear examples showing the need for such an approach for the understanding of the soil fabric. Moreover, this paper is one of the few publications dealing principally with the mineralogical and micromorphological aspects of salt neoformations in soils, a subject of utmost importance for the arid and semiarid saline soils, which cover a large part of the earth.

Paper 25, by R. Miedema, A. G. Jongman, and S. Slager, was selected as an example of the way micromorphology can be applied in the study of mineral neoformations and transformations in soils. The authors used Brewer's (1964) morphological approach, which facilitated a detailed analysis of the spatial and chronological relationships between the pedogenetic minerals (pyrite, jarosite, gypsum, ferri-hydrate) and other soil features, such as pores and organic material. It is clear that several microenvironments may coexist only a few millimeters apart in the soil material, a fact too frequently neglected by many soil scientists, who believe only in the results of mineralogical and chemical analysis of randomized bulk samples. Micromorphological concepts are here discussed in conjunction with mineralogical and geochemical considerations.

REFERENCES

Bal, L., 1975a, Carbonate in soil: A theoretical consideration, and proposal for its fabric analysis. 1. Crystic, calcic and fibrous plasmic fabric. *Netherlands Jour. Agric. Sci.* **23**:18–35.

Bal, L., 1975b, Carbonate in soil: A theoretical consideration, and proposal for its fabric analysis. 2. Crystal tubes, intercalary crystals, K fabric, *Netherlands Jour. Agric. Sci.* **23**:163–176.

Beaudou, A. G., and Y. Chatelin, 1979, La pédoplasmation dans certains sols ferrallitiques rouges de savane en Afrique Centrale, Cah. ORSTOM, sér. Pédologie, vol. 17, pp. 3–8.

Brewer, R., 1964, *Fabric and Mineral Analysis of Soils*, J. Wiley & Sons, London, New York and Sydney, 470p.

Carrol, D., 1970, *Rock Weathering*, Plenum Press, New York, 203p.

Delvigne, J., 1965, *Pédogenèse en zone tropicale*, ORSTOM, Paris.

Kubiena, W. L., 1948, *Entwicklungslehre des Bodens*, Springer-Verlag, Vienna, 215p.

Kubiena, W. L., 1953, *The Soils of Europe*, Thomas Murby & Co., London, 314p.

Loughnan, F. C., 1969, *Chemical Weathering of the Silicate Minerals*, Elsevier Publ., New York, 154p.

Parfenova, E. I., E. F. Mochalova, N. A. Titova, 1964, Micromorphology and chemism of humus-clay new-formations in gray forest soils, in *Soil Micromorphology*, A. Jongerius, ed., Elsevier, Amsterdam, p. 201–212.

Romashkevich, A. I., 1965, Micromorphological features of processes leading to the formation of red earths and the red weathering crust of the Black Sea coast of Caucasus, *Soviet Soil Sci.* **4**:407–415.

Sehgal, J. L. and G. Stoops, 1972, Pedogenic calcite accumulation in arid and semi-arid regions of the Indo-Gangetic alluvial plain of erstwhile Punjab (India)—their morphology and origin, *Geoderma* **8**:59–72.

Wieder, M., and D. H. Yaalon, 1974, Effect of matrix composition on carbonate nodule crystallization, *Geoderma* **11**:95–121.

Die taxonomische Bedeutung der Art und Ausbildung von Eisenoxydhydratmineralien in Tropenböden

Von *W. Kubiena*[*])

Abteilung für Bodenkunde und Forstökologie der Bundesforschungsanstalt für Forst- und Holzwirtschaft Hamburg-Reinbek

Böden werden als umweltbedingte Naturbildungen durch folgende ihrer Eigentümlichkeiten am stärksten charakterisiert: durch ihr Bodenprofil, durch die Art ihrer Humusbildung und — wie wir auf Grund unserer jahrzehntelangen Erfahrungen mit vollem Recht sagen können — durch ihre Mikromorphologie. Wenn wir aus diesen Aufzählungen die Mikromorphologie herausgreifen und uns fragen, wieso gerade sie zu einem so überaus charakterisierenden Element der verschiedenen Böden geworden ist, so läßt sich folgendes antworten: jede Formbildung im Boden ist in erster Linie abhängig 1. von der Stoffart, 2. von der Ausbildungsart der Stoffe, 3. von der Vergesellschaftung und der Affinität bestimmter Stoffarten zueinander, 4. von der Dynamik und Kinematik (d. h. dem Kräftespiel und den Bewegungsvorgängen) in bestimmten Stoffgefügen.

Alle angeführten Begebenheiten der Formbildung stehen in enger Abhängigkeit zum Stofflichen, auch die in Punkt 4 angeführte Dynamik und Kinematik. Dies zeigt, wie sehr Gefügeuntersuchungen stets im Zusammenhang mit der Kenntnis der stofflichen Zusammensetzung des Bodens durchgeführt werden müssen. Die Entwicklung der Bodenkunde hat gezeigt, daß die alleinige Anwendung der chemischen Analyse zur stofflichen Charakterisierung des so außerordentlich heterogenen Bodens nicht ausreicht. Die Böden unterscheiden sich, wie *Pallmann* dies ausdrückte, voneinander nicht durch ihren Gesamtchemismus, sondern durch ihre Gemengteile. In der einfachen und raschen Ermittlung der Gemengteile liegt darum die Zukunft der Bodenkunde. Sie läßt uns heute bereits auf Schritt und Tritt nicht nur die Art des Einflusses des Stofflichen auf die Gefügeteile überprüfen und spezifizieren, sie zeigt uns auch, daß bestimmten Gemengteilen erhebliche taxonomische Bedeutung zukommt.

Beides soll an einem besonderen Beispiel, den Mineralien des Eisenoxydhydrates in Tropenböden, gezeigt werden. Die Untersuchungen wurden an einem umfangreichen, während mehrerer Afrikareisen aufgesammelten Material durchgeführt. Neben den Gefügeuntersuchungen wurden von Dr. *Schmidt-Lorenz*, z. T. auch von Dr. *Erkwoh*, zahlreiche Bauschanalysen durchgeführt, die laufend durch Röntgen- und DTA-Untersuchungen ergänzt werden. Das gesamte Untersuchungsmaterial wird zusammengefaßt an anderer Stelle veröffentlicht werden. Ferner konnte ich mich auf frühere mit anderen Fachkollegen gewonnene Rönt-

---
[*]) Prof. Dr. *W. Kubiena*, Reinbek/Hamburg, Schloß

gen- und DTA-Untersuchungen von Böden stützen, die von mir mikromorphologisch untersucht wurden (s. Schrifttum 1, 3, 4, 5, 6).

Während im gemäßigten Klima in terrestrischen Böden Eisenoxydhydrat in überwiegend amorpher Form vorkommt, neigt es in Tropenböden leicht zu Auskristallisierung. Auch ist im gemäßigten Klima die Formenmannigfaltigkeit gering (meist handelt es sich um flockige Ausscheidungen von amorphem Eisenoxydhydrat), während sie in Tropenböden auch bei rein terrestrischen Bildungen in typischer Weise zunimmt. Kristallisiertes Eisenoxydhydrat kommt in gemäßigtem Klima fast nur in semiterrestrischen Böden vor, die mit den entsprechenden tropischen Varianten (z. B. Pseudogleye) große Ähnlichkeit zeigen.

Das bildungsfähigste Bodengefüge ist das Braunlehmgefüge (Abb. 1). In ihm ist Eisenoxydhydrat in diffusibler Form vorhanden. Wir haben auch eine dichte, für Diffusion stark und gleichmäßig leitfähige Grundmasse vor uns, in der das als Peptisator weitaus stärkste Schutzkolloid des Bodens, die wasserhaltige kolloidale Kieselsäure, in hohem Grade wirksam ist. Braunlehmgefüge bilden sich vorzugsweise im immerfeuchten tropischen bis subtropischen Klima bei

Abbildung 1:
Braunlehm unter äquatorialem Regenwald auf Granit, Abogonsu, Span. Guinea. In der dichten, eigelben Grundmasse (*Munsell* 10 YR 8/8) überwiegend peptisiertes amorphes Eisenoxydhydrat
Braunlehm under tropical rain forest on granite, Abogonsu, Spanish Guinea. In the dense egg-yellow coloured matrix (*Munsell* 10 YR 8/8) peptized amorphous iron hydroxide predominates

Abbildung 2:
Erdiger Braunlehm auf Basalt. Pico de Sta. Isabel aus 2440 m. Fernando Póo. Stark aufgelockertes Gefüge. Überwiegend geflocktes, amorphes Eisenoxydhydrat von dunkelbrauner Farbe
(*Munsell* 5 YR 4/3 bis 3/4)
Earthy braunlehm on basalt, Pico de Sta. Isabel, from 2440 m, Fernando Póo. Fabric considerably losened. Flocculated amorphous iron hydroxide predominates, producing a dark brown coloration of the soil mass
(*Munsell* 5 YR 4/3 to 3/4)

Abbildung 3:
Subtrop. Pseudogley, Li Quo Si, China (Probenahme durch Dr. *St. Kowalinski*). In einer dichten, hellockergrauen (5 Y 8/3), enteisenten Grundmasse ockergelbe Flecken (2,5 Y 8/8) von amorphem Eisenoxydhydrat und dunkle, orangerote (10 R 5/8 bis 4/8) Ausscheidungsnester von vorw. Goethit. In der Mitte Leitbahndurchschnitt mit Füllung von hellgrauem, konzentrisch gelagertem Tonplasma
Subtropical pseudogley, Li Quo Si, China (sample taken by Dr. *St. Kowalinski*). In a dense, light ochregray (5 Y 8/3), de-ironized matrix yellowish stains (2.5 Y 8/8) of amorphous iron hydroxide and orange coloured (10 R 5/8 to 4/8) crystal aggregates of preponderately goethite. In the center a cross-section of a conducting channel filled concentrically by a light gray clay plasma

Abbildung 4:
Tropischer Pseudogley auf Basalt mit leichter Laterisierung, Kamerunberg (Probenahme durch Dr. *W. Dohmke*). In einer hellockergrauen (5 Y 8/3) Grundmasse ockergelbe Flecken von amorphem Eisenoxydhydrat mit dunklen, orangeroten Ausscheidungsnestern von vorwiegend Goethit
Aufnahmen von Dr. *R. Schmidt-Lorenz*
Tropical pseudoglay on basalt (slightly laterized; sample taken by Dr. *W. Domke*, Kamerun Mountain). In a light ochregray groundmass intensely ochre coloured stains of amorphous iron hydroxide and orange coloured precipitations of preponderately goethite

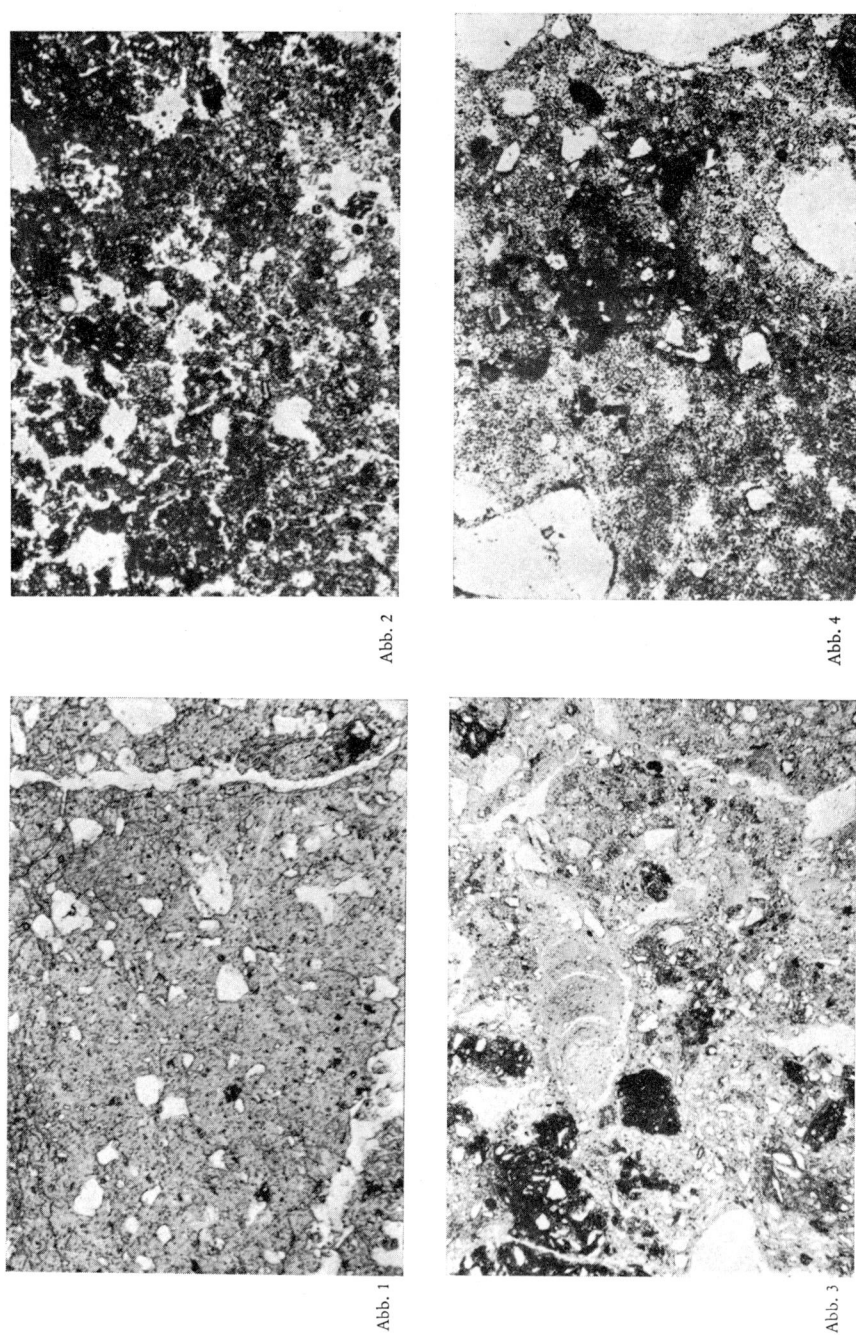

geringen Temperatur- und Feuchtigkeitsschwankungen. Die hohe Beweglichkeit und feine Zerteilung des amorphen Eisenoxydhydrates im Braunlehm bewirkt seine leichte Umwandlung und Verlagerung. Die Möglichkeiten der Umwandlung, bezüglich derer kein anderes Bodengefüge größere Mannigfaltigkeit zeigt, sind gegeben durch: bloße Flockung (braune Vererdung), durch Rubefizierung, rote Vererdung, Laterisierung, lateritische Schlackenbildung, Pseudovergleyung und Vergleyung. Im Braunlehmstadium selbst ist als typische Umwandlungsform die Bildung von runden, tiefbraunen Konkretionen mit glatter Oberfläche von mikroskopisch kleinen bis zu kindskopfgroßen Ausbildungen bekannt, die sich in erster Linie aus amorphem Eisenoxydhydrat zusammensetzen, obwohl ihre Umwandlung zu kristallinem Eisenoxydhydrat weit leichter vor sich geht als bei dem geflockten Eisenoxydhydrat, das sich durch die braune Vererdung abscheidet.

Die braune Vererdung tritt dort auf, wo sich das Klima dem Charakter der gemäßigten und kühlen Breiten nähert, bzw. dort, wo starke Temperaturschwankungen (Abkühlung und Wiedererwärmung) und Feuchtigkeitsschwankungen auftreten. In solchen Lagen bleibt das Braunlehmgefüge nicht stabil. Das peptisierte, amorphe Eisenoxydhydrat wandelt sich zu mehr oder minder irreversibel geflockten Formen um. Das so entstandene Gefüge des **erdigen Braunlehms** (Abb. 2) ist locker, hohlraumreich, mit stabiler Aggregatbildung und zeigt zumeist deutliche Auswaschungsverluste an kolloidaler Kieselsäure (ähnlich wie bei vielen stark vererdeten Rotlehmen und Lateriten). Es wird dem Gefüge der mitteleuropäischen Braunerde ähnlich. Das geflockte amorphe Eisenoxydhydrat ist wie dort außerordentlich stabil und wandelt sich schwer in

Abbildung 5:
Rotlehm, Mount Moriah. Natal. Gefügeplasma von frei schwebenden feinsten Kristalliten von Goethit und Hämatit durchsetzt
Rotlehm on dolerite, Mount Moriah, Natal. Matrix interspersed by finest crystallites of goethite and hematite

Abbildung 6:
Laterit bei Kano, Nigeria. Ehemalige Braunlehm-Grundmasse von zusammenhängenden Kristallaggregaten von vorwiegend Hämatit und Goethit (nadelige Ausbildung in unterer Bildmitte) durchsetzt
Laterite, near Kano, Nigeria. Former braunlehm matrix interspersed by crystal aggregates preponderately of hematite and goethite (at the lower center development of needles) fused together to coherent formations

Abbildung 7:
Laterit, Chiribosa, Ghana. (Probenahme durch Dr. R. Hamilton.) Auflichtaufnahme. Im Oberteil dichte hämatitisierte Konkretion. In der gelben Grundmasse vorwiegend Ausscheidungen von Goethit
Laterite, Chiribose, Ghana (sample taken by Dr. R. Hamilton). Vertical light, on the upper part dense hematitisized concretion. In the yellow matrix predominately precipitations of goethite

Abbildung 8:
Laterit, Chiribosa, Ghana. Aufnahme der gleichen Partie mit Durchlicht bei x Nicols
Laterite, Chiribosa, Ghana. Photography of the same preparation taken by transparent light and x Nicols

Aufnahmen von Dr. R. Schmidt-Lorenz

Abbildungen Nr. 5, 6, 7, 8 zu Beitrag Kubiena

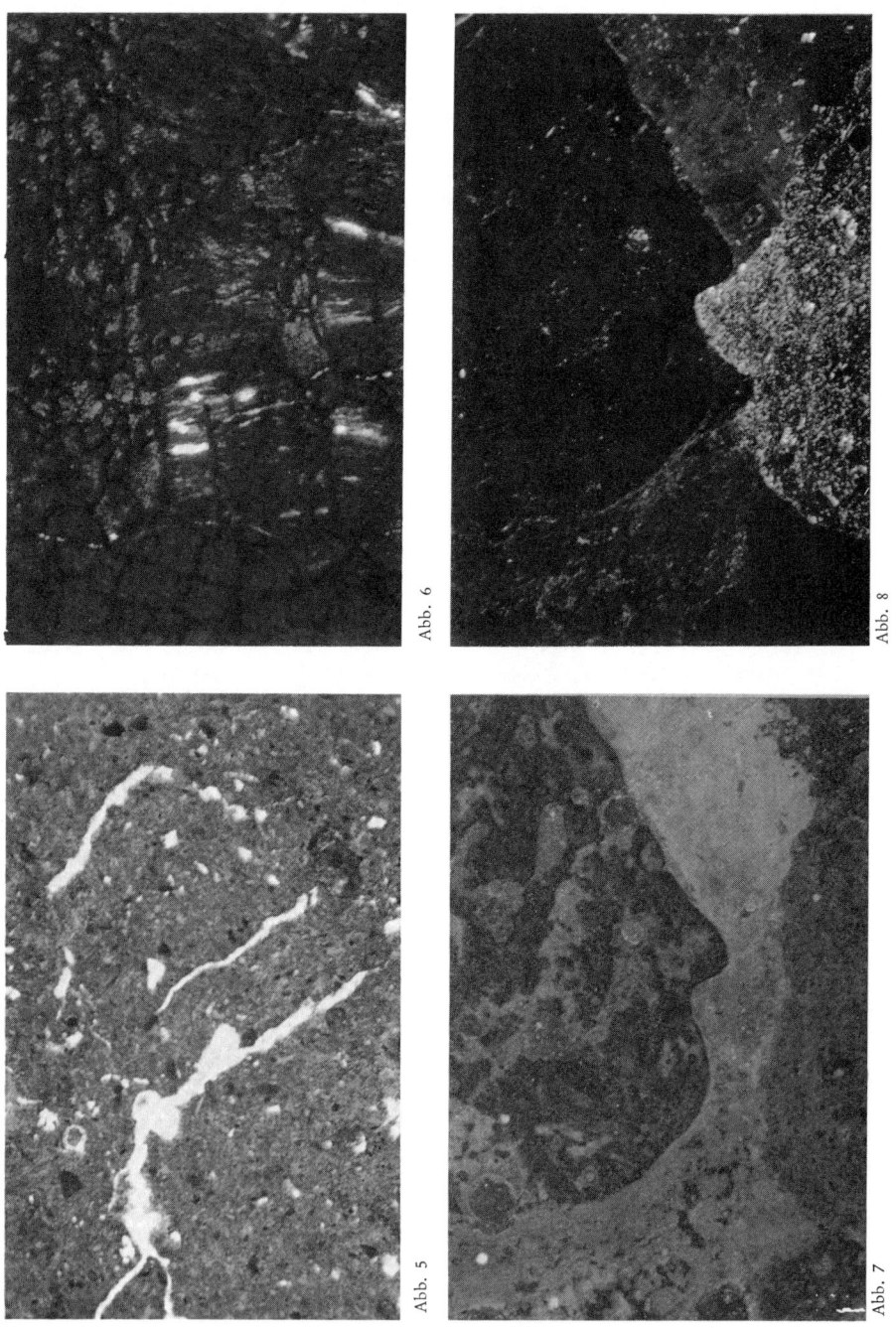

[*Editors' Note*: These photographs are reproduced in color in the original.]

kristalline Formen um. In den Tropen entstehen solche Gefügeumwandlungen überall in Gebirgslagen — in der Zone des äquatorialen Regenwaldes auf Basalt in Seehöhen oberhalb 600—700 m, in Zonen mit ausgeprägten Trockenzeiten können sie noch tiefer hinabreichen. Eutrophe erdige Braunlehme gehören zu den fruchtbarsten Tropenböden.

Eine gewisse Ähnlichkeit mit der braunen Vererdung zeigt die rote Vererdung. Auch bei ihr tritt Auflockerung und Zunahme des Hohlraumgehaltes ein. Sie entsteht in wechselfeuchten, tropischen und subtropischen Klimagebieten mit heißen Trockenzeiten. Ihr geht die sogenannte Rubefizierung voraus. Diese beginnt (im Unterschied zu der Laterisierung) in den oberen Bodenhorizonten und greift erst allmählich auf den (B)/C-Horizont über. Gefügekundlich läßt sich in einer Braunlehmgrundmasse (nur eine solche ist zur Rubefizierung fähig) das Auftreten von schwebenden feinsten Kristalliten von Goethit und Hämatit beobachten, die das Gefüge mehr oder minder gleichmäßig durchsetzen und den Boden grellrot färben (Rotlehmgefüge, Abb. 5). Im weiteren Verlaufe kann Flockung und Auswaschung von Braunlehmgrundmasse eintreten (Roterdebildung). Die entstehenden Gefügetypen sind einfach, wenig formenreich und variieren lediglich nach der Intensität und Dichte der Ausscheidung von Eisenoxyd-Kristalliten und nach dem Grade der Vererdung.

Im Gegensatz hierzu bewirkt die Laterisierung einen großen Formenreichtum der Gefügebildung. Zur Laterisierung sind sowohl Braunlehm- als auch Rotlehmgefüge fähig. Während bei der Rubefizierung das Substrat einem ständigen Wechsel von Durchfeuchtung und starker Austrocknung ausgesetzt ist, bleibt es bei der Laterisierung bei gleichem Temperaturgang dauernd und gleichmäßig durchfeuchtet. Während die Rubefizierung in den oberen Bodenhorizonten vor sich geht, vollzieht sich die Laterisierung vorzugsweise in tieferen Bodenhorizonten und Bodenablagerungen und läßt sich als diagenetischer Vorgang auch in alten Sedimenten von Braunlehm- oder Rotlehmcharakter in mitunter erheblicher Tiefe feststellen. Das Auskristallisieren des Eisenoxydhydrats erfolgt in verhältnismäßig großen Kriställchen und Kristallaggregaten, die zu großem Teil auch bei Normalvergrößerung (Objektiv 10 ×) im Dünnschliff sichtbar werden (Abb. 6 und 7). Die Laterisierung erfolgt in verschiedenen Phasen (die mikroskopisch leicht erkannt und unterschieden werden können). Ihnen entsprechen verschiedene Bildungsbedingungen, die nicht nur sehr verschiedene Ausbildungsarten, sondern auch die Entstehung verschiedener Mineralarten zur Folge haben. Unter den größeren Kristallindividuen findet man vorzugsweise Goethit und Lepidokrokit; primärer Hämatit bildet zumeist kleinere schwebende Kriställchen (größer als bei der Rubefizierung), die sich aber zu dichten opaken Aggregaten aneinanderlagern können. In Partien mit stärkerer Vernässung entstehen auch Neubildungen von Magnetit und Maghemit. So können bei der Laterisierung fast alle Mineralarten des Eisenoxydhydrates in sehr mannigfaltiger Ausbildung vorkommen, mit Ausnahme des geflockten amorphen Eisenoxydhydrats.

Tabelle der häufigsten Tropenböden und der sie charakterisierenden Eisenoxydhydratmineralien

Representation of the most frequently tropic soils and its significant iron hydroxide minerals

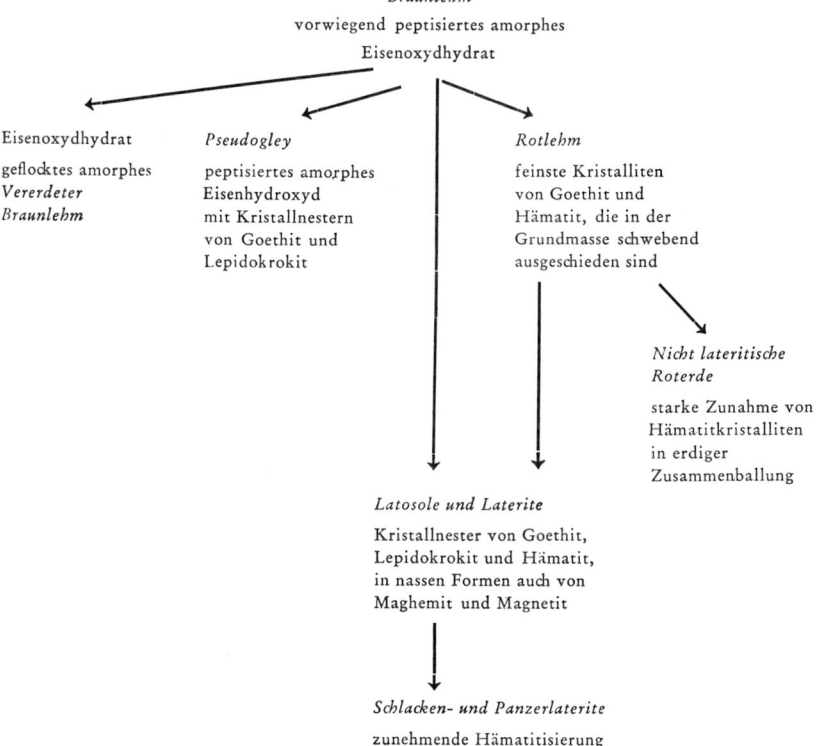

Bei der lateritischen Schlackenbildung, die durch Freilegung und starke irreversible Austrocknung weicher Lateritschichten ausgelöst wird, tritt eine engere Aneinanderlagerung der Kristalle und Kristallaggregate der Eisenmineralien ein, gleichzeitig geht eine weitgehende Umwandlung der wasserhaltigen Mineralien in wasserfreien Hämatit vor sich. Die Schlacken- und Panzerbildung wird im wechselfeuchten Klima stärker gefördert als im Gebiete des äquatorialen Regenwaldgürtels, obwohl die Laterisierung an sich hier mit größerer Vollkommenheit und Mannigfaltigkeit verläuft.

Die Pseudovergleyung kommt nicht nur im gemäßigten Klima, sondern auch in den Tropen vor und läßt weitgehend analoge Bodenformen entstehen (Abb. 3 und 4). Ihre Bildung vollzieht sich auch hier durch Staunässe aus Braunlehmplasmen. Durch Eisenverlagerung entsteht ein Fleckengefüge, das sich aus eisenarmen und eisenreichen Partien zusammensetzt. In den eisenreichen Flecken tritt sowohl in den Tropen als auch im gemäßigten Klima eine Auskristallisierung des amorphen Eisenoxydhydrats und die Bildung von bei Normalvergrößerung meist deutlich unterscheidbaren Kristallen und Kristallaggregaten von Goethit oder Lepidokrokit ein. Der Vorgang hat eine gewisse Ähnlichkeit mit der Bildung der Fleckenschichten bei der Laterisierung.

Die bisherigen Befunde sind in der auf Seite 6 wiedergegebenen Tabelle zusammengefaßt, auf der durch Pfeillinien auch die Entwicklungstendenzen ausgedrückt werden.

Auch der Stagnogley ist in den Tropen, besonders im äquatorialen Regenwaldgürtel, an Standorten mit dauernder Vernässung und starker Reduktionskraft des Bodenwassers nicht selten. Wie im gemäßigten Klima treten auch hier schwärzliche Fleckenbildungen in den Vordergrund. Es scheint, daß Magnetit und Maghemitbildung dabei stark begünstigt ist, doch liegt noch zu wenig Untersuchungsmaterial vor, das zu verallgemeinernden Schlüssen berechtigt.

Zusammenfassung

Die Mikromorphologie, die zu den kennzeichnendsten Merkmalen eines Bodens gehört, steht in engem Zusammenhang mit der chemischen Natur seiner Bauelemente. Viele Bauelemente, besonders die neugebildeten Minerale, haben taxonomische Bedeutung. Dies gilt vor allem für die Minerale des Eisenoxydhydrats, ihrer Formbildung, der Art ihrer Entwicklung, ihrer Rolle bei der Entstehung bestimmter Gefügetypen und ihrer Vergesellschaftung.

In Tropenböden zeigt das Vorwiegen von peptisiertem, amorphem Eisenoxydhydrat Braunlehm an, Vorwiegen von geflocktem, amorphem Eisenoxydhydrat erdigen Braunlehm; Flecken von peptisiertem, amorphem Oxydhydrat und Kristallaggregate von Goethit und Lepidokrokit in einer hellen enteisenden Grundmasse deuten auf Pseudogley; feinste, in der Grundmasse schwebende Kristallite von Goethit und Hämatit charakterisieren den Rotlehm (fortschreitende Hämatitisierung); ihre Zusammenballung in Flocken die Roterdebildung, während sich in Lateriten zusammengewachsene, grobkristalline Aggregatkomplexe von Goethit (z. T. Lepidokrokit) und Hämatit (in bestimmten Fällen auch von Maghemit und Magnetit) zeigen, die in weiterer Entwicklung entscheidenden Einfluß auf die Schlackenbildung nehmen.

Schrifttum

(1) *Albareda, J. Ma.*, Aleixandre, u. Sanchez Calvo, Ma.: An. Edafol. IV, 11, 543—563 (1955). — (2) *Alexander, L. T.*, Gady, J. G., Whittig, L. D., u. Dever, R. T.: VI. Intern. Bodenk. Kongr. Paris, Bd. E, V 11 (1952). — (3) *Alia, M. T.*, u. Muñoz Taboadela, M.: An. Edafol. XI, 1, 1.32 (1952). — (4) *Hoyos, A.*, u. Alias, L.: VI. Intern. Bodenk. Kongr. Paris, Bd. E, V 62 (1956). — (5) *Kubiena, W.*: Int. Bodenk. Kongr. Leopoldville, Vol. IV, pp 11—84 (1954). — (6) *Kubiena, W.* (mit Untersuchungen der Ton- und Eisenoxydhydratminerale von *R. C. Mackenzie* und *W. A. Mitchell*): Erdkunde IX, 2, 125—132 (1955). — (7) *Kubiena, W.*: VI. Int. Bodenk. Kongr. Paris, Bd. E, V 39 (1956). — (8) *Kubiena, W.*: X 46, p. 64—75 (1958). — (9) *Kubiena, W.*: Untersuchungen über die Dynamik und Systematik der Tropenböden (mit besonderer Berücksichtigung der Braunlehm- und Rotlehmvarianten). Bericht DFG. Vervielfältigt. Manuskript. (1958). — (10) *Raymond, R. E.*: Amer. J. Sci. 240 (1942). — (11) *Scheffer, F.*, Welte, E., u. Ludwieg, F.: Chem. d. Erde, 19, 51 (1957).

The Taxonomic Value of the Type and Formation of Iron Hydroxide Minerals in Tropic Soils

by W. *Kubiena*

The micromorphology of a soil which can be regarded as one of its most characterizing features is closely related to the chemical composition of its constituents. Many constituents, primarily newly formed mineral formations, have inportant taxonomic value. This is particularly true with the iron hydroxide minerals, shown up by the presence of particular forms and associations, a particular development and their rôle in the formation of characteristic fabrics.

In tropical soils preponderance of peptized amorphous iron hydroxide indicates braunlehm, preponderance of flocculated amorphous iron hydroxide earthy braunlehm, mottles of peptized amorphous iron hydroxide and crystal aggregates of goethite and lepidocrocite in a more or less de-ironized groundmass are characteristics of pseudogley; finest crystallites of goethite and hematite indicate rotlehm, their flocculation (rote Vererdung) is produced in Roterde, while in laterites coarse crystal aggregates of goethit (partly lepidocrocite) and hematite (in some cases also maghemite and magnetite) are found which tend to fuse together and take decisive influence on the later scoria formation.

SUMMARY*

Whereas iron oxyhydrates occur mainly in an amorphous state in the terrestrial temperate soils, they tend to crystallize in tropical soils, where they also show a large variation of forms.

The Braunlehmfabric is the one that is most apt to be transformed to other fabrics. Iron hydroxydes are present in a diffuse state and protected by colloidal silicic acid. They are formed in constantly humid tropical or subtropical environments. The iron is mobile and may form deep brown nodules with a smooth surface. "Braune Vererdung" occurs in more temperate and cooler zones with strong variations in temperature and humidity (mountains). The peptized amorphous iron is transformed to a more or less irreversible flocculated form, which is very stable. The fabric is loose porous, with stable aggregates.

Also in the "rote Vererdung" the material becomes looser and more porous; it forms in tropical and subtropical regions with hot dry seasons. Very fine crystallites of goethite and hematite are formed in the Braunlehmmass and give a strong red color (rubefaction). During progressing development flocculation and leaching of the Braunlehm groundmass may occur (formation of Roterde).

Braunlehm- and Rotlehm fabrics can be laterized in constantly and evenly moist and warm conditions, starting from the deeper horizons. Laterization takes place in different phases, corresponding to different conditions of formation, resulting not only in different morphology, but also different mineral types (goethite, hematite, lepidocrocite, magnetite, and maghemite). When soft laterites are exposed to strong irreversible drying, they are transformed to scoria, whereby a heavy accumulation of crystals and crystal aggregates of iron minerals takes place, together with a transformation to hematite.

In pseudogley formation, the translocation of iron gives rise to a flecked fabric. Iron hydroxydes of the Braunlehmplasma crystallize as relative large crystals of goethite and lepidocrocite.

*This summary was prepared by the volume editors.

CHARACTERISTIC FEATURES OF CERTAIN USSR SOILS IN THIN SECTIONS

E. I. Parfenova and E. A. Yarilova

[*Editors' Note:* In the original, material precedes and follows this excerpt.]

In the Soviet Union, the study of soil micromorphology has as yet been very little applied. It is made use of only by individual pedologists specializing in soil mineralogy. There is no doubt, however, that the method deserves wide introduction into the everyday work of pedologists. The obstacle seems to be absence of a Russian language manual, as well as the preconceived notion that such studies require special mineralogical training. In fact, the micromorphological method employs the same concepts as the macromorphological method, but is applied to very small areas of soil. The mineralogical information in the possession of every pedologist is quite sufficient for mastering the method. Obviously, a deeper knowledge of mineralogy will make such studies more comprehensive.

The essential aspects of soil formation which may be studied by the micromorphological method will be described below, on the basis of the practical experience gained by the authors.

On commencing his work with thin sections every investigator becomes aware of the extremely diverse nature of the structure of the microareas of soil. This diversity indicates the different processes occurring in the different microareas, together constituting the entire process of soil formation characteristic of the given conditions. Since the fabric of the microareas is closely related to the physical, physicochemical, chemical and other properties of the soil, it is important to study these relationships. For instance, differences in the concentrations of bicarbonate solutions in soil affect the nature of calcite crystals. In Figure 28, one of the two neighboring pores is surrounded by microcrystalline calcite precipitated from solutions of high concentration, while the other pore is surrounded by coarsely crystalline calcite precipitated from dilute solutions.

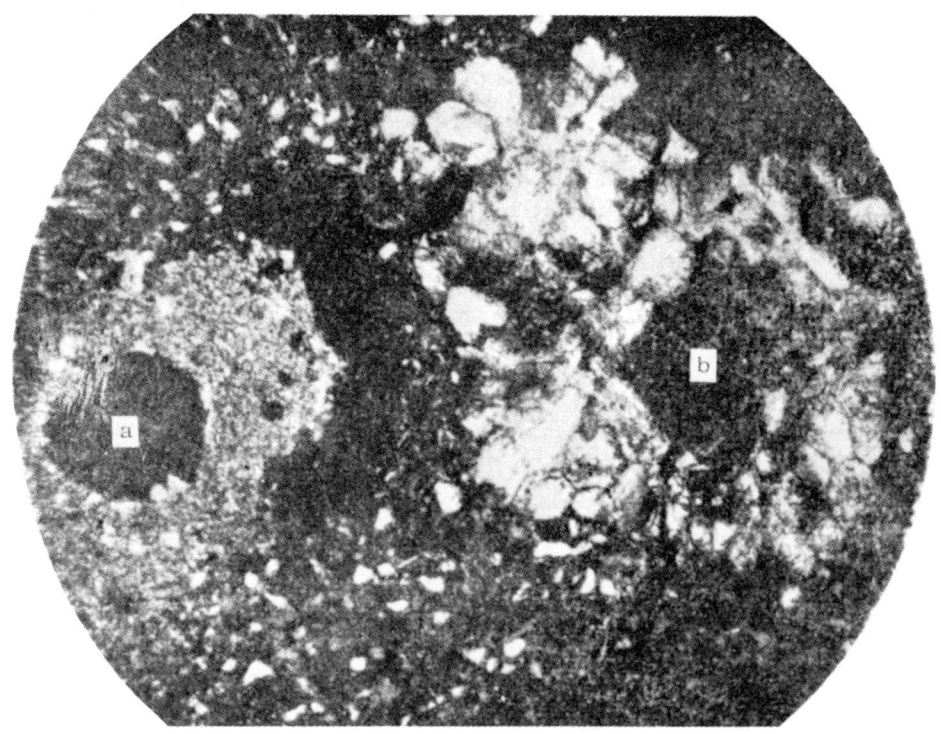

FIGURE 28. Calcite separations in neighboring pores

a — microcrystalline; b — coarsely crystalline. × 360.

Characteristic Features

Another example is the presence of two neighboring ferruginous concretions of compact and annular structures, shown in Figure 29. The different structures indicate that the different soil microareas contained different amounts of free iron oxides at the time when the concretions were formed.

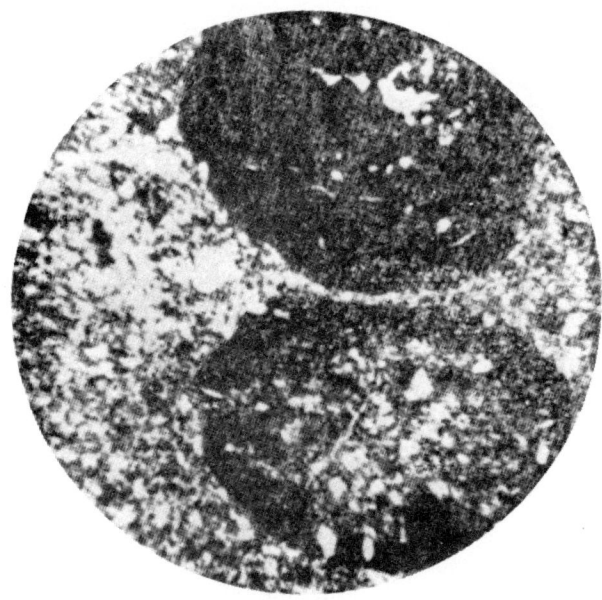

FIGURE 29. Different forms of the iron separations in adjacent microareas. ×360

Thin sections show very small amounts of certain substances which cannot be detected morphologically or by simple chemical tests in field investigations. For instance, the upper boundary of the carbonate horizon is usually established by effervescence with hydrochloric acid. Yet its first signs are distinguishable under the microscope above this boundary also, in the form of sparsely distributed individual calcite crystals or their small aggregations. Calcite microcrystals disseminated over the profile are seen in noneffervescing gray forest soils. An interesting phenomenon was observed in a thin section from podzol litter in the Arkhangelsk Region. Despite the strong leaching of these soils, the plant residues contained accumulations of whewellite, which apparently had just separated from the plant tissue but had not yet become decomposed (Figure 30, a and b). Other objects which can be observed under the microscope but are invisible to the naked eye include gypsum crystals, small accumulations of iron oxides, etc. It is only by the micromorphological method that observation can be made of the earliest signs of the illuvial process, manifested by the translocation of clay material and its deposition on the form of thin films on pore walls.

Details of the soil fabric which may be discerned in thin sections will now be more closely examined.

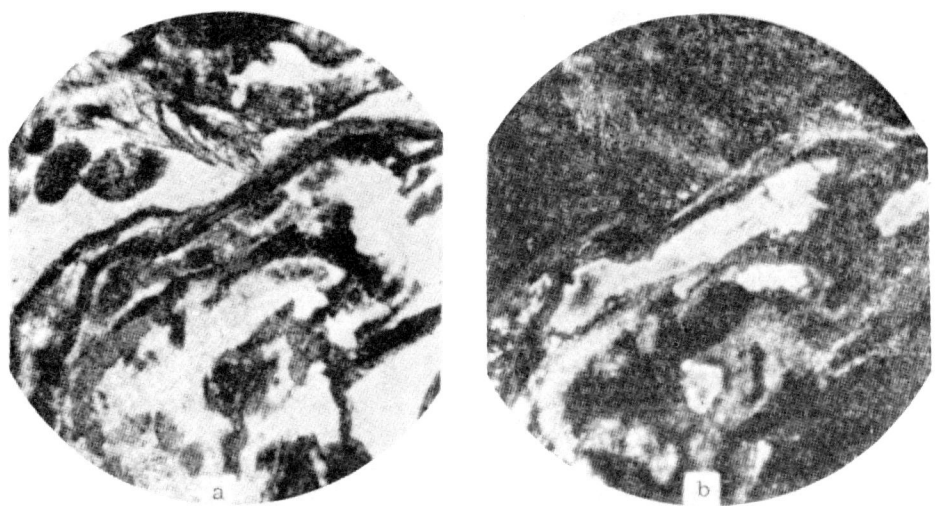

FIGURE 30. Whewellite which has separated from plant tissue

a — in transmitted light (light gray areas); b — with crossed nicols (light areas). × 100.

Mechanical Composition

Field investigations of the mechanical composition of soil by fingering it will provide only a rough idea of the coarseness or fineness of the particles. Mechanical analysis will yield the amounts of each of the different fractions. Viewing the thin sections under the microscope does not provide a substitute for mechanical analysis, but it allows observation of the shape and size of grains larger than $1\,\mu$, determination of the relative arrangement of particles of different sizes, and their approximate quantitative ratios, and, in addition, tentative determination of the constituent minerals. Figure 31,a, represents a sandy soil made up of grains of similar size with the predominance of angular and slightly rounded grains surrounded by thin films of clayey birefringent substance. It can be seen that the minerals consist principally of quartz, microcline and other feldspars, partially altered. Figure 31,b, represents a loamy soil with a nonuniform distribution of coarse primary minerals in the fine material. Figures 31,c, and d, clearly display the differences between heavy loam and clayey soil compositions. In the first example, a larger number of fine fragments of primary minerals can be seen, which are almost completely absent from the second. Thin sections also make it possible to distinguish binary deposits when these cannot be distinguished in the field. For instance, field investigation of the morphology of a podzolic heavy-loamy soil on the Karelian Isthmus revealed a very slight increase in coarseness in the upper 45 cm layer, which provided no grounds for suspecting the presence of a binary deposit. Under the microscope, however, the difference in the structure of layers lying above and below the 45 cm level became apparent. In the upper layer the clayey material contained nonuniformly distributed unsorted fine fragments of primary minerals, which were almost absent from the lower horizon.

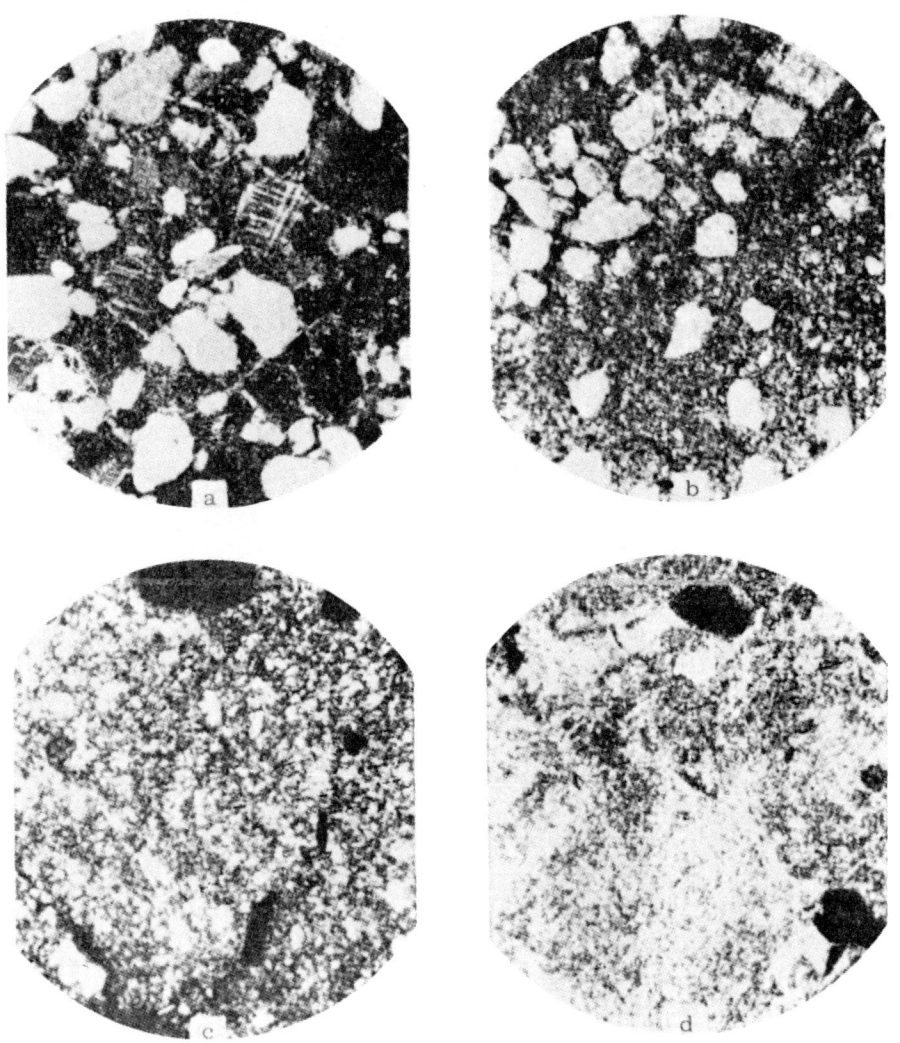

FIGURE 31. Mechanical composition of soils as seen in thin sections

a — sandy soil; b — sandy loam; c — heavy loam; d — clayey soil. × 100.

Aggregation and Porosity

Aggregation and porosity determine the nature of the soil fabric as regards distribution of the solid soil material and pore space. The solid soil material may be aggregated and nonaggregated, and the aggregates in the thin section may be completely or partially separated by the pore space.

Thin sections reveal the considerable differences in the internal structure of microaggregates. The separate granular aggregates in the chernozem are complex, consisting of smaller microaggregates which are in turn composed of even finer microaggregates (Figure 32, a). If the finest aggregates are termed primary, in combination they form aggregates of the second, third, and higher orders. Aggregation of this kind is related to the predominance of humic acids in the organic matter and the saturation of the absorbing complex, principally in calcium. Moreover, chernozems also contain aggregates of rounded-oval shape (Figure 32, b), which are apparently excretions of earthworms, not yet decomposed.

Light-gray forest soils of lower calcium saturation and with a different composition of humus contain microaggregates of a different nature, which are not composite but are in fact relatively large primary aggregates (Figure 32, c). The microaggregates in dark-gray soils are usually composite. In the horizon underlying solonchak-like solonets the microaggregates have the dimensions of sand particles (Figure 32, d), and this horizon is designated the "sand" horizon; here, aggregates are pushed apart by the crystallization of salts (calcium and sodium sulfates).

Micropores in soils differ widely in shape, dimension and arrangement. As regards origin, the following types of pores may be distinguished:

1. Biogenic pores. Root channels, burrows of small insects and animals, and closed chambers for the development of larvae. Such pores are rounded or rounded-oval in cross section (Figure 33, a).

2. Pores related to the aggregation of soil particles. According to Kachinskii (1947), there is a distinction between the "inter"-aggregate porosity and the "intra"-aggregate porosity, the former being porosity inside the crumbs. Up to 60% of pores have a diameter of less than 1 μ and therefore escape observation under the microscope. Interaggregate pores appear as a highly ramified network (see Figure 32, a, and b); their walls are mostly uneven because of the projecting mineral grains and fine primary aggregates.

3. Crack pores formed by changes in temperature and moisture. These are present between the microstructural soil units, mainly with vertical and horizontal directions, but also as the horizontal cracks arranged in parallel and responsible for the microlamellar fabric of surface horizons in podzolic and gray forest soils, solonetses, solods, etc. (Figure 33, b). In flushed soils the walls of pores and cracks have a smooth surface, produced by the deposition of a fine birefringent clay substance; this surface may be flat or undulating (Figure 33, c). Pores in carbonate horizons are often sharply outlined by the precipitated microcrystalline calcite (Figure 33, d).

4. Pores formed by the liberation of gas bubbles, of more or less rounded contours (Figure 33, e), 1-1.5 mm in size, visible to the naked eye and especially numerous in surface crusts of saline soils.

Thin sections are used by certain investigators to describe the porosity of soils and its quantitative estimation (Pol'skii, 1950, 1952, 1955).

Characteristic Features

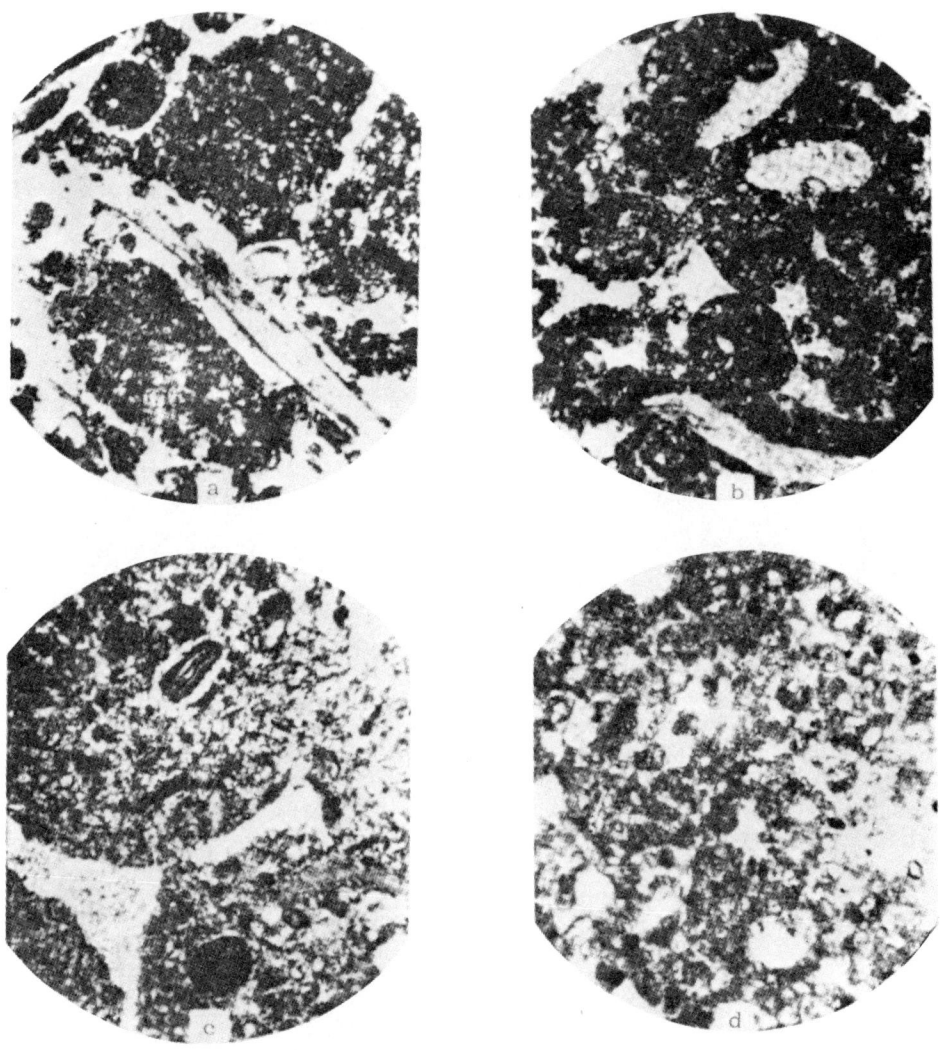

FIGURE 32. Soil microaggregates

a — angular in chernozem; b — rounded in chernozem; c — in gray forest soil; d — "sand" in solonets. × 100.

Organic Residues and Humus

Thin sections display sections through plant rootlets (Figure 34, a) and pieces of dead vegetable tissue in different stages of decomposition, in which the cellular structure is still retained or has already become lost. Their color ranges from light straw yellow to brown and almost black; some tissues are charred (Figure 34, b). Certain pores with root residues

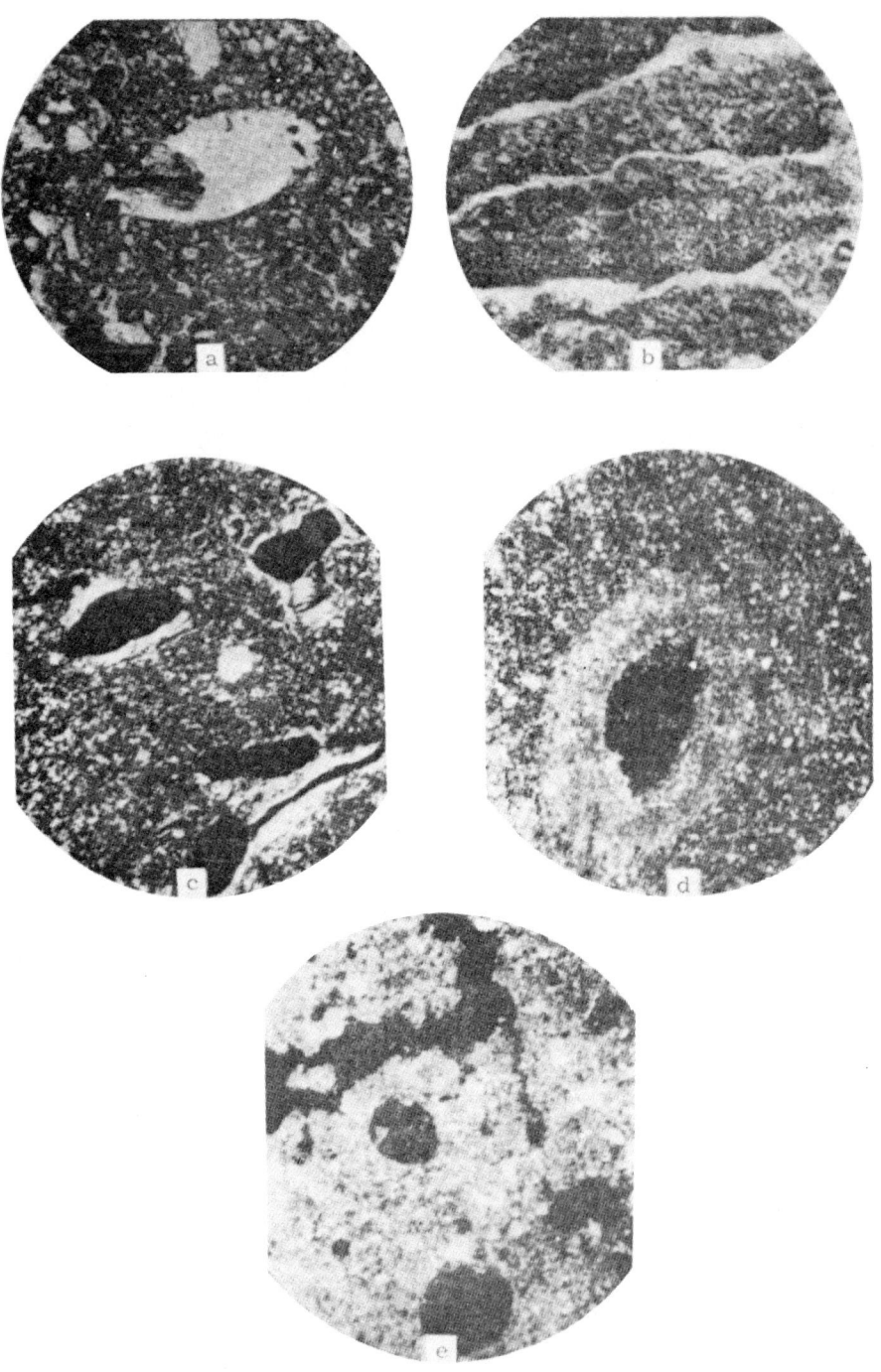

FIGURE 33. Pores

a — root channels, ×360; b — horizontal in solonets; c — covered by collomorphic clay, with smooth surface; d — covered by calcite; e — formed by gas bubbles. ×100.

Characteristic Features

of different degrees of preservation also contain excretions of soil-inhabiting animals. Other frequently occurring elements are fungi hyphae (Figure 34, c), mushroom spores (individual or in sporangia), spores of other plants, pollen, siliceous skeletons of diatomaceous algae, spicules of sponges, and colonies of comparatively large microorganisms. Remains of small animals are sometimes observed.

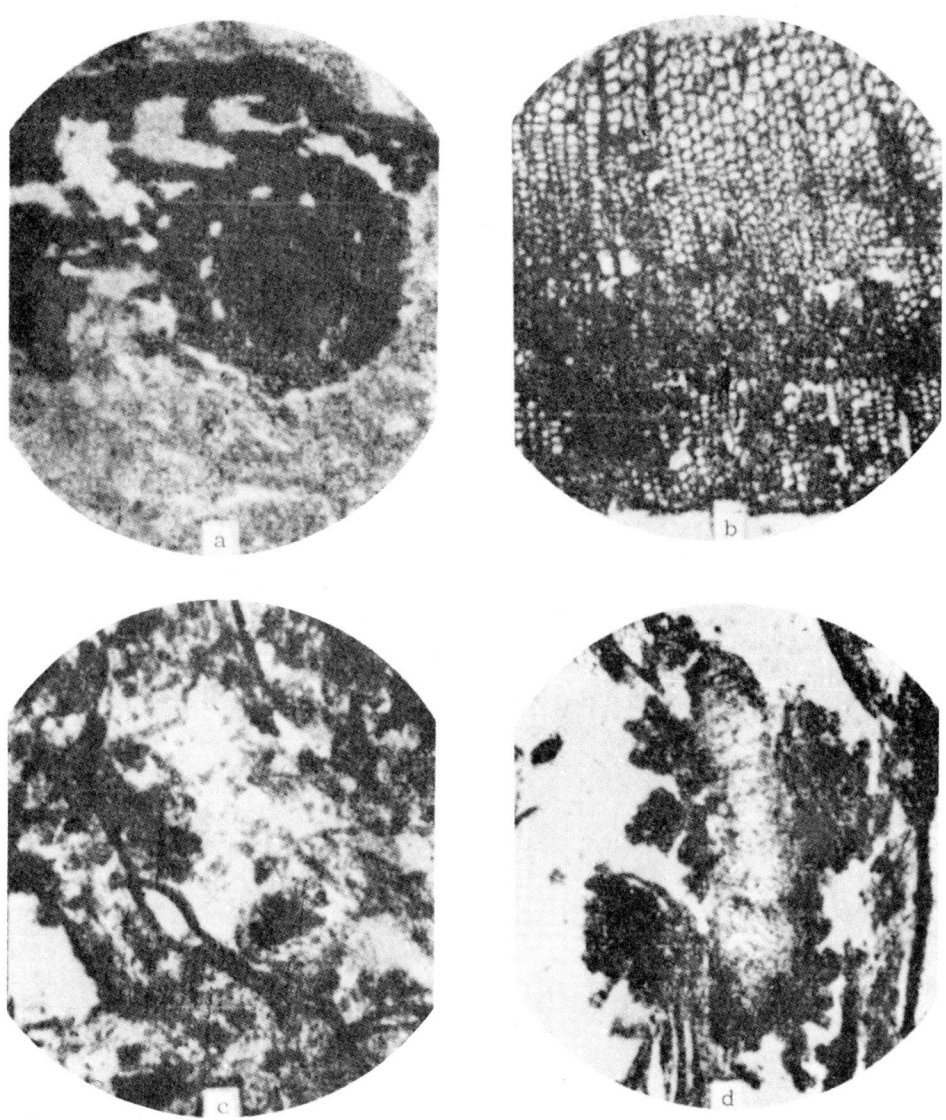

FIGURE 34. Organic residues in soil

a — section through a rootlet; b — charred leaf tissue; c — fungi hyphae; d — mull humus amongst plant residues. × 100.

The strongly altered plant residues, turned brown, represent raw humus, principally a characteristic of undeveloped gravelly, peaty, and podzolic soils. Among such residues in the litter layer of a podzol, the appearance of amorphous lumps of fine mull-humic substance of an intense brown-black color (Figure 34, d) is also observed. The distribution of humus in horizon A_1 of these soils is nonuniform; it is disseminated, often in the form of small punctate accumulations. In sod-podzolic, gray forest, and other similar soils, the plant residues are still numerous but are more strongly decomposed, and contain larger amounts of dark amorphous humus accumulations. It is possible to see the translocation of finely dispersed humus into the illuvial horizon where it becomes accumulated together with the clay substance. In certain instances, humus penetrates to a considerable depth. In chernozems, dark amorphous humus of the mull type also predominates in the upper horizons, completely coloring the soil. Here all the humus is present in the coagulated state, and no translocation is observed over the profile.

Salt Minerals in Soils

Carbonates. Micromorphological observation has established the existence of different types of carbonate accumulations in soil:
a) concentrations around pores (Figure 35, a); b) dense separations appearing as specks in section; c) pseudomorphs of calcite after gypsum (Feofarova, 1958); d) more or less uniform cementation of soil by crystalline calcite (Figure 34, b); e) calcified rootlets (Figure 35, c); f) calcite in earthworm excretions (Ponomareva, 1953). The most common form of carbonates is the microcrystalline calcite, mainly of irregularly rounded form not over 3 or 4 μ in size.

Larger crystals of carbonates from 0.01 to 0.10 or 0.15 mm are mainly represented by calcite, and, to a considerably lesser extent, by dolomite. Calcite crystals are of various shapes, usually irregular with uneven faces (Figure 35, d); twinned grains are often encountered; sometimes large crystals are aggregates. Soils contain grains of a radial or fanlike structure which could be aragonite (Feofarova, 1950); this, however, is subject to checking, especially since Dobrovol'skii (1960) mentioned that this mineral, which is detected by Meigen's reaction only as traces in the composition of pseudomicelium, occurs very rarely. Large individual calcite crystals are usually disseminated throughout the soil, but they sometimes form accumulations filling the pores (Figure 36, a) or fringing their walls, occasionally with a regular orientation along them (Figure 36, b). Individual dolomite crystals are usually rhombohedral (or approach the rhombohedral), often with curved faces. Acicular calcite crystals — lublinite — are precipitated in certain soils (chernozems, chestnut soils), growing out of the walls into the pores and forming matted accumulations (Figure 36, c, d).

Small fragments of mollusk shells are fairly common in soils. They do not extinguish under crossed nicols because they are formed by intersecting aragonite fibers.

Characteristic Features

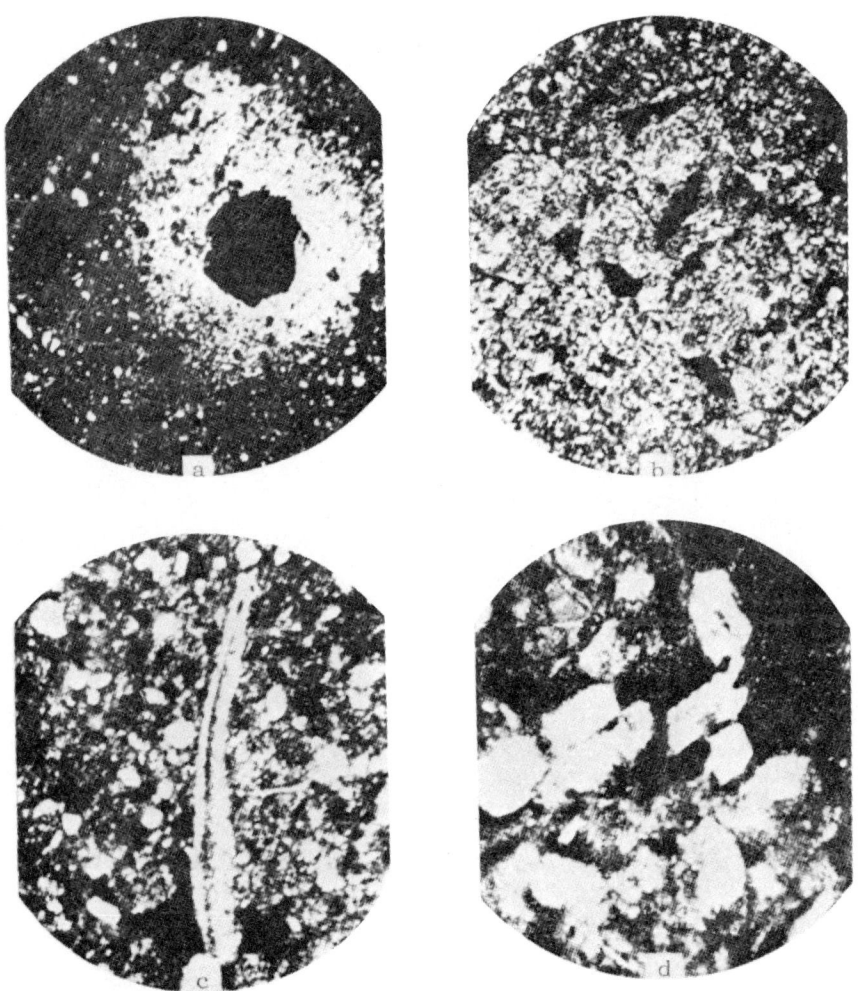

FIGURE 35. Forms of calcite

a — concentration of microcrystalline calcite around a pore, × 100; b — uniform cementation of soil by calcite, × 100; c — calcified rootlet, × 360; d — large disseminated crystals of calcite and dolomite, × 360.

As regards origin, carbonates in soils can be subdivided into those inherited from the parent rock (unsuitably designated as "primary" by certain authors) and those in the soil itself ("secondary"). Obviously, all carbonates in soils developed on magmatic rocks belong to the second category. Soils developed on sedimentary rocks and continental (detrital) deposits contain carbonates of both categories. However, it is not always possible to distinguish between the two. Beyond doubt, shell fragments occurring in soils of arid regions are inherited, while in wet habitats the shells may also belong to contemporary mollusks living in the region.

Limestone fragments containing microfauna are also inherited. Coarsely crystalline calcite, if separated out in pores, is of soil origin, as are disseminated large calcite grains occurring in soils developed on noncarbonate rocks. In the case of a carbonate deposit the distinction is very arbitrary and the nonsoil origin can be more or less reliably established only for fragmentary grains. We cannot agree with N.G. Minashina's classification (1960) of crystals of calcium carbonates in loess (fragmentary calcite) as "primary" crystals because their dimensions are similar to those of the predominant part of the primary minerals and are uniformly disseminated throughout the deposit. The similarity in the dimensions of the crystals of carbonate and primary minerals cannot serve as proof that they are native to the parent rock, since they cannot have retained their dimensions and shape in transport to the same extent as primary

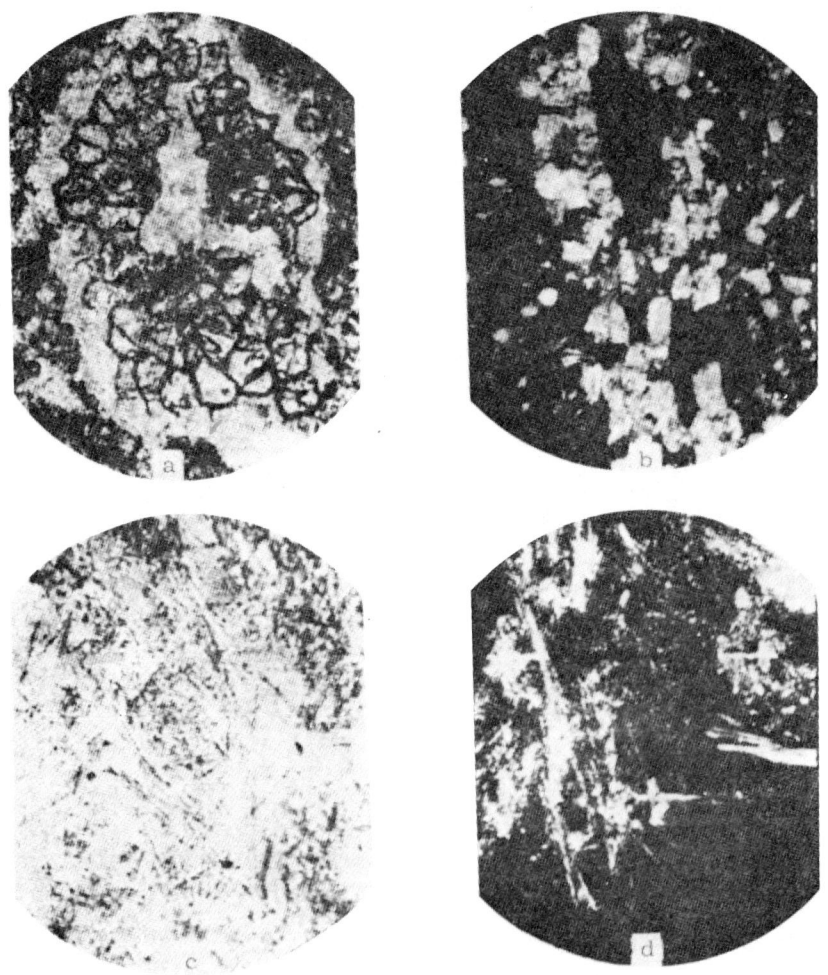

FIGURE 36. Forms of calcite

a, b — coarsely crystalline calcite in a pore; c — lublinite; d — the same; crossed nicols, × 360.

minerals (quartz, feldspars, etc.); because of their lower stability, they should have been comminuted to a greater degree. Likewise, their uniform distribution is no indication that the crystals necessarily belong to the parent rock.

Microcrystalline calcite disseminated in soil can be both inherited and of soil origin, and it is impossible to distinguish between the two forms. Its concentration in pores and cracks is undoubtedly related to the type of soil formation. Dense microaccumulations of calcite, appearing as specks in sections, would seem to be mainly of soil origin since similar but larger accumulations of white calcareous inclusions (nodules) known as "white eye" ["beloglazka"] form a clearly delimited horizon of the illuviation of carbonates in many soils.

Gypsum occurs in soils of dry steppes, with the circulation of solutions containing appreciable amounts of the SO_4^{2-} ion. Especially high gypsum concentrations are found in saline soils where gypsum sometimes even appears on the surface. Ordinarily, gypsum separates in cracks and pores, forming accumulations of monoclinic tabular crystals, often of pseudorhombohedral habit (Figure 37, a), as well as crystals of irregular shape with slightly rounded faces (Figure 37, b); corroded grains are also encountered. Sometimes, individual gypsum crystals are disseminated in the soil. In dense accumulations, the strongly deformed crystals are in close contact with one another, producing a mosaic pattern due to their nonsimultaneous extinction.

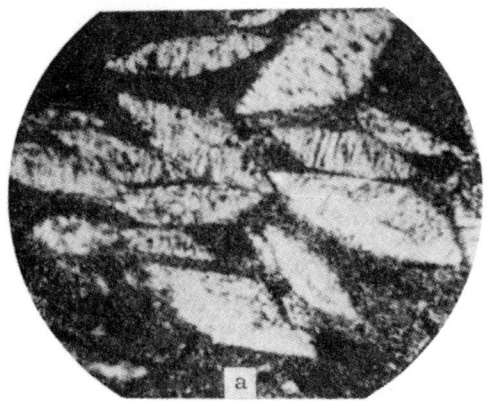
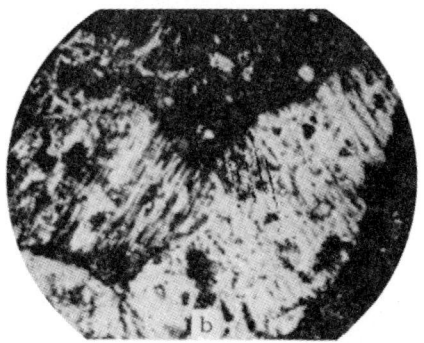

FIGURE 37. Gypsum in soils

a — rhombohedral crystals; b — crystals with rounded faces.

The dimensions of gypsum microcrystals vary over a broad range, from the finest (a few microns) to 1 or 2 mm and larger, which may be seen in the thin section by the naked eye.

As a rule, smaller crystals occur in upper horizons.

It should be noted that in thin sections we are not dealing with gypsum containing two water molecules, but with a hemihydrate of the formula $CaSO_4 \cdot 0.5\ H_2O$, to which gypsum is converted by the boiling of the sample in resin before grinding. The hemihydrate crystals have a fibrous structure and higher indexes of refraction and birefringence.

Whewellite (calcium oxalate) can be observed inside vegetable tissues in thin sections.

Apparently, separations of other anhydrous or slightly hydrated salts can also be observed in thin sections, but little has been published on this subject.

Ferruginous New Formations

Ferruginous new formations are very widespread and are found in larger or smaller amounts in all soils subject to alternating oxidation and reduction processes. These new formations have various forms, although they may be similar in different soils; microphotographs of the most characteristic forms are represented in Figure 38. Accumulations of iron oxides are present in upper horizons; they are compact when occurring in conjunction with humus (Figure 38, a) or loose (Figure 38, b); sometimes their shape is irregular and sometimes more or less rounded. In transmitted light they appear brown-black, while in reflected light their color is nonuniform, being reddish brown in spots of iron accumulation, black in the presence of humus, and grayish black earthy or with metallic luster in the presence of manganese. Often, dense rounded micro-ortsteins display appreciable concentric banding (Figure 38, c). Ortsteins are looser in deeper-lying horizons; in them, the concentric layers occur at greater distances from one another and include particles of birefringent clayey substance, sometimes humus. Sections through ortsteins show that some have a dense nucleus (Figure 38, d), while others are denser on the periphery (Figure 38, e). Soil horizons with a small amount of free iron oxides form so-called diffusion rings, which are iron concretions in the form of very thin wavy irregular closed lines, a single line or a series of concentrically arranged lines (Figure 38, f). Other forms of the separation of iron hydroxide are its depositions in pores and around rootlets (Figure 38, g).

Glei soils exhibit light-gray microareas containing reduced iron, neighboring on microareas with precipitated oxidized iron of a brighter yellowish-brown color (appearing brownish-red in reflected light). Ordinarily, the dark areas are not very sharply delimited and are of an irregular shape, often with denser floccular concretions inside. Sometimes, minute floccular iron oxide is disseminated in individual microareas and stands out sharply because of the dark-brown color (appearing bright red in reflected light). In iron-rich soils, iron concentrations may be very abundant over a considerable area of a thin section.

The general shape, appearance, and amount of ferruginous new formations are very significant for the identification of soils and the determination of processes related to the liberation, translocation and concentration of iron hydroxides in soils.

Characteristic Features

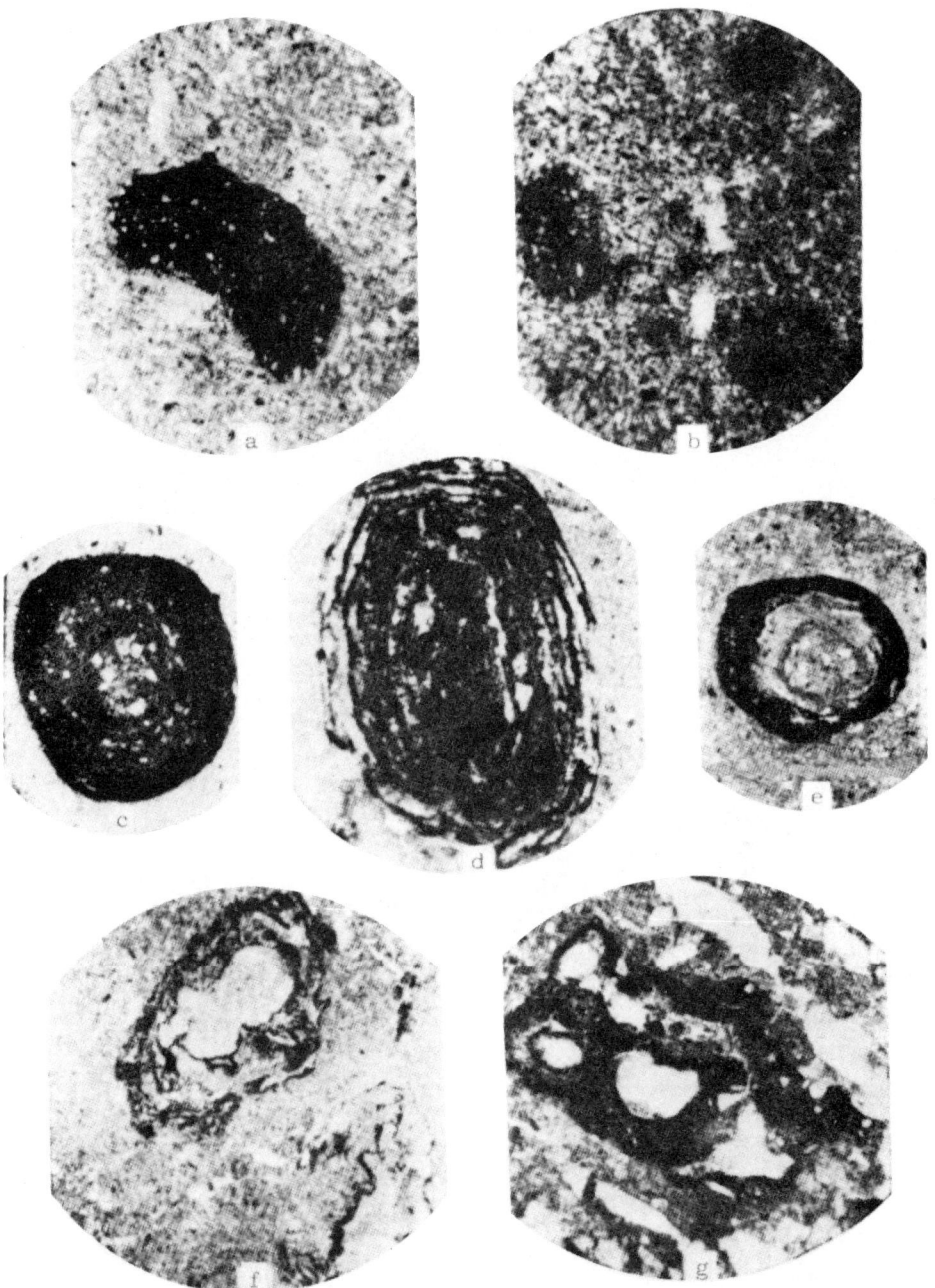

FIGURE 38. Forms of iron separations

micro-ortsteins: a — compact (with humus); b — loose; c — dense concentric constitution; d — with a dense nucleus; e — denser on the periphery; f — diffusion rings; g — dark accumulations around pores and rootlets. × 100.

Separation and Orientation of Clay Matter

Important diagnostic features are the nature of orientation and form of deposition of the clay matter. In soils from which the illuvial process is absent, the clay matter either is not oriented at all or its orientation is very poor and is observed in the form of the finest birefringent bands and scales or fine matted fibers (Figure 39, a), and is related to the shift of

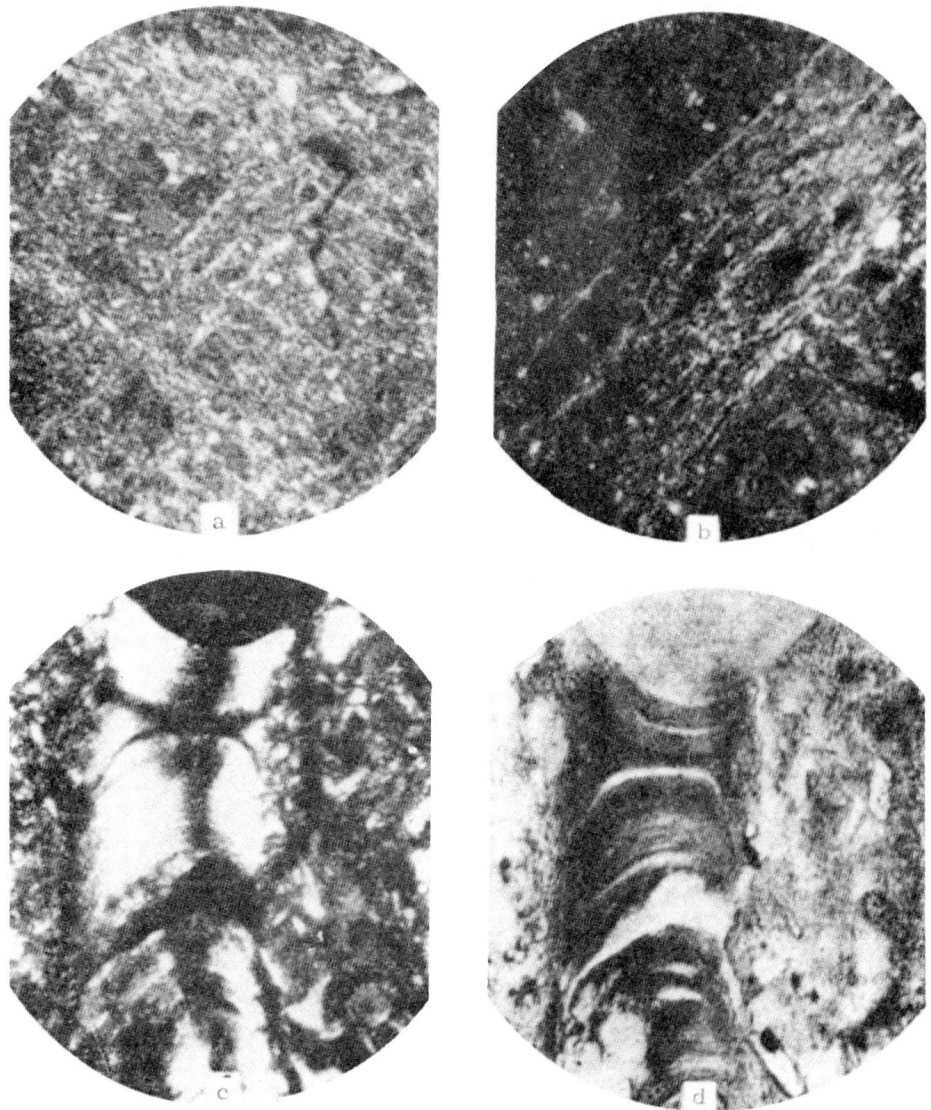

FIGURE 39. Forms of oriented clay in soil

a — matted; b — striated; c — oriented in the form of a pseudocrystal; d — the same, with crossed nicols, ×100.

clay particles inside the microareas only; bands also surround individual minerals or clay microaggregates. The initial signs of illuviation are reflected by a striated arrangement of the birefringent clay bands predominantly in one direction (Figure 39, b), and by the appearance of more noticeable films of mobile clay around the walls of micropores and cracks, as well as around minerals and microaggregates.

When the illuvial process is intensive (in podzolic, sod-podzolic, and gray forest soils, as well as in krasnozems,* laterites, and many other soils), transparent collomorphic** clay matter is abundant, especially in the horizon B. It separates out in the pores in the form of scaly layers, fringing their walls, filling the cracks, forming striated streaks in the soil mass, and filling all micropores (Figure 39, c and d). The color of this substance depends on the presence of iron in the crystal lattice and as an admixture of free oxides, varying from the golden yellow characteristic of the majority of soils (tervalent iron in the lattice and an admixture of hydroxides, apparently goethite) to vivid red in krasnozems and laterites (with an admixture of dehydrated iron forms, probably hematite) and nearly colorless greenish gray in glei soils in which iron is present in its bivalent form. Sinters of such material usually display inclusions differing from the colloidal matrix, such material being: a) silty and larger particles nonuniformly distributed or forming more or less distinctly expressed intercalations; b) finely pulverized punctate humus, in the presence of which the clay matter becomes less transparent and acquires the dark brown color characteristic, for instance, of gray forest soils, especially in their second humus horizon (Figure 40, a); c) residues of decomposed vegetable tissues. Translocation of the silty (and even coarser) fraction as a whole is often observed in horizon A_2 and in the horizon transitional to horizon B; in such a fraction colloids constitute a very small part.

In field investigations the attention of the authors was attracted to the presence in horizon B of distinctly expressed pseudomorphs of dark brown clay substance after thin tree roots (1 cm and less), with their shape and ramification distinctly preserved. Thin sections prepared from such "roots" are generally seen to consist of a birefringent – but, in spots, a nonbirefringent – collomorphic substance of the same habit as that described above. The section displays the fibrous structure of clay repeating the structure of a living root.

Sometimes the preparations enabled observation of the preserved remains of vegetable tissues and phytoliths arranged along the edges of the section in the place of the former bark (Figure 40, b); this undoubtedly indicates the formation of pseudomorphs after vegetable tissue.

Pseudomorphs of this type were observed in many sections along our route, which traversed territories of gray forest soils (from Moscow to the Cis-Ural uplands). They were especially distinct when the lower soil horizons contained carbonates (more often in the form of dense calcareous nodules ("zhuravchiks")) and where the water table was not too deep. It appears to the authors that the formation of pseudomorphs is related to the penetration of soil solutions into the dead residues of rootlets. Initially,

---

* [Red soils.]
** The collomorphic structure of a finely dispersed substance is the form of its precipitation from colloidal solutions as lamellar, scaly, rounded, irregularly winding sinters filling pores and cracks in rocks and soils.

when the tissues were sufficiently fresh, they could be penetrated only by ionic solutions, but later, as the rootlets became decomposed, they may also have been penetrated by sols. These sols gave rise to (synthesized) a clay substance in different stages of crystallization, judging from the nonuniform distribution of the birefringent microareas. It is highly probable that the formation of the clay substance proceeded with the participation of elements liberated on the spot by the mineralization of the roots. The dark color of pseudomorphs is caused by the fixation of humus by calcium from the upwelling carbonate solutions.

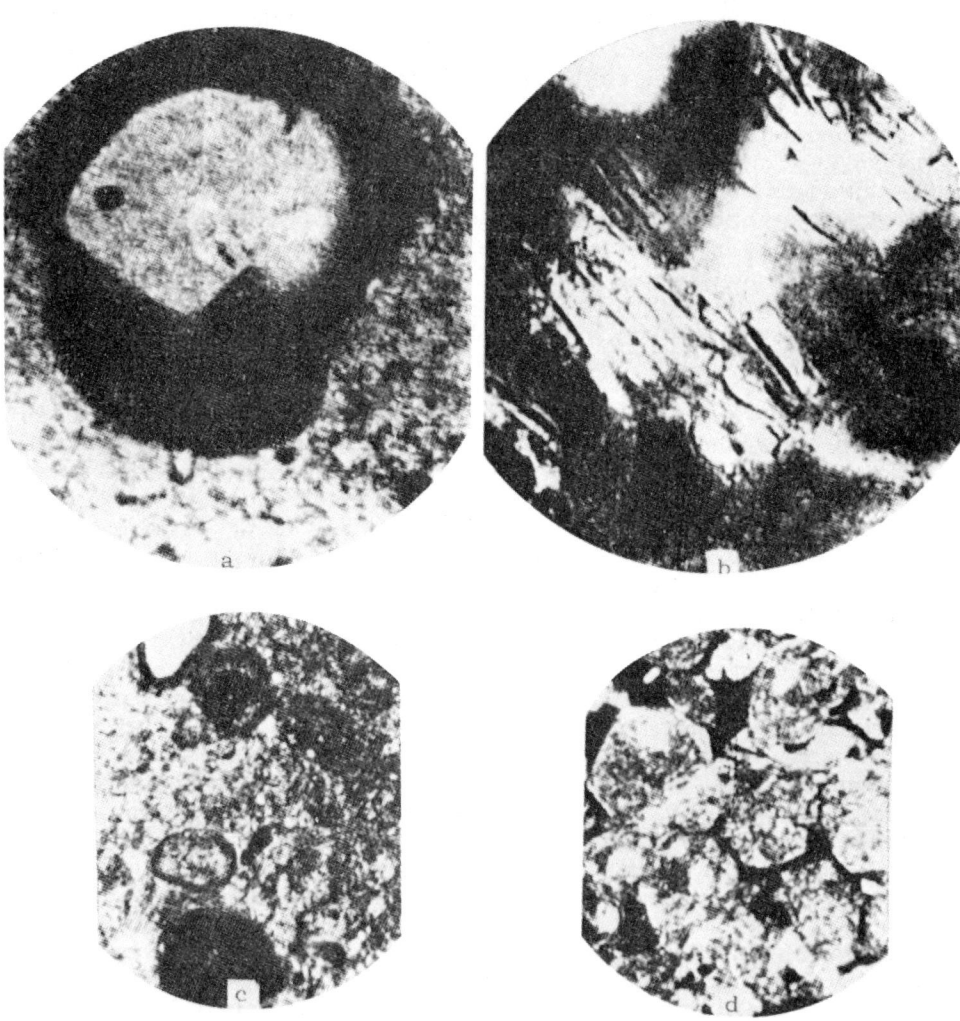

FIGURE 40. Forms of oriented clay in soils

a — collomorphic clay with humus, ×360; b — pseudomorphs of collomorphic clay after a rootlet, ×360; c — rounded compactions not separated from the soil mass, ×100; d — the same, separated from the soil mass, ×100.

Yet another form of the separation of clay matter is represented by concretions. They were observed in gray and brown forest soils under the illuvial horizon, with a somewhat higher content of absorbed calcium. The concretions were compact, rounded-oval compactions or looser formations of a concentric lamellar structure which were not separated from the surrounding soil mass (Figure 40, c). This form of the coagulation of clay should probably be explained by the presence of calcium. Dense rounded concretions separated from the soil mass were encountered in horizon D of permafrost soils, the permafrost apparently being responsible for their separation (Figure 40, e).

REFERENCES

Dobrovol'skii, V. V. (1960). Mineralogy of Supergene Quaternary Deposits in the Forest-Steppe of Central Russia. Byulleten'MOIP, Otdelenie Geologii, 35, no. 4.

Feofarova, I. I. (1950). Aragonite in Soils. Trudy Pochvennogo Instituta AN SSSR, 54.

Feofarova, I. I. (1958a). Sulfates in Saline Soils. Trudy Pochvennogo Instituta AN SSSR, 54.

Kachinskii, N. A. (1947). Soil Structure, Some of its Moisture Properties, and Differential Porosity. Pochvovedenie, 6.

Minashina, N. G. (1960). Micromorphological Investigation of Loess and of its Alteration During Soil Formation. Doklady sovetskikh pochvovedov k VII Mezhdunarodnumu Kongressu v SShA. Izdatel'stvo AN SSR.

Pol'skii, M. N. 1950. Role of Earthworms in the Structuration of Sod-Podzolic Soils. Pochvovedenie, 8.

Pol'skii, M. N. (1952). The Study of the Porosity and Microstructure of Soil Aggregates in Polished Thin Sections. Pochvovedenie, 4.

Pol'skii, M. N. (1955). Some New Ways of Investigating the Porosity and Structure of Soil. Pochvovedenie, 5.

Ponomareva, S. I. (1953). Influence of the Life Activity of Earthworms on the Formation of a Stable Structure in Sod-Podzolic Soil. Trudy Pochvennogo Instituta imeni V. V. Dokuchaeva, 41.

23

Copyright © 1968 by the International Society of Soil Science
Reprinted from *Internat. Congr. Soil Sci. 9th Trans.* **4**:343–351 (1968)

PEDOGENIC ALTERATION OF HIGHLY WEATHERED PARENT MATERIALS

K. W. Flach, J. G. Cady and W. D. Nettleton

Soil Conservation Service, U.S. Department of Agriculture, Riverside, California and Hyattsville, Maryland

Introduction

In many soils on old land surfaces, most commonly in tropical and warm temperate regions, weathering is along a sharp front at great depth. In humic environments secondary minerals formed are hydrous oxides, 1:1 layer phyllosilicates and subordinate amounts of nonexpanding 2:1 layer phyllosilicates. In subhumid and arid environments (and in a few soils in humid climates) weathering along a similar front may lead to formation of montmorillonite. Either type of weathering takes place with little volume change and the weathering product assumes a replica structure of the original rock (Cady 1960). This material, saprolite, is usually considered to be the C horizon of the overlying soil. This paper is concerned with changes in morphology and physical and chemical properties that take place in the transition from saprolite to subsoil (B) horizons of almost identical mineralogical composition.

Methods

The concepts presented here are based on studies of many soils from various parts of the United States, but only two soils, an Oxisol with kaolinitic mineralogy and a Vertisol with montmorillonitic mineralogy (Table 1), are discussed in detail. The Oxisol is a Tropeptic Haplorthox from the uplands of Puerto Rico. It formed on a chloritized andesite flow breccia that has weathered to a depth of at least 10 m at the sampling site. The mean annual precipitation exceeds 2000 mm and there is no pronounced dry season; the mean annual temperature is 24·5°C with only small variations throughout the year. The profile consists of a thin A horizon and a deep brown and red (oxic) B2 horizon that grades through a B3 to saprolite, the C horizon. The Vertisol is a Chromic Pelloxerert from the peninsular range of Southern California. The soil is in a region with a mediterranean climate; the mean annual rainfall is about 380 mm, all of which falls during the winter; the mean annual temperature is 15·5°C. The profile consists of a deep dark gray A horizon, a thin transitional B (cambic) horizon, and a saprolite C horizon. The saprolite formed from a coarse textured tonalite (quartz diorite). Although there is no assurance that the two soil profiles formed from saprolite in place there is evidence indicating that B and C horizons formed from similar parent materials.

Surface area was determined by ethylene glycol retention (Bower et al. 1959). All other measurements were made by standard Soil Survey Laboratory methods (Soil Conservation Service 1967) as follows:

Particle size distribution analysis—pipette method, using sodium hexametaphosphate and reciprocal shaking (12 hr); Cation exchange capacity (CEC)—NH_4OAc (at pH 7), $NaOAc$ (at pH 8·2); organic carbon—wet combustion; extractable (free) iron—buffered dithionite at room temperature; bulk density—displacement of plastic-coated soil fragments equilibrated at 1/3 bar tension and oven dried; linear extensibility

$$(LE) = \left[\left(\frac{\text{bulk density (oven dry)}}{\text{bulk density (1/3 bar)}}\right)^{1/3} - 1\right] \times 100;$$

water retention—at 1/3 bar with undisturbed clods and pressure cooker, and at 15 bar with crushed samples and pressure membrane apparatus; mineralogy—X-ray diffraction of oriented slides and of box mounts of the whole soil ground to 100 mesh, and DTA following equilibration of the samples over magnesium nitrate, heating rate 10°/min, peak heights compared with standard kaolinite-aluminium oxide mixtures.

TABLE 1

PROFILE DESCRIPTIONS*

OXISOL; Tropeptic Haplorthox; clayey, oxidic, isohyperthermic (Soil Survey Staff 1967)

Ap   0-13 cm; dark yellowish brown (7·5YR 5/6) clay; weak fine subangular blocky structure; friable, slightly sticky; abrupt smooth boundary.

B2   13-107 cm; strong brown (7·5YR 5/6) clay; moderate very fine subangular blocky structure, grading to dark red (2·5YR 3/6) and angular blocky structure below 58 cm; smooth pressure faces; friable, slightly sticky; common fine pores; clear smooth boundary.

B3   107-260 cm; dusky red (10R 3/4) silty clay loam; moderate very fine angular blocky structure, rough ped faces; friable, slightly sticky; few, very fine pores; gradual smooth boundary.

C   260-640 cm+; 50% dusky red (10R 3/4), 30% strong brown (7·5YR 5/8), 20% white (10YR 8/2) loam; massive; friable and plastic.

VERTISOL; Chromic Pelloxerert, montmorillonitic, thermic (Soil Survey Staff 1967) 1967)

Ap   0-10 cm; very dark gray (10YR 3/1) clay, moderate medium granular structure; very hard, firm, sticky and plastic; abrupt smooth boundary.

A12 10-56 cm; similar to Ap but with moderate, medium blocky structure and few slickensides.

B2ca 56-71 cm; very dark grayish brown (10YR 3/2) clay; moderate medium blocky structure; very hard, very firm, sticky and plastic; calcareous; some rock structure; clear wavy boundary.

C   71-145 cm+; pale yellow fine gravelly loamy sand; massive, quartz diorite saprolite; very hard, firm, nonsticky and nonplastic; calcareous.

* Drastically shortened from field descriptions; subhorizons are not described individually but are used for presentation of data in Tables 2 and 3.

DISCUSSION

To the observer in the field the transition from the saprolite to the subsoil horizon of the Oxisol consists of the disappearance of rock structure, the merging of red, brown and white components, the formation of blocky soil structure, and the change in texture from loam to clay.

The macromorphological changes are reflected in the micromorphology. The C at 260 cm (Fig. 1a) consists of alternating bands of anisotropic, colourless, and of opaque, red material. In the colourless bands kaolinite books ranging from 0·01 mm to 0·13 mm in diameter can be clearly identified (crystic plasmic fabric, Brewer 1964). The opaque bands appear bright red (10R 4/6) in incident light and consist of goethite and amorphous hydrated iron oxides. In the B3 horizon the bulk of the matrix has been altered to a weakly translucent and essentially isotropic reddish brown material, but in places alternate bands of red, nearly opaque material and brown, birefringent material resembling the kaolinite books and the red iron bands of the C horizon are detectable. In the lower B2 horizon, at 90 cm (Fig. 1b), remnant structures of the saprolite are restricted to a

Fig. 1.—Photomicrographs

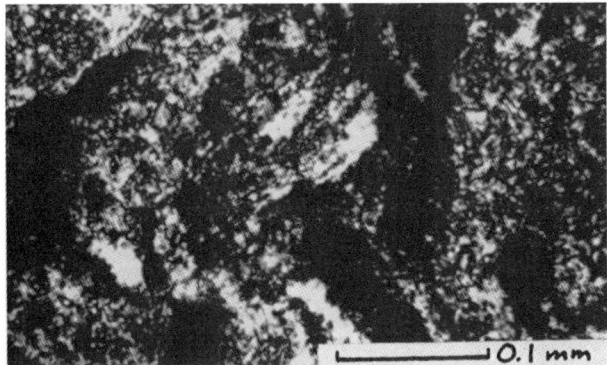

1a   Oxisol C1 horizon, 260-335 cm.

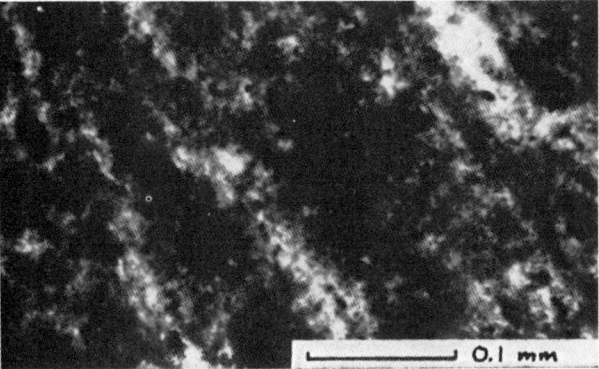

1b   Oxisol B25 horizon, 81-107 cm.

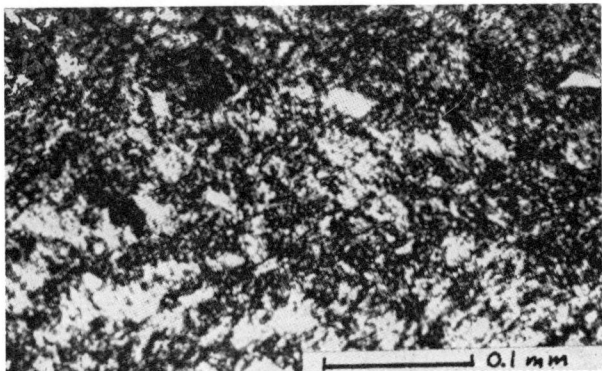

1c   Oxisol B22 horizon, 25-41 cm.

1d   Vertisol C1 horizon, 71-114 cm.

few strong brown, moderately oriented (lattisepic plasmic fabric, Brewer 1964) spots in a very uniform, weakly oriented (vosepic plasmic fabric, Brewer 1964), red matrix. Ped peripheries show weak stress orientation that is reflected in the smooth ped faces that had been observed in the field. Further up (at 60 cm) and progressing toward the surface the matrix becomes more uniform and more highly oriented (bimasepic or bimaomnisepic, Brewer 1964) (Fig. 1c).

The transformation from saprolite to oxic horizon, hence, consists of two major types of morphological rearrangements. The contrasting components of the saprolite first break down and combine into a uniform, nearly unoriented matrix in the lower B horizon. In the upper B2 horizon this matrix assumes a new orientation pattern in response to stresses generated within the soil. One of the measurable changes associated with the transition from saprolite to the oxic horizon is the increase in measured clay from 25% in the lower C to 83% in the upper B horizon (Table 2).

TABLE 2

PHYSICAL AND CHEMICAL PROPERTIES

Horiz.	Depth (cm)	Sand 2–0.05 mm (%)	Silt 0.05–0.002 mm (%)	Clay <0.002 mm (%)	Organic carbon (%)	Extractable iron (%)	CEC m-equiv. /100 g	Water retention 1/3 bar (%)	Water retention 15 bar* (%)	Surface area m²/g	Bulk density 1/3 bar g/cc	LE** (%)
Oxisol												
Ap	0–13	5.9	18.0	76.1	5.8	9.7	16.1[a]	39.9	31.2 (39.7)	125	0.92	13
B21	13–25	3.3	14.0	82.7	1.9	12.7	9.8[a]	43.0	36.5	n.d.	1.13	6
B22	25–41	1.5	18.9	79.6	1.6	13.8	8.8[a]	44.2	39.8 (44.2)	111	1.06	7
B23	41–58	1.5	24.9	73.6	1.3	15.7	7.5[a]	43.0	38.2 (44.1)	108	1.08	6
B24	58–81	2.1	28.0	69.9	1.0	16.1	7.6[a]	41.7	37.4	100	1.13	5
B25	81–107	4.2	33.2	62.6	0.4	16.5	7.7[a]	35.3	32.0 (39.4)	83	1.29	3
B31	107–145	8.6	46.3	45.1	0.1	16.1	8.4[a]	30.8	27.5	78	1.46	2
B32	145–183	6.9	49.8	43.3	0.1	14.3	9.0[a]	30.3	26.8 (34.7)	79	1.44	2
B33	183–260	14.6	44.5	40.9	0.1	15.6	8.4[a]	n.d.	27.2	81	n.d.	n.d.
C1	260–335	15.5	54.0	30.5	0.1	14.0	10.0[a]	n.d.	24.5 (35.9)	91	n.d.	n.d.
C3	427–518	13.0	60.0	27.0	0.01	12.2	11.3[a]	n.d.	24.2 (31.9)	84	n.d.	n.d.
C5	610–640	15.1	59.7	25.2	0.01	11.5	11.7[a]	n.d.	21.0 (33.6)	86	n.d.	n.d.
Vertisol												
Ap	0–10	38.5	25.5	36.0	1.59	0.6	34.7[b]	n.d.	15.6	200	n.d.	n.d.
A12	10–56	33.7	28.3	38.0	0.52	0.6	42.4[b]	25.2	17.1		1.50	5.6
B2ca	56–71	50.7	22.9	26.4	0.28	0.5	53.3[b]	23.7	17.2		1.53	3.8
C1	71–114	79.6	11.1	9.3	0.01	0.5	65.7[b]	19.4	15.4	260	1.59	3.1
C2	114–145	81.5	11.3	7.2	—	0.5	64.0[b]	15.5	14.3		1.73	2.5

* Values in parentheses after grinding.
** Linear extensibility.
a *NH₄OAc* method.
b *NAOAc* method.

Since the CEC, the total surface area, and the total kaolinite content change relatively little this clay increase must reflect increased dispersibility of the clay rather than accumulation through formation of phyllosilicates or illuviation. This is confirmed by the shift of kaolinite from the sand to the silt and the clay fractions (Table 3). Other soil properties that depend on particle size and orientation also change. Shrink-swell capacity as measured by linear extensibility increases by a factor of three and 15 bar water retention increases nearly two-fold. Lund (1959) and others have shown that 15 bar water retention (wilting point) is related to clay content.

TABLE 3

MINERALOGY OF SIZE FRACTIONS OF THE OXISOL

Horizon	Kaolinite						Other minerals
	Percent of fractions[1]			Percent of total kaolinite[2]			Whole soil[3]
	Sand	Silt	Clay	Sand	Silt	Clay	
Ap	n.d.	0	47	n.d.	0	100	Qtz, Vm, Go
B22	n.d.	28	54	n.d.	11	89	Qtz, Vm, Go
B23	n.d.	44	47	n.d.	24	76	Qtz, Go, He
B24 & B25	n.d.	54	54	n.d.	32	68	Qtz, Go, He
B31	n.d.	62	55	n.d.	54	46	Qtz, He, Go
B32	29	59	61	3	51	46	He, Go
B33	31	57	64	8	45	47	He, Go
C1	31	51	64	9	53	38	He, Go
C3 & C5	42	58	62	10	61	29	He, Go

[1] Percentages based on 560°C endotherm and kaolinite-aluminium oxide reference.

[2] $\dfrac{\% \text{ Kaolinite} \times \% \text{ fraction}}{\Sigma(\% \text{ Kaolinite} \times \% \text{ fraction})_{s,\,si,\,c}} \times 100$

[3] In order of importance based on size of X-ray peak: Go = goethite; He = haematite; Qtz = quartz; Vm = vermiculite

Fifteen bar water retention has also been reported to be related to surface area (Abrol et al. 1966) although this relationship should only be expected to hold for soils of similar mineralogy. The increase between saprolite and the B horizon while surface area and mineralogical composition remain nearly constant would indicate that fabric rearrangement and increasing dispersibility and fineness of phyllosilicates also influence water retention. This was confirmed by measuring 15 bar water retention after the samples had been thoroughly ground in a mortar. Grinding increased 15 bar water retention most in the C3 (60%) and least in the B21 horizon (11%), with intermediate increases in horizons with intermediate dispersibility.

Morphological changes associated with the formation of the B horizon in the Vertisol consist of a slight darkening of the soil material, the formation of blocky structure, and a large change in texture. Thin sections show that the C horizon consists largely of pseudomorphs after feldspar, biotite, and hornblende. The plagioclase pseudomorphs are weakly birefringent (mosepic fabric) and some of the grains are outlined by highly birefringent, apparently stress-oriented material (Fig. 1d). Significant portions of apparently unaltered material remain in most of the primary mineral grains. In the B2ca horizon about 30% of the material has lost the outlines of the parent mineral and has a stress oriented (masepic, Brewer 1964) soil fabric. All skeleton grains in the A horizon appear unweathered. They are imbedded in a moderately stress-oriented (vo-skel-masepic, Brewer 1964) groundmass.

Measured clay percentages are reasonably well related to field estimates of texture. What had been designated fine gravelly loamy sand in the field contains 7% clay and what had been called clay contains 38% clay. Yet, the C horizon has a higher CEC and a slightly larger surface area than the A horizon. Assuming the A horizon to be completely dispersed and the same surface/% clay ratio in both horizons, the C contains 47% "clay", only 15% of which is readily dispersible. X-ray analyses of the clay fractions and, in the C horizon, of ground plagioclase pseudomorphs indicate montmorillonite as the only significant phyllosilicate in both horizons. The charge density in both horizons is the same. As in the Oxisol, the 15 bar water/surface water ratio increases slightly and linear extensibility increases four-fold in the transition from saprolite to soil.

Conclusions

Well established concepts of physical and chemical weathering assign to physical weathering the initial comminution of fresh rock without mineralogical alteration and to chemical weathering the subsequent mineralogical and chemical changes leading to the formation of clay minerals. Evidence presented here suggests a second, primarily physical process that effects the transformation of weathered rock (saprolite) to soil B horizons. The authors proposed the term *pedoplasmation*, the formation of soil plasma, for this process.

In this process the physical properties of the weathering products change significantly. The effect of pedoplasmation on the dispersibility of clay, on water retention, and on shrink-swell capacity has been demonstrated in this paper, but it is likely that other rheological properties are also affected. Mechanical disturbance, notably shrinking and swelling on wetting and drying, root action, and soil fauna are likely agents in pedoplasmation *in situ* but local alluvial transport cannot be ruled out. Pedoplasmation may be the dominant process causing the formation of soil horizons from weathered rock in Oxisols, Inceptisols and Vertisols. It is a contributing process in the formation of many argillic horizons.

References

Abrol, J. P., Khosla, B. K. (1966)—Surface area—a rapid measure of wilting point of soils. *Nature, Lond.* **212**, 1392.
Bower, C. A., Goertzen, J. V. (1959)—Surface area of soils and clay by an equilibrium ethylene glycol method. *Soil Sci.* **87**, 289-292.
Brewer, R. (1964)—"Fabric and Mineral Analysis of Soils." (John Wiley and Sons, New York.)

Cady, J. G. (1960)—Mineral occurrence in relation to soil profile differentiation. *Trans. 7th Congr.Int.Soil Sci.Soc.*, Madison **4**, 418-424.

Lund, Z. F. (1959)—Available water holding capacity of alluvial soils in Louisiana. *Proc. Soil Sci.Soc.Am.* **23**, 1-3.

Soil Conservation Service (1967)—Soil Survey Laboratory methods and procedures. Soil Survey Investigations Report No. 1, Soil Conservation Service, U.S. Department of Agriculture.

Soil Survey Staff (1960)—Soil classification, a comprehensive system, 7th Approximation, 1967 supplement (Soil Conservation Service, U.S. Department of Agriculture).

Summary

In many soils on old land surfaces, weathering is at great depth along a sharp front. The weathering product, consisting of 2:1 or 1:1 phyllosilicate minerals and hydrous oxides, and resistant minerals, retains a replica structure of the original rock. This weathering product, saprolite, has the chemical and mineralogical properties of phyllosilicate minerals but resembles silt and sand in mechanical properties, texture, and shrink-swell properties. Under the influence of pedogenic processes of horizon differentiation, the replica rock structure is destroyed and a soil B horizon having soil structure is formed. In the transition from saprolite to soil B horizon the texture, water retention, and shrink-swell properties change significantly, although mineralogical composition, surface area and cation exchange capacity may remain nearly constant. This transition is demonstrated with the formation of a kaolinitic oxic (latosolic B) horizon, and a montmorillonitic cambic (structural B) horizon. Shrinking and swelling of the soil upon drying and wetting, root action and, possibly, local alluvial transport are likely forces in this transition.

Résumé

Dans beaucoup de sols sur de vieilles surfaces de la terre l'altération se fait à une grande profondeur sur un front bien défini. Le produit d'altération qui consiste de 2:1 ou 1:1 de minéraux phyllosilicats et d'oxides hydrates, et de minéraux résistants retiennent une structure replique du rocher original. Ce produit d'altération, le saprolite, a les propriétés chimiques et minéralogiques de minéraux phyllosilicats mais ressemble au limon très fin et au sable dans ses propriétés mécaniques, sa texture, et sa capacité de retrait. Sous l'influence de procédés pédogéniques de différenciation des horizons, la structure replique du rocher est détruite et un horizon B du sol avec une structure de sol est formé. Dans la transition de saprolite à un horizon B la texture, la rétention de l'eau, et les propriétés de retrait changent significativement quoique la composition, l'étendue da la surface, et le CEC soient capables de rester constants. Cette transition est démontrée par la formation d'un horizon oxic-kaolinitique (latosolique B), et un horizon cambic montmorillonitique (structurel B). Le rétrécissement et le gonflement du sol avec dessication et humectation, l'action de racines et, peut-être, le transport local alluvial sont des forces probables dans cette transition.

ZUSAMMENFASSUNG

In vielen Böden alter Landformen die Verwitterung erfolgt in grosser Tiefe und an einer scharfen Front. Verwitterungsprodukte, wie die 2:1 und 1:1 Tonminerale und Oxyhydratminerale, und widerstandsfähige Minerale haben eine Struktur die die Struktur des Muttergesteins widerspiegelt. Dies Verwitterungsprodukt, Saprolit, hat die chemischen und mineralogischen Eigenschaften der Tonminerale aber es ähnelt Schluff und Sand in mechanischen Eigenschaften, Bodenart und Schrumpf-Quell Eigenschaften. Unter dem Einfluss bodenbildender Umschichtungsvorgänge wird die gesteinsartige Struktur des Saprolit zerstört und ein Unterboden mit typischer Bodenstruktur gebildet. In der Umschichtung des Saprolit zum Unterboden ändert sich sowohl die Bodenart, als auch das Wasserhaltungs und Schwellungsvermögen des Bodens nachdrücklich obwohl die mineralogische Zusammensetzung, die spezifische Tonoberfläche und die Umtauschkapazität fast gleich bleiben. Diese Umschichtungsvorgänge sind an Hand der Bildung eines kaolinitreichen "oxic" (Latosol B) Horizontes und eines montmorillonitreichen "cambic" (Struktur B) Horizontes aufgezeigt. Schrumpfung und Quellung des Bodens durch Trocknen und Wasseraufnahme, Wurzelwachstum und möglicherweise lokale alluviale Umlagerung, sind als mögliche Vorgänge in der Bildung dieser Unterbodenhorizonte anzusehen.

COMBINED STAGE-BY-STAGE MORPHOLOGICAL, MINERALOGICAL AND CHEMICAL STUDY OF THE COMPOSITION AND ORGANIZATION OF SALINE SOILS

T.V. TURSINA, I.A. YAMNOVA, and S.A. SHOBA, V.V. Dokuchayev Soil Institute

A combined morphological study by stages of the makeup of the soil mass and salts by several modern techniques has made it possible to identify the minerals of readily soluble salts, to trace the effect of various salt minerals on the microstructure of the plasma and on the general makeup of the soil mass, to identify the diversity of forms of salt crystallization, to establish the relation between them and crystallization conditions, and to refine data on the content of readily soluble salts obtained by extraction with water.

Various techniques have been developed and are widely used now for the morphological study of soils at different levels: macro-, meso-, micro-, and submicromorphological. These methods are normally used in various combinations, one of them being the basic method, as a rule [3, 13]. A complex of methods, usually macro-, meso-, and micromorphological, is much less frequently used [2, 15].

In the study reported here, we proceeded from the following considerations: (i) that a maximum effect can be obtained by using the entire set of morphological methods; (ii) that the investigations must be performed successively, by stages, beginning at the macromorphological and ending at the submicromorphological level; and (iii) that morphological methods must be supplemented by mineralogical and chemical methods at all levels.

A set of morphological methods was used to study saline soils, including the forms of new salt formations, the mineralogy of readily and poorly soluble salts, and the effect of the salts on the makeup of the soil mass. At the present time readily soluble salts in soils are described mainly on the basis of water-extract data by associating the cations and anions into hypothetical salts. There are much fewer direct data on the mineral composition of salt formations in the soil literature.

Feofarova [6-10] gave the first basic information on the mineralogy of salts found in soils. She suggested various approaches for separating the salts, dividing them into fractions by specific weights using a mixture of nonaqueous liquids (bromoform and xylene), and obtaining salt minerals from a water extract or from a mixture of salt solutions of given concentration and of definite chemical composition after evaporation. Using this method, I.I. Feofarova identified four mineral forms of readily soluble salts: halite, thenardite, glauberite, and astrakhanite.

In recent years S. Stoops (of Belgium) and colleagues from other countries have been working most actively in this field. They have published several papers on the micromorphology of the saline soils of Egypt, Iraq, and India [12, 14-16]. These papers are concerned mainly with the diversity of forms of poorly soluble salts (gypsum and carbonates). Only occasional mention is made of the occurrence of readily soluble salts (halite and thenardite) in some surface horizons of the soil profile [14, 16].

We used the methods suggested by I.I. Feofarova and G. Stoops as well as the whole set of morphological methods, together with microchemical, mineralogical [11] and crystallographic [1, 4] methods and modern equipment. At the same time we developed a technique for preparing special thin sections so as to preserve the integrity of the microstructure of loose saline horizons and the crystals of readily soluble salts. The thin sections were prepared by low-temperature boiling with balsam or without boiling by saturating the sample in vacuo and under pressure.

Proceeding from the aforementioned considerations, we developed the following program of investigation.

Stage I. Detailed macromorphological study of all salt concentrations in the field at the natural moisture content; determination of the nature of the inhomogeneity of the soil mass with detailed description of the makeup of all horizons along the profile and of structurally inhomogeneous areas within horizons; detailed sampling for further study (samples of undisturbed makeup and samples with their natural moisture content); photographing or sketching the macrostructural elements in the profile; and performing qualitative chemical tests for major cations and anions to determine the composition of large salt concentrations (in crusts, druses, spots, etc.).

Stage II. Mesomorphological study of samples under a binocular microscope (in a range of magnifications from 8-10 to 7-100x) and determination of the details of the makeup at the subhorizon structural level; identification of salt formations with determination of their size, shape, and their confinedness to certain mesostructural elements. At this stage one can separate and identify individual crystals, druses, aggregates, and accumulations of salts 0.025 mm or more in diameter. Here great emphasis is given to the study of the makeup of the pore space and to the surface of peds and associated new formations. In addition, salt formations are prepared for further study by the immersion method, and, if necessary, the salts are fractionated using water and nonaqueous liquids for subsequent identification in immersion liquids. At this stage samples are selected for the preparation of thin sections, taking into account structural inhomogeneity at the mesomorphological level.

Stage III. The soil mass and salts are studies at the micromorphological level in specially prepared thin sections over a range of magnifications from 70-100x to 200-500x. A thin slice of soil is studied in thin sections to obtain additional information on the nature of microaggregation, the relative position of the plasma and skeleton, the makeup of the pore space, and the nature of salt aggregates and their relative position with respect to the main matrix, peds, and pores. One can also identify individual salt minerals optically. The chemical and mineral compositions of the salts are refined in uncovered thin sections using microchemical reactions and immersion liquids. In addition, the mineralogical nature of the salts separated during mesomorphological investigation is further refined.

Stage IV. Submicroscopic study using scanning electron microscopes and various kinds of microanalyzers. The range of magnifications extends from 200-400x to several thousands. We used a scanning electron microscope and "Comebax" and "Kevex-5000" microanalyzers. At these magnifications one can describe the microstructural details of the plasma in the salt and carbonate horizons, and the minutest aggregates (domains) of clay particles in horizons that differ in degree of salinization and chemistry. One can also determine the distribution of the minutest and finest salt crystals with respect to microstructural elements (aggregates, pores, mineral grains, plant residues, and new formations), and the chemistry (at the qualitative level) at any given point and the individual mineralogical nature of the salts on the basis of crystal morphology.

It was assumed that each stage of this study program will: (i) answer the unresolved questions arising in the preceding stages; (ii) answer questions arising in a given stage, and (iii) raise questions to be answered in later stages. As a result it will become possible to investigate an object at all levels of structural organization using a set of correlated morphological, morpho-chemical, and morpho-mineralogical methods, and to determine on this basis the nature and character of soil salinization as well as the character and genesis of the salt formations themselves.

Two very saline hydromorphic Solonchaks from Mongolia differing in the type of salinization (sulfate and sulfate-chloride) were studied. The samples were kindly provided by Ye.I. Pankova. The soils had developed on calcareous loesslike loams with weakly mineralized groundwater of sodium sulfate or calcium-sodium sulfate composition.

Salt accumulation had the following characteristics.

1) Most of the salts were concentrated in the upper horizons (20-25 cm) (Table 1), where they accounted for 80-90% of the total salt content in the upper 2-m layer.

2) The salt content decreased sharply down the profile from 60% (in the case of sulfate salinization) or 25% (in the case of sulfate-chloride salinization) in the upper layer (0-20 cm) to 2-4% at a depth of 20 to 50 cm and to a fraction of a per cent below 1 m.

3) A salt crust had formed on the surface, where the chloride content increased sharply, being 3-30 times greater than in the most saline horizon.

4) Both profiles had a classic type of salt distribution with the following downward sequence of salinization zones: chloride, sodium-sulfate, gypsum, and carbonate.

5) With the groundwater at the same level and having an equally low mineral content (3.5-5.0 g/liter), the studied soils differed noticeably in the salt content of the most saline horizons, the degree of salinization being very high in both cases.

Stage I of the field macromorphological investigation showed that the soil profile was noticeably differentiated (Fig. 1A) and that this differentiation was associated mainly with the salt content and with the effect of the salts on the makeup. Horizons with a maximum salt concentration were identified, and the compactness of their makeup did not always correlate with their salt content. For example, the 10- to 20-cm layer with a sodium sulfate content of 60% in the field had an extremely compact makeup because of its high content of vitreous salts, while the overlying horizon with the same high salinization of the same chemistry (Na_2SO_4), was identified as the most friable and aggregated in the profile. The reason for such a different makeup of the two salinized horizons could not be determined at this stage.

The field description of the profile revealed an insignificant amount of new formations associated with poorly soluble salts, namely carbonates and gypsum. It was known, however, that the soil develops over shallow groundwater containing calcium carbonates and sulfates. Chemical analyses

Table 1

Description of the salinization of hydromorphic Solonchaks

Depth, cm	Solid residue, %	Alkalinity CO$_3$	Alkalinity HCO$_3$	Cl	SO$_4$	Ca	Mg	Na	K	CaSO$_4 \cdot$ 2H$_2$O	Carbonate CO$_2$
					% per 100 g of soil						
Sulfate-chloride Solonchak, Profile 4											
0—2	18.60	0.02	0.09	7.24	5.46	0.24	0.22	6.44	0.12	N.d.	1.20
2—7	14.30	None	0.03	3.73	5.12	0.47	0.09	4.14	0.15	1.70	0.90
7—12	25.40	»	0.02	7.75	8.11	0.37	0.06	8.42	0.09	10.30	0.90
12—25	4.60	»	0.02	0.90	1.90	0.25	0.46	1.12	0.05	7.80	1.50
25—50	3.10	»	0.02	0.60	1.38	0.27	0.23	0.64	0.08	8.80	2.70
50—85	1.40	»	0.04	0.31	0.58	0.03	0.007	0.48	0.02	0.70	4.30
85—100	0.20	»	0.06	0.03	0.06	0.004	0.001	0.06	0.002	0.05	1.00
100—135	0.20	»	0.05	0.02	0.05	0.004	0.001	0.04	0.001	0.07	3.80
135—150	0.10	»	0.06	0.02	0.03	0.01	0.003	0.03	0.02	0.20	4.80
150—200	0.30	»	0.05	0.07	0.10	0.01	0.004	0.06	0.03	0.30	6.70
200—250	0.20	»	0.05	0.04	0.07	0.01	0.003	0.04	0.01	0.40	5.50
Groundwater, g per liter	3.50	0.02	0.43	0.63	1.36	0.14	0.08	0.92	0.03	Not det.	
Sulfate Solonchak, Profile 46											
0—3	5.60	0.003	0.06	1.15	2.45	0.50	0.06	1.16	0.08	12.80	3.70
3—10	51.90	0.03	0.35	2.27	31.20	1.14	0.05	14.64	0.80	16.00	2.40
10—20	62.80	0.02	0.10	0.57	40.90	0.51	0.04	18.87	0.30	19.70	2.20
20—40	2.10	None	0.04	0.19	1.14	0.23	0.03	0.27	0.16	33.10	3.40
40—80	1.90	»	0.01	0.20	1.09	0.25	0.02	0.23	0.13	25.10	4.50
80—90	1.60	»	0.02	0.12	0.95	0.25	0.02	0.16	0.07	39.00	7.50
90—140	0.60	»	0.04	0.10	0.24	0.03	0.004	0.11	0.07	0.50	10.30
150—190	0.20	»	0.01	0.04	0.12	0.01	0.02	0.02	0.02	0.30	9.10
Groundwater, g per liter	5.40	0.01	0.16	0.40	3.05	0.45	0.10	1.04	0.11	Not det.	

also speak of a substantial content of these salts: 2.5-10% carbonates and 13-39% gypsum. The reason for the discrepancy between the results of chemical analysis and field descriptions also remained obscure.

Mesomorphological stage II helped answer the raised questions in first approximation, and also provided more detailed information on the nature of large salt formations. It was found that the essentially different makeup of the most saline horizons is associated with differences in the mineral composition of the salts: when the makeup was compact or even monolithic, sodium sulfate was present in the form of mirabilite, while the loose makeup of the overlying horizon was determined by another mineral form of sodium sulfate, thenardite (anhydrous sodium sulfate). The underlying mirabilite horizon is probably continuously recharged from the water table by capillaries. The overlying horizon dries out, and the sodium sulfate by losing water and converting to another mineral form radically changes the structure and compactness of the soil mass. The thenardite horizon has a finely aggregated structure similar to the pseudosandy structure of the subsolonetzic horizons of Solonetzes in the Transvolga region and Western Kazakhstan. There are two types of aggregates: purely salt aggregates, which are relatively few, and clay-salt aggregates, which make up the main mass of the horizon. Compared with the underlying mirabilite horizon, the thenardite horizon appears visually at the mesomorphological level to be considerably less rich in pure salt new formations (Fig. 1B). The reason for this difference will be revealed in the next stages of the investigation.

In the laboratory all the mirabilite converted to thenardite as the samples dried. Mirabilite converts very rapidly to thenardite. After this conversion, the consistence

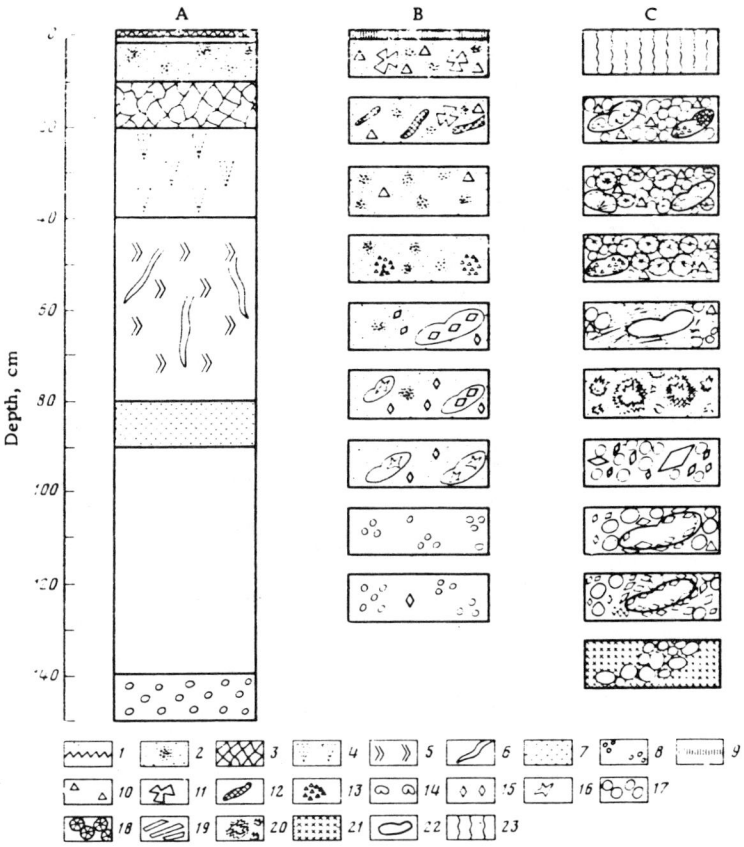

Fig. 1. Morphological makeup of the soil-salt mass in a sulfatic hydromorphic Solonchak studied in successive stages.

A) macromorphological level; B) mesomorphological level; C) micromorphological level; 1) crust; 2) mealy salt accumulations; 3) very compact vitreous layers; 4) salt film appearing upon drying of profile wall; 5) beige-colored salt veins (probably gypsum); 6) plant roots; 7) scattered small salt accumulations; 8) carbonate concretions; 9) dense halite crust; 10) large primary mineral grains and rock fragments; 11) skeletal halite aggregates; 12) salt pseudomorphoses along plant tissues; 13) nests of microcrystalline gypsum; 14) halite-filled pores; 15) gypsym crystals; 16) gypsum druses; 17) clay-carbonate aggregates; 18) thenardite-glauberite aggregates covered with clay; 19) thenardite crystals; 20) chrysanthemum-like aggregates of artificial thenardite; 21) completely carbonate-saturated zones; 22) pores; 23) weakly crystallized halite mass.

The vertical position of the structural schemes in B and C corresponds to the position of the horizons in A.

303

of the soil mass of the horizon becomes only very slightly viscous and loose. When dry, the makeup and structure of this horizon differs significantly from the natural thenardite horizon. At the mesomorphological level it appears visually to consist of pure salt, to consist only of salt aggregates (Fig. 2a), while the overlying thenardite horizon with the same high salt content (50%) appears to contain few new salt formations. The difference between the structure of the natural and artificial thenardite horizons was established in the subsequent stages of the study.

New carbonate formations were not found in the upper horizons at the mesomorphological level. A high content of clastic (primary) forms of carbonates was established. The content of gypsum along the profile, on the other hand, proved to be quite high and varied. It was established that the reason for the difference between field descriptions and the results of chemical analysis is the relatively small size of crystals gypsum as well as their confinement in certain horizons exclusively to the internal mass of peds. However, gypsum was not found in the 10- to 20-cm horizon, even at the mesomorphological level, although analysis showed that its content in that horizon was about 20%. This question was clarified in the next stage of investigation.

Investigation at the mesomorphological level also revealed other important features of the organization of the soil mass. It was found that: (i) The surface crust consists of two parts, a hard upper part of pure halite salts, and a lower part consisting of a loose, well aggregated soil-salt mass. (ii) The salts most often organize into pure salt accumulations, but may also form part of clay-salt aggregates, (iii) The structure of pure salt aggregates varies, viz., the halite aggregates are usually large (0.1 to 1-2 mm in diameter) with loose packing of large halite crystals, while thenardite aggregates may vary in size and very small thenardite crystals are much more densely packed. (iv) The upper horizons contain many slightly decomposed plant residues on which salt accumulates actively. (v) Various types of soil-mass aggregation were identified and the nature of porosity was found to differ in salinized horizons and horizons with a low salt content. The aggregates of salinized horizons, which contain many salt and clay-salt aggregates, are considerably larger and porosity is very varied in these horizons. The horizons below the salt horizons were also aggregated, but the packing of the small aggregates in them is much denser.

Studies at the micromorphological level (Stage III) answered some questions raised in the preceding stages of investigation and revealed some new structural features of the soil-salt mass. The difference between the analytical determination of gypsum (20%) and the actual very low content of gypsum grains observed under the binocular microscope in the 10- to 20-cm horizon is attributable to the presence in this horizon of glauberite, identified from its characteristic grain shape (prismatic or acicular) and from the index of refraction during immersion testing, and representing the double salt $Na_2SO_4 \cdot CaSO_4$ (Fig. 1C). Thus the chemically determined gypsum is actually a component of a double salt, the mineral glauberite.

Gypsum is easily identified in the horizons below the salt horizon. It is confined mainly to the interior of the structural units and takes various shapes: columnar, rhombic, lenticular, and spindle. Individual crystals occur more often and are idiomorphic (see Fig. 3a). With proximity to the water table, gypsum formations are found more and more frequently in the zones directly adjacent to the small pores and in the pores themselves. Starting at a depth of 50 cm cutans form along pores; they consist of individual fine gypsum grains with their long edges parallel to the pore walls. Farther down the profile these gypsum cutans (or gypsans, in the international nomenclature) transform into compact continuous, sometimes multilayer crusts on the pores (Fig. 3b). The gypsum grains in these gypsans are recognized as hypidiomorphic or allotriomorphic (i.e., partly having

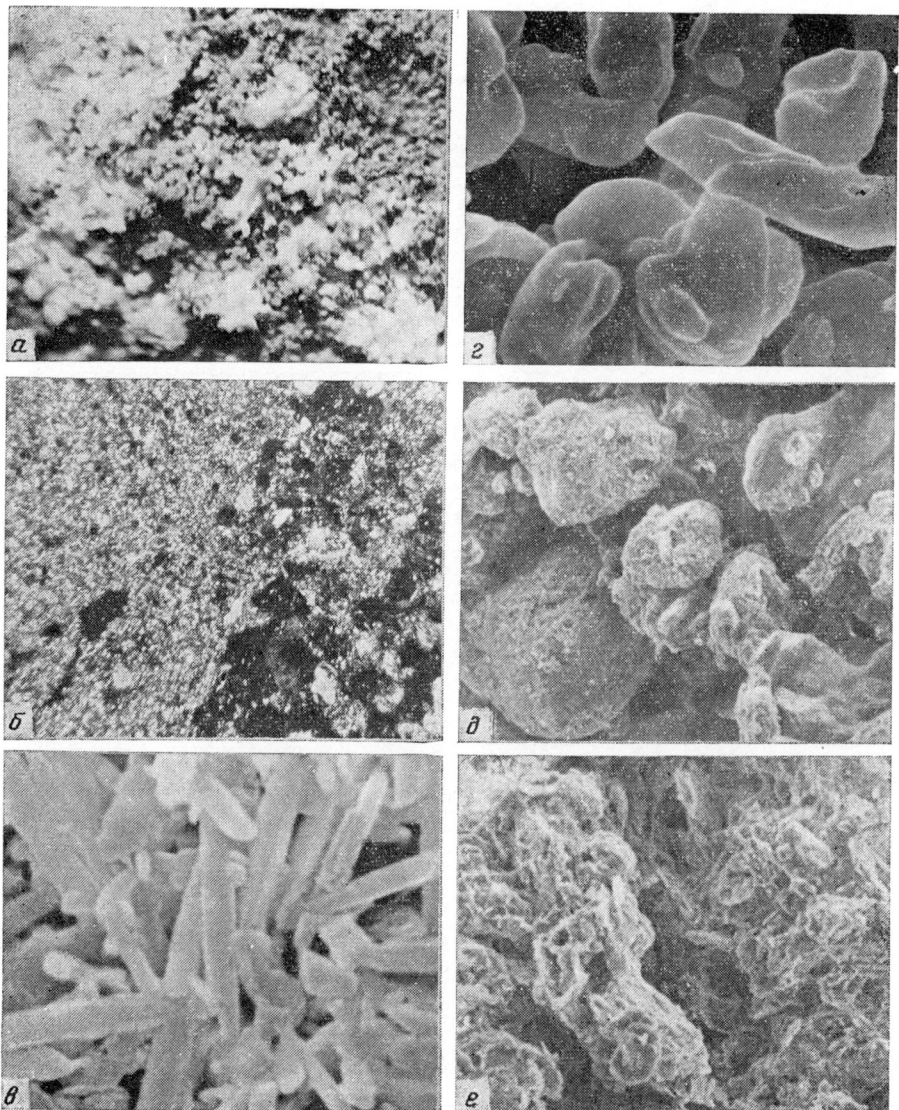

Fig. 2. Morphology of the plasma and salts:

a) large salt aggregates of former mirabilite horizon, 25x (binocular microscope); b) loesslike aggregation of horizons below the salt horizon (on the right) and compact makeup in the region with a high carbonate content (on the left). Crossed Nicols, 64x (polarizing microscope); c) acicular thenardite crystals formed under natural conditions, 6000x (SEM); d) tabular thenardite crystals formed in the laboratory as a result of conversion of mirabilite to thenardite, 5000x (SEM); e) spherical plasma aggregates and large halite crystals, 120x (SEM); f) vermiform aggregation of plasma in thenardite horizon, 200x (SEM).

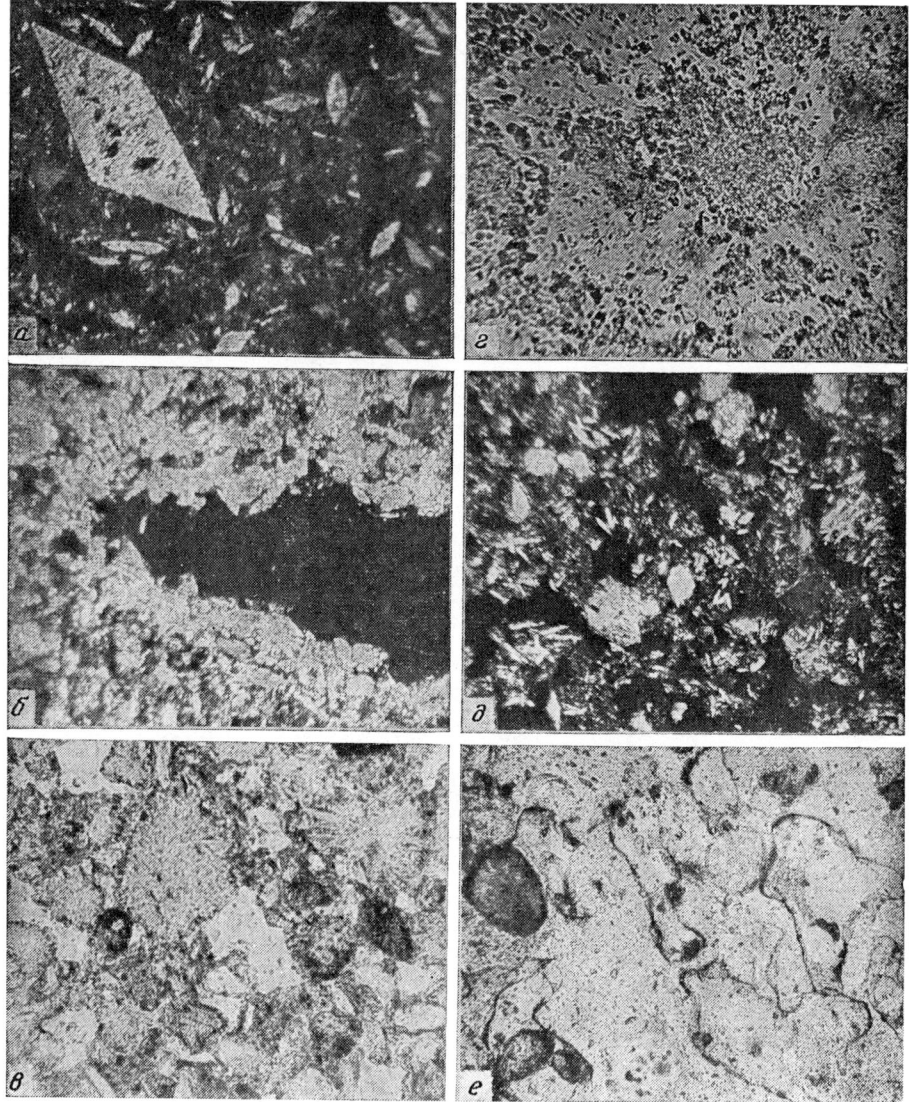

Fig. 3. Characteristics of the micromorphological structure.

a) scattered idiomorphic gypsum crystals in the main mass of the horizon below the salt horizon, crossed Nicols, 64x; b) compact covering of hypidiomorphic gypsum crystals along pores in the zone of excessive capillary wetting, crossed Nicols, 64x; c) strong aggregates of acicular thenardite covered with thin clay films and forming the base of the pseudosandy horizon, parallel Nicols, 64x; d) fragile chrysanthemum-like aggregates of a former mirabilite horizon, parallel Nicols, 160x; e) accumulations of small acicular thenardite crystals in fine pores, crossed Nicols, 64x; f) nests of irregularly shaped halite crystals along large pores, parallel Nicols, 64x.

their own faces and partly bounded by the faces of adjacent grains). This gypsum, concentrated in the lower soil horizons almost exclusively in the gypsans, can be regarded as hydrogenic, having formed relatively recently as a result of precipitation from solutions rising from the groundwater. The concentric structure of the gypsans, with clear separation of unequigranular zone-layers can serve most certainly as evidence of the hydrogenic origin of the gypsum.

Another form of gypsum scattered through the main mass as single crystals probably forms as a result of decalcification of near-lying areas of clay-carbonate plasma, clearly identified in thin sections. We find similar evidence of the formation of scattered idiomorphic gypsum in Feofarova's papers [10].

Another question concerning the low content of visible new salt formations in the so-called pseudosandy horizon, containing about 50% of analytically determined salts, has been resolved. Analysis of the microstructure in thin sections revealed that the salts form relatively small aggregates (0.07-0.42 mm in diameter), of mainly round or almost round shape. All the salt aggregates consisting of acicular or acicular-rhomboidal salt crystals are covered with thin clay films (see Fig. 3c). It is this film that masks the salts, as it were, and gives the impression that the horizon is much less saline than the adjacent mirabilite horizon.

One can also answer another question, that of the differences in the forms of crystallization of natural and artificial thenardite obtained from mirabilite. The first produces stable small salt aggregates of well-crystallized acicular crystals and in this case the clay plasma organizes in the form of films around these aggregates and of bridges between aggregates. As a result, the soil-salt mass acquires a fairly strong, unusual openwork structure.

In contrast to the pseudosandy horizon, the former mirabilite horizon has an extremely weak structure and generally poorly defined aggregation. As it converts to thenardite mirabilite produces very fragile chrysanthemum-like aggregates, which rarely persist, so that the main salt mass consists of very small individual crystals, which are much smaller than the crystals of the overlying horizon (Fig. 3d). The clay plasma in the structureless thenardite horizon occurs in the form of clay aggregates, disorderly aggregated in certain areas. Therefore the main mass of this horizon looks in some places like a loose salt mass with occasional chrysanthemum-like aggregates, and in other places like a succession of pure salt areas with areas in which groups of clay aggregates are disorderly scattered among salt aggregates and fine salt crystals (of thenardite, glauberite, and gypsum).

In addition to answering the questions posed earlier, the investigations at the micromorphological level revealed the following important microstructural features.

1. The aggregation of the horizons below the salt horizon is fairly distinct and is determined by the degree of preservation of the microstructure inherited from the loessified calcareous loam. The middle part of the profile has a distinct typical loesslike aggregation (Fig. 2b, right), which gradually deteriorates in the lower part of the profile as the content of carbonates increases: the size of the aggregates increases, they become more closely packed, and porosity decreases (Fig. 2b, left).

2. In spite of the large amount of salts and the aforementioned high content of carbonates, the plasma has a considerable optical orientation in some cases in the form of bright optically oriented films around primary mineral grains at a depth of 3-10 cm or partial optical orientation in the clay aggregates of the former mirabilite horizon. In this case the optical orientation of the plasma is probably associated exclusively with its ordered structure produced by pressure during the development of new salt formations and of salt horizons. The influence of other possible factors on

the ordering of the plasma microstructure, such as illuviation and alternate wetting and drying, is extremely limited here or totally impossible.

3. A difference was shown to exist in the crystallization of halite and thenardite with respect to microstructural elements; namely, it was determined that thenardite crystallizes along small pores (Fig. 3e), and halite along large pores (Fig. 3f). It was also concluded that halite plays a much more active role in porosity and in the modification of the initial microstructure because of the stronger separating effect of the large halite crystals than of the small thenardite crystals.

Stage IV, submicromorphological stage of investigation, using a scanning electron microscope, provided answers to some questions that arose in the preceding stages, as well as to general questions concerning the structure of the plasma in various salt horizons and the identification of the individual nature of the mineral salts. It was found, for example, that under natural growth conditions thenardite crystals have mainly the shape of long prismoidal needles (Fig. 2c), while during recrystallization from mirabilite their appearance approximates that of tabular crystals (Fig. 2d). The effect of salts on the structure of the plasma is manifest in the formation of an aggregational structure with variously shaped aggregates packed in various degrees. If the halite content is high, the plasma organizes into separate spherical aggregates (Fig. 2e), while a high content of thenardite produces a vermiform plasma aggregation (Fig. 2f).

Speaking very generally it can be said that as the content of salts decreases, the packing of the aggregates becomes denser and the size of the aggregates themselves increases.

The relative position and the interrelation of the plasma and salts were also examined. It was found that most often the plasma aggregates and new salt formations occur separately. One can find both pure salt zones with occasional clay aggregates and pure plasma zones with single salt crystals. The plasma and salts rarely form a more or less homogeneous type of microstructure. This type of microstructure was observed only once during the formation of a pseudo-sandy horizon. As regards the interrelation between salts one can state that the salts crystallize separately in zones without mixing with each other. Growth of the crystals of one mineral into another has not been observed for the given type of salinization.

Using the scanning electron microscope and "Comebax" and "Kevex-5000" type microanalyzers, we identified salt minerals also from descriptions of their crystallographic appearance and determination of the chemical composition of individual salt crystals. In addition, the form of crystallization of salt minerals was shown to vary greatly, depending on moisture conditions, salt concentration, and the chemical composition of the groundwater.

Summarizing the results of the stepwise structural analysis of the profile of Solonchaks, it can be stated very generally that the microstructure of the upper horizons, rich in readily soluble salts, is determined entirely by the amount of salts and their individual nature, and this influence is manifest through the specific aggregation of the plasma and through the effect of the salt crystals themselves. Different salt minerals are responsible for different types of plasma aggregation, on the one hand, while on the other hand, having different rates of growth and different crystals sizes, they affect porosity in different ways and deform the initial makeup of the horizons differently. The microstructure of the lower horizons, where the content of readily soluble salts decreases to 1% or less, is determined by the inherited microstructure of the calcareous loesslike loams as well as by the development of new carbonate and gypsum formations.

Thus, by using the developed program of complex stage-by-stage investigation we found the following salt minerals in the studied Solonchaks: halite, mirabilite, thenardite, glauberite, and gypsum, as well as carbonates and borates. These salts were identified by all four morphological methods and chemical, optical-chemical (electron-microscope and X-ray spectral) and mineralogical methods, each of which in a given case may be the critical, controlling method of identification. In the soils studied salts occur more often in the form of aggregates, druses, and compact and loose accumulations, rather than as individual crystals.

Halite most often produces openwork aggregates or druses ranging from 0.1 mm to 1-2 mm in diameter and is confined to the large pores of the upper horizons. The crystals in these openwork aggregates are rarely cubic and usually of irregular shape. One or two faces of the crystals are well developed, while the surface of the other faces is often corroded (Fig. 4A, a). Another common form of halite in the upper horizons consists of rods of different shapes and sizes that frequently connect openwork aggregates (Fig. 4A, b). The rods and the openwork aggregates are the largest salt formations, except, of course, the salt crusts that form on the soil surface. The crusts in these Solonchaks consist exclusively of halite, even when soil salinization is of the sodium-sulfate type (Profile 46). Even though it is only 0.2-0.7 mm thick, the crust forms a compact and fairly strong cover. The halite in this crust consists of a monolithic vitreous mass with a few individual crystals. The surface of the crust is often dusted with fine thenardite grains and has rounded erosion holes (Fig. 4A, c).

The next type of halites densely fills small single pores. Their shape is determined entirely by the shape of the pores. In addition, halite occurs in the form of individual crystals in the main mass or in the pores. Their shape is usually irregular, but sometimes it is quite perfect. Halites often forms pseudomorphoses on plant tissues in the upper horizons, crystallizing along preserved cell spaces or filling the empty shells of plant residues with a compact undifferentiated mass.

Thenardite, like halite, produces a variety of new formations, the shape and structure of which are closely associated with definite genetic horizons, i.e., with definite crystallization conditions. Thenardite is found as a dusting of fine elongated and irregular crystals on the surface of the hard halite crust. In the loose lower part of the surface crust, which contains numerous large primary mineral grains and rock fragments, thenardite forms hairs on the surface of coarser fragments (Fig. 4B, g), recognizable at high magnification as an accumulation of fine, well-faceted, and very enlongated crystals (Fig. 2, c).

The main mass of the pseudosandy horizon (Profile 46, depth 3-20 cm) consists of loosely packed aggregates of acicular (Fig. 2c) and acicular-rhomboidal (Fig. 4e) thenardite crystals. All these aggregates have a surface clay cover with angular and blade-shaped projections, outwardly resembling coral (Fig. 3c). The former mirabilite horizon consists almost entirely of fragile chrysanthemum-like thenardite aggregates or fragments of these aggregates. These are salt aggregates 0.05 to 0.1 mm in diameter, consisting of flexible elongated, often nearly acicular fine thenardite crystals. The main mass of the former mirabilite horizon (Fig. 3d) consists of fragments of these aggregates and of individual crystals of varying sizes and usually of elongated irregular shape. Less often the same horizon contains larger single elongated thenardite crystals. Pseudomorphoses of thenardite on plant tissues are quite frequent (Fig. 4B, f).

Glauberite was identified in the most saline horizon (Profile 46) where salinization was determined analytically to be almost exclusively of the sodium sulfate type and where the amount of gypsum was found to be about 20%. Actually, however, gypsum crystals occur singly in this horizon, while glauberite has been identified in the form of separate grains.

Borates were determined in the three upper horizons of Profile 46 by microchemical reactions. We were unable so far to refine the mineralogical characteristics of borates because of the high concentration of thenardite and the low concentration of borates in this horizon.

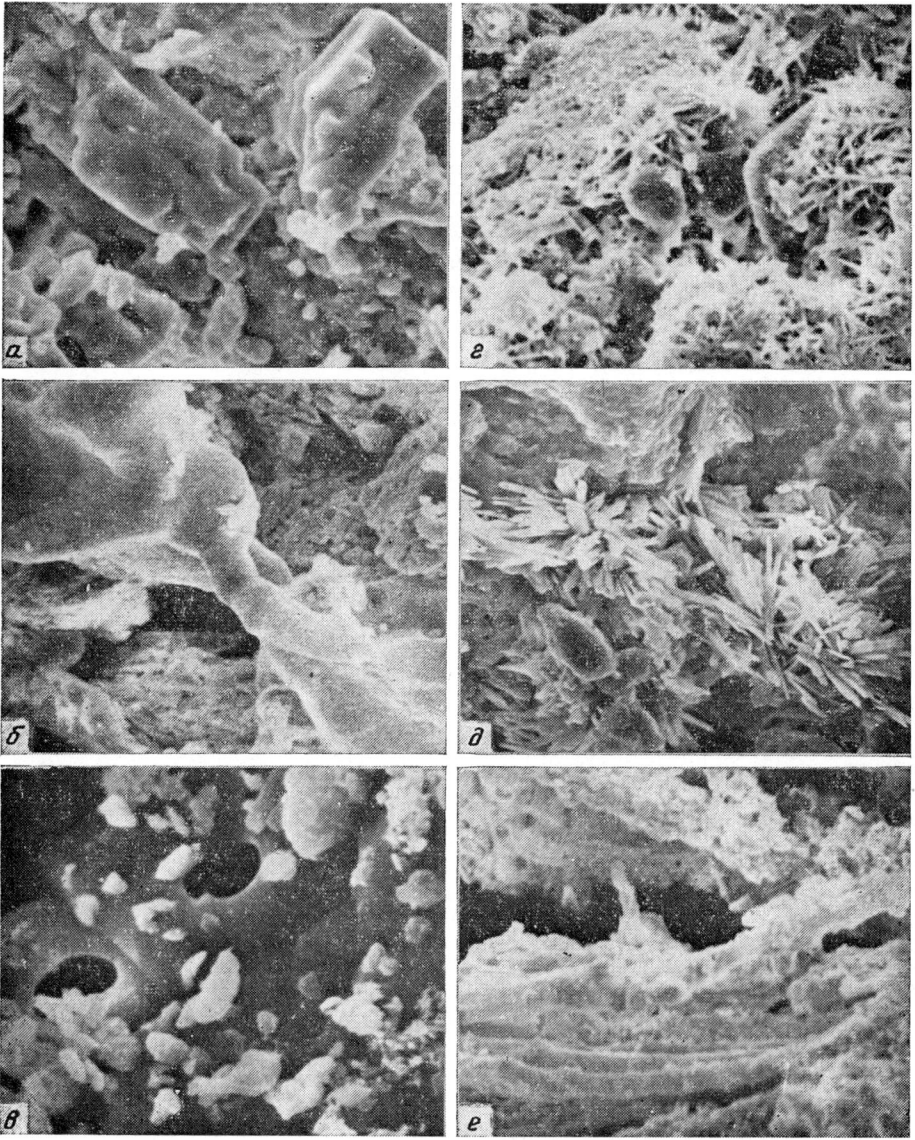

Fig. 4. Forms of halite (left) and thenardite (right) minerals under a scanning electron microscope:

a) minerals of irregular shape forming the base of openwork aggregates, 600x; b) rods connecting openwork aggregates, 300x; c) surface crust with round pores and powdering of small salt crystals, 200x; d) hairlike covering on the surface of mineral grains, 600x; e) acicular-rhomboidal thenardite crystals, 700x; f) pseudomorphoses of thenardite on plant tissues, 200x.

Gypsum was found in the upper horizons as individual crystals scattered through the main mass of material. The shape of the crystals was quite diverse: rhombic, spindle, and columnar (Fig. 3a). The crystals were idiomorphic, as a rule. Gypsum in the lower horizons is more often confined to pores, forming gypsans, single or multiple-layer gypsum covers along pores (Fig. 3b). In addition, pores were sometimes completely or disorderly filled with gypsum crystals of hypidiomorphic or allotriomorphic shape, and there were sometimes roselike aggregates of gypsum crystals. The gypsum crystals varied from 0.05 to 0.5 mm in diameter. Finely crystalline gypsum was also identified in the upper crust horizon in the form of pockets along small pores.

Carbonates occur in the upper part of the profile as variously shaped fragmental (primary) carbonates and as a constituent part of the clay-carbonate plasma. The primary fragmental carbonates disappear in the middle part of the profile. No new carbonate formations were found anywhere in the profile. Small carbonate nodules and very large hard concretions were found only in the direct vicinity of the water table beginning at a depth of 140 cm.

CONCLUSIONS

1. Use of a set of successively coordinated macro-, meso-, micro-, and submicromorphological methods in combination with chemical methods makes it possible to ask and answer questions that cannot arise when only one of the methods or an incomplete, limited set of morphological methods is used. The advantage of this combined and stagewise approach to the study of soils consists in that continuous information is obtained on the structure of the object, beginning with visual studies and ending with studies at the highest magnifications (several thousand times). In this case it is possible to describe the soil profile at all levels of its structural organization: profile, horizon, subhorizon, intraped, and even plasma levels.

2. Use of the complex stage-by-stage approach to the study of Mongolian hydromorphic Solonchaks made it possible to establish the following major patterns in the makeup of the soil-salt mass: (i) the concentration of salts alters significantly the makeup and structure of the soil profile, as well as the microstructure of the entire soil mass; (ii) the individual mineralogical nature of the salts affects the development of a specific structure and all the properties of the soil mass, regardless of salt concentration; (iii) with a mixed type of salinization individual salt minerals crystallize separately and do not mix with one another nor with the clay plasma; and (iv) the salts, being a very mobile system, alter the form of new formations as the soil moisture content changes, and some salts transform from one mineral form to another.

3. The form of salt crystallization and the type of new salt formations as well as their localization in particular structural elements make it possible to draw genetic conclusions not only concerning salt accumulation processes but also the soil formation process as a whole.

4. The method used revealed discrepanies between the hypothetical salts obtained by associating cations and anions (water extract data) and the results of chemical determination of gypsum, on the one hand, and the actual mineral forms of salts determined in the soil by direct methods, on the other hand. The development of a method for correcting hypothetical salts on the basis of soils of different types of salinization is on the agenda.

Received July 16, 1979

BIBLIOGRAPHY

1. BETEKHTIN, A.G. 1951. Kurs mineralogii (Textbook of mineralogy). Moscow, Gosgeolizdat.

2. GERASIMOVA, M.I. 1978. Meso- and micromorphology of the Pale Yellow Sod-Podzolic soils of the Valday Hills. Pochvovedeniye, No. 10.

3. DOBROVOL'SKIY, G.V. and S.A. SHOBA. 1972. Micromorphological study of a secondary Podzolic soil using a scanning electron microscope. Pochvovedeniye, No. 7.

4. PREOBRAZHENSKIY, I.A. and S.G. SARKISYAN. 1954. Mineraly osadochnykh porod (The minerals of sedimentary rocks). Moscow, Gostoptekhizdat.

5. TARGUL'YAN, V.O., T.A. SOKOLOVA, A.G. BIRINA, A.V. KULIKOV, and L.K. TSELISHCHEVA. 1974. Organizatsiya, sostav i genezis dernovo-palevo-podzolistoy pochvy na pokrovnykh suglinkakh (The organization, composition, and genesis of a Pale Yellow Sod-Podzolic soil on mantle loams). Moscow, Nauka.

6. FEOFAROVA, I.I. 1940. Mineralogical determination of water-soluble minerals in saline soils. Pochvovedeniye, No. 12.

7. FEOFAROVA, I.I. 1950. Aragonite in soils. Tr. Pochv. inst. im. V.V. Dokuchayeva, Vol. 34.

8. FEOFAROVA, I.I. 1950. Calcite pseudomorphoses on gypsum in soils. Tr. Pochv. inst. im. V.V. Dokuchayeva, Vol. 34.

9. FEOFAROVA, I.I. 1958. Microscopic determination of carbonates in saline soils. Tr. Pochv. inst. im. V.V. Dokuchayeva, Vol. 53.

10. FEOFAROVA, I.I. 1958. Sulfates in saline soils. Tr. Pochv. inst. im. V.V. Dokuchayeva, Vol. 53.

11. YARZHEMSKIY, Ya.Ya. 1966. Mikroskopicheskoye izucheniye galogennykh porod (Microscopic study of halogenic rocks). Novosibirsk, Nauka.

12. BARZANJI, S. and G. STOOPS. 1974. Fabric and mineralogy of gypsum accumulations in some soils of Iraq. Works of the 10th Intern. Congr. of Soil Sci., v. 7.

13. BISDOM, E.B.A., S. HENSTRA, A. JONGERIUS, and F. THIEL. 1975. Energy-dispersive X-ray analysis on thin sections and unimpregnated soil material. Neth. J. Agric. Sci., No. 23.

14. HANNA, F.S. and G.J. STOOPS. 1976. Contribution to the micromorphology of some saline soils of the North Nile Delta in Egypt. Pedologie, 26(1).

15. SEHGAL, I.L. and G. STOOPS. 1972. Pedogenic calcite accumulation in arid and semi-arid regions of the Indo-Gangetic alluvial plain of erstwhile Punjub (India). Their morphology and origin. Geoderma, No. 8.

16. STOOPS, G., H. ESWARAN, and A. ABTAHI. 1977. Scanning electron microscopy of authigenic sulphate minerals in soils. Proc. of the 5th Intern. Working meeting on Soil Micromorphology. Granada, Spain.

MICROMORPHOLOGICAL OBSERVATIONS ON PYRITE AND ITS OXIDATION PRODUCTS IN FOUR HOLOCENE ALLUVIAL SOILS IN THE NETHERLANDS

R. MIEDEMA, A. G. JONGMANS and S. SLAGER

INTRODUCTION

The accumulation of sulphides, and in particular of pyrite, is quite common in marine and estuarine deposits all over the world (Moorman, 1963). The oxidation of pyrite has been described in various publications with special reference to chemical and microbiological processes (Quispel et al., 1952; Harmsen, 1954; Van Breemen, 1972). Some micromorphological aspects of pyrite accumulations were described by Eswaran (1967), Rickard (1970) and Slager et al., (1970). Apart from the above mentioned publications, the micromorphology of pyrite and its oxidation products has received little attention. The authors studied these micromorphological aspects of 26 thin sections (8 x 15 cm) from four profiles, situated in the Western part of the Netherlands.

MATERIALS AND METHODS

The investigated profiles (Mijdrecht 2, Haarlemmermeer 1, Zuidplas 1 and Purmerend 2) are situated in inland polders. These polders were reclaimed from artificial or natural lakes, which resulted from the excavation of the peat toplayer or natural abrasion of the peat. This peat was overlying Subboreal tidal flat deposits sedimented as soft muds in a reed vegetation, in which high amounts of pyrite accumulated. The profiles developed in these deposits, which are now lying at the surface of the inland polders. Artificial drainage gave rise to oxidation of pyrite in these profiles.

The Dutch climate belongs to the maritime climates of the temperate zone. Mean annual temperature is $9.0°$ C. The precipitation is rather evenly distributed and on the average totals 765 mm annually. From May through July the evapotranspiration exceeds the rainfall.

The thin sections have been prepared according to the method described by Jongerius and Heintzberger (1963). The terminology is mainly according to Brewer (1964). The investigations were carried out on a Leitz Ortholux microscope using transmitted plane or polarized light and incident light. The magnifications used varied from 50 to 300 times. Samples of these profiles were also studied on a Scanning Electron Microscope, using magnifications up to 25,000 times, by Dr. H. Eswaran (Geological Institute, State University, Ghent).

RESULTS

Micromorphology

Profile: Mijdrecht 2. Typic Sulfaquept (Soil Taxonomy, 1970). The profile is situated on the transition from a creek ridge to a former reed marsh. It is imperfectly drained and the groundwater level fluctuates between 40 and 120 cm below the soil surface. The altitude is about 5.5 m below mean sea level.

Horizon	Depth	Description
Ap	0- 28 cm	2.5Y 5/2, dry; silty clay loam; strong subangular blocky; common fine channels; few, fine, faint, yellowish brown (10YR 5/6) mottles; clear and wavy on:
II O	28-28/33 cm	5YR 3/2, dry; clayey peat; moderate subangular blocky; common fine channels; clear and broken on:
II B21g	28/33-45/50 cm	10YR 4/1, moist; silty clay; strong subangular blocky; common fine and few large channels; common, coarse, distinct, dark brown (7.5YR 3/2) mottles associated with reed remains; gradual and wavy on:

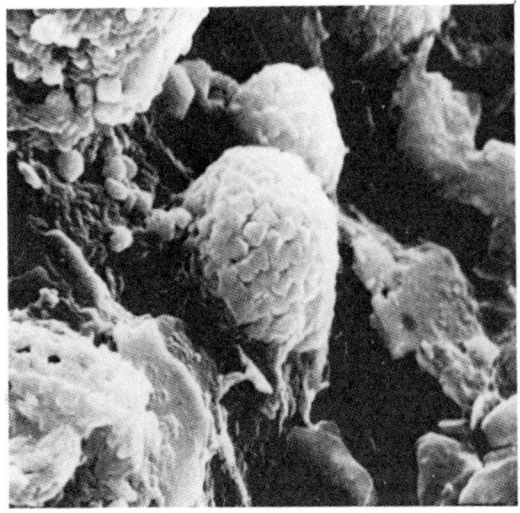

(2)

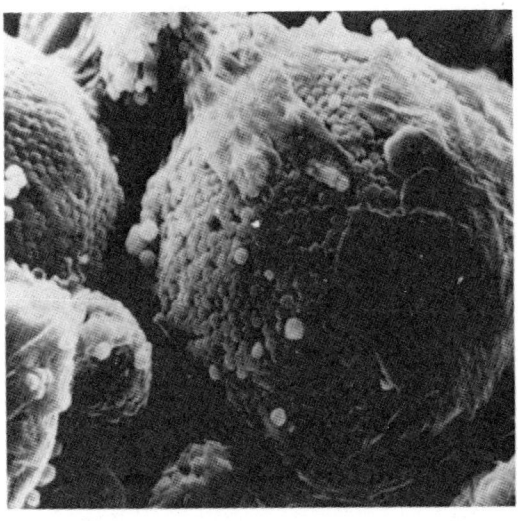

(3)

Figure 2. SEM photo of pyrite framboid consisting of microcrysts. The coating along the framboid consists most likely of organic matter (x2400).

Figure 3. SEM photo of pyrite framboid consisting of bipyramidal microcrysts. A coating along the framboid is absent (x6000).

(Photo's courtesy of Dr. H. Eswaren, Ghent).

II B22g	45/50-63/78 cm	10YR 5/1, wet; silty clay loam; moderate smooth prismatic; common fine and large channels; common, coarse, prominent, yellow (2.5Y 8/6) mottles, locally associated with dark brown (10YR 3/3) mottles; in channels reddish brown (5YR 4/4), slightly hardened reed remains; gradual and tongued on:
II B3g	63/78-105/115 cm	5Y 4/1, wet; silty clay; weak prismatic, changing with depth into macro-structureless; many large channels; common, coarse, prominent, pale yellow (2.5Y 8/3) mottles with increasing depth substituted by dark brown (10YR 3/3) mottles; diffuse and tongued on:
II Cg	+ 105/115 cm	5Y 4/1, wet; silty clay; macro-structureless; abundant large channels.

The profile in general consists of clay minerals with an admixture of organic matter in the topsoil. The skeleton grains (quartz and some micas, glauconites and feldspars) are mainly randomly distributed and from 2 to 50 microns in size. In the topsoil some coarse skeleton grains occur. Calcium carbonate is absent throughout this profile. Biogenic voids are numerous throughout the profile with diameters up to 1 cm. They often still contain reed remains. Some planes were encountered in the upper part of the profile.

A summary of the micromorphological phenomena concerning pyrite and its oxidation products is given in Table 1. Pyrite mainly occurs as framboids (10-15 microns in diameter) in clusters, associated with reed remains (photo 1). The internal fabric of the framboids is shown on photos 2 and 3. Besides the pyrite framboids, clusters of intercalary pyrite crystals occur, which have an angular shape in the thin section and were found to have an octahedral shape under the Scanning Electron Microscope. The boundary between pyritic and non-pyritic soil material lies at 70 cm below the soil surface.

Most of the jarosite in this profile occurs as channel neo- and quasicutans. The latter were always found to be associated with neo-ferrans (photo 4). The highest amount of neo- and quasijarositans was observed in the 44-75 cm zone. Some jarosite occurs in voids, mainly as irregularly shaped bodies which generally contain skeleton grains and reed remains. Another part of the jarosite in voids occurs as framboids with the same diameter (10-15 microns) as the pyrite framboids. With decreasing depth all the jarosite tends to become stained by ferri-hydroxides. The maximum jarosite content occurs in the 60-75 cm zone; jarosite is absent above 44 cm and below 91 cm. Channel neoferrans are the predominant form in which the amorphous ferri-hydroxides occur. Iron impregnated reed remains commonly occur throughout the oxidized zone. Further throughout the oxidized zone some ferri-hydroxide spheres occur which have the same diameter (10-15 microns) as the jarosite and pyrite framboids. The total amount of ferri-hydroxoides increases with decreasing depth within the oxidized zone.

In the 35-75 cm zone some well-crystallized goethite (X-ray identified) was found to occur associated with amorphous ferri-hydroxides which impregnated reed remains (photo 5). Some intercalary gypsum crystals along large channels were found from 94-109 cm.

Profile: Haarlemmermeer 1. Sulfic Haplaquept (Soil Taxonomy, 1970). The profile is situated in an almost flat position. It is imperfectly to moderately well-drained. The groundwater level was lowered in 1967 by artificial drainage and fluctuates between 60 and 120 cm below the soil surface. The altitude is about 5 m below mean sea level.

Ap   0-40 cm   10YR 3/2.5, dry; humid silty

TABLE 1. Some important micromorphological observations of Profile Mijdrecht II.

TABLE 2. Some important micromorphological observations of Profile Haarlemmermeer I.

		clay loam; strong subangular blocky; common fine channels; abrupt and smooth on:
II B21g	40-50/70 cm	10YR 5/1, moist; silty clay; moderate compound prismatic subdivided into weak subangular blocky; common fine and large channels; on ped faces common, coarse, distinct, dark grayish brown (10YR 4/2) mottles; along large channels abundant, distinct, vertically elongated, slightly hard ferri-hydroxide coatings associated with abundant, prominent, yellow (2.5Y 8/6) mottles; in large channels soft irregularly shaped jarosite nodules; gradual and tongued on:
II B22g	50/70-94/100 cm	10YR 5/1, wet; (silty) clay; weak prismatic; many large channels; in the upper part of the horizon common, coarse, distinct, strong brown (7.5YR 4/6) mottles; on walls of channels of dark reddish brown (2.5YR 3/4) coatings often associated with jarosite; along channels common, yellow (2.5Y 8/6) mottles; gradual and tongued on:
II C1g	94/100-175 cm	10Y 4/1, wet; clay; macrostructureless; many large channels; clear and smooth on:
II C2g	+ 175 cm	10YR 5/1; silty clay loam; calcareous.

This profile in general consists of clay minerals and

skeleton grains (quartz and micas, glauconites and feldspars) of sizes between 2 and 50 microns. In the topsoil the skeleton grains occur in a random distribution pattern, which downwards gradually changes into a banded distribution pattern. Biogenic voids, often with reed remains, are present throughout the profile, while planes occur mainly in the upper part of the profile.

A summary of the micromorphological phenomena concerning pyrite and its oxidation products is given in Table 2. Nearly all the pyrite in this profile occurs as framboids (10-50 microns in diameter) in clusters, associated with reed remains. A minor part of the pyrite was observed as angular crystals (5-15 microns in diameter). The boundary between pyritic and non-pyritic soil material lies at about 85 cm below the soil surface.

The jarosite occurs in this profile mainly as neo- and quasicutans along channels. The quasijarositans were always found to be associated with neoferrans. Especially in the lower part of the oxidized zone jarosite framboids regularly occur in voids (photo 6). They have the same diameter (10-50 microns) as the pyrite framboids and they contain sometimes (in the 94-109 cm layer) still a nucleus of pyrite (photo 7). The detailed structure of jarosite framboids is shown on photo 8. In the middle part of the oxidized zone irregularly shaped bodies of jarosite were found in voids. These jarosite bodies generally contain skeleton grains and reed remains. The above-mentioned morphological types of jarosite tend to become increasingly stained by ferri-hydroxides with decreasing depth. The highest amount of jarosite occurs between 35 and 70 cm below the soil surface. Jarosite is absent above 25 cm. The lower boundary of the jarosite occurrence lies below 110 cm.

The amorphous ferri-hydroxides occur regularly throughout the soil as channel neoferrans and as iron impregnated reed remains. Between 15 and 90 cm ferri-hydroxide spheres occur with the same diameter as the pyrite framboids (10-50 microns). The content of ferri-hydroxides remains relatively constant with depth. Locally amidst the ferri-hydroxides some goethite (X-ray identified) was observed in the reed remains. Between 85 and 110 cm gypsum was found to some extent as intercalary crystals adjacent to channels and as crystal tubes filling channels.

Profile: Zuidplas I. Typic Sulfaquept (Soil Taxonomy, 1970). The profile is situated in the lowest part of a weakly undulating reed marsh. It is imperfectly drained and the groundwater level fluctuates between 40 and 100 cm. Its altitude is about 5.5 m below mean sea level.

Macromorphology

A11	0-26 cm	10YR 3/3, moist; silty clay; moderate subangular blocky; fine channels; common, fine, brown (10YR 5/7) mottles, often along channels; clear and smooth on:
A12g	26-33/38 cm	7.5YR 2/2, moist; silty clay; moderate subangular blocky, fine channels; common, fine, brown (7.5YR 4/2) mottles; abrupt and wavy on:
B2g	33/38-70/80 cm	10YR 5/1.5, moist to wet; silty clay; moderate subangular blocky, changing with depth into weak prismatic; large channels; common, coarse, yellow (2.5Y 8/6) mottles associated with common, coarse, yellowish brown (10YR 5/8) mottles occurring along channels; some channels partly filled with coarse, yellow (2.5Y 8/6) nodules; clear and tongued on:
C1g	70/80-95/105 cm	10Y 4/1, wet; silty clay; macro-structureless; large channels; common, coarse, yellow-

C2g   95/105-170 cm   ish red (5YR 4/6-3/4) mottles along channels; gypsum crystals; clear and tongued on: 10Y 4/1, wet; silty clay; macro-structureless; large channels.

This profile consists of clay minerals with an admixture of organic matter in the topsoil. The skeleton grains (quartz and some micas, glauconites and feldspars) have sizes between 2 and 50 microns and are generally randomly distributed except for the 47-79 cm layer where the distribution is mainly clustered and banded. In the 92-117 cm zone calcium carbonate skeleton grains were observed in appreciable amounts. Biogenic voids with diameters up to a few millimeters are abundant throughout the profile and normally contain reed remains. Some planes are present to depths of 62 cm.

A summary of the micromorphological phenomena concerning pyrite and its oxidation products is given in Table 3. Nearly all the pyrite in this profile occurs as framboids (10-50 microns in diameter) in clusters, associated with reed remains. Locally some angular crystals (5-15 microns in diameter) of pyrite were found. The boundary between pyritic and non-pyritic soil material lies at 67 cm below the soil surface.

Most of the jarosite in this profile occurs as neo- and quasicutans along channels. The quasijarositans are always associated with neoferrans. Some of the jarosite occurs as framboids, which have the same diameter as the pyrite framboids (10-50 microns), and only occurs in voids. They generally contain skeleton grains and reed remains. The three above mentioned morphological types of jarosite tend to become increasingly stained by ferri-hydroxides with decreasing depth. The highest amount of the jarosite occurs between 47 and 62 cm below the soil surface. Jarosite is absent above 26 cm and below 97 cm. Most of the amorphous ferri-hydroxides observed in this profile occur as neoferrans along channels. Some of the ferri-hydroxides occur as spherical bodies having the same diameter as the pyrite and jarosite framboids (10-50 microns).

PYRITE IN THE NETHERLANDS

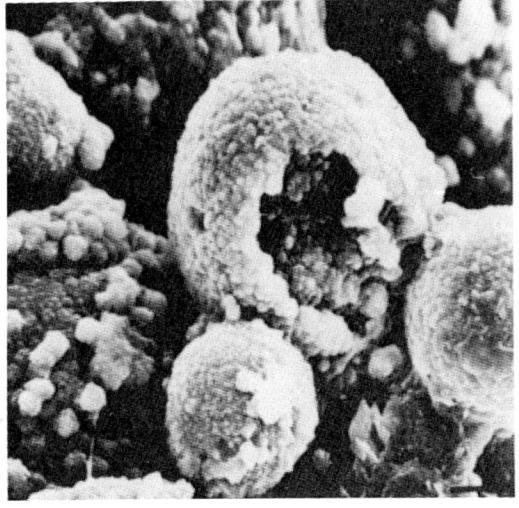

(8)

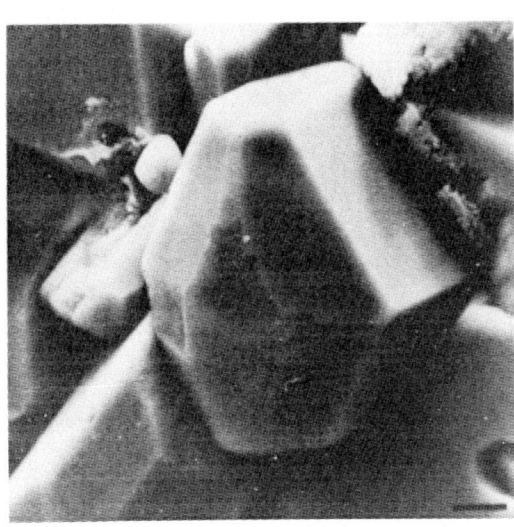

(10)

Figure 8. Scanning Electron Microscope photo of a jarosite framboid consisting of microcrysts (x2350).
Figure 10. Scanning Electron Microscope photo, showing the monoclinic prismatic shape of a gypsum crystal (x1000).
(Photo's courtesy of Dr. H. Eswaren, Ghent).

In the lowest part of the oxidized zone (81-117 cm) some of the spheres contain a nucleus of pyrite. The total amount of ferri-hydroxides increases within the oxidized zone with decreasing depth. The amount of neoferrans shows the same trend; the amount of spherical bodies and iron impregnated reed remains stays rather constant within the oxidized zone.

Apart from the amorphous ferri-hydroxides goethite was observed between 26-62 cm. It occurs in channels, associated with the ferri-hydroxides which impregnated reed remains. Gypsum crystals were observed in appreciable amounts in the 82-117 cm layer. They occur as crystal tubes within channels and as intercalary crystals along channels. Photo 9 shows the shape of the crystals as observed in thin sections; photo 10 shows the crystal shape as observed under the Scanning Electron Microscope.

Profile: Purmerend 2. Typic Haplaquoll (Soil Taxonomy, 1970). The profile is situated in a flat position. It is imperfectly drained. The groundwater level fluctuates between 25 and 150 cm below the soil surface. The profile is situated at 3 m below mean sea level.

A1	0-24/30 cm	10YR 2.5/1, dry; silty clay; strong compound prismatic subdivided into moderate angular blocky; many large and fine channels; diffuse and tongued on:
AB(g)	24/30-40/53 cm	5Y 4/1, dry; silty clay; moderate compound prismatic subdivided into moderate subangular blocky; many large and fine channels; few, fine, faint, strong brown (7.5YR 5/8) mottles; diffuse and tongued on:
B21g	40/53-80 cm	5Y 5/1, moist; clay; disturbed stratification and weak compound prismatic subdivided in-

		to weak subangular blocky; many large and fine channels; common, medium, distinct, dark reddish brown (5YR 3/4) mottles; diffuse and smooth on:
B22g	80-102/108 cm	5Y 5/1, moist; silty clay; disturbed stratification and sponge structure; many large and fine channels; common, coarse, prominent, yellowish red (5YR 4/8) mottles, sometimes along channels; diffuse and wavy on:
B3g	102/108-119/125 cm	5Y 5/1, wet; silty clay; disturbed stratification; few large and fine channels; common, coarse, distinct, (dark) olive (5Y 4/3) mottles; clear and wavy on:
C1g	119/125-133/143 cm	coarsely spotted, 2.5GY 3.5/1 and 7.5GY 2.5/1 (Japanese color chart), wet; clay; partly disturbed stratification; few fine channels; abrupt and wavy on:
C2g	133/143-180 cm	7.5GY 2.5/1 (Japanese color chart), wet; clay; nearly undisturbed stratification; few fine channels.

The part of this profile, which is of interest in connection with pyrite and its oxidation products (i.e. from 68-178 cm), consists of a mixture of clay minerals and much finely divided primary calcium carbonate (asepic plasmic fabric tending to crystic). The skeleton grains (quartz with some micas, chalcedonies, glauconites and feldspars) with sizes between 2 and 50 microns show a banded and clustered distribution pattern. Biorelicts (Eswaran, 1967) in the form of shell frag-

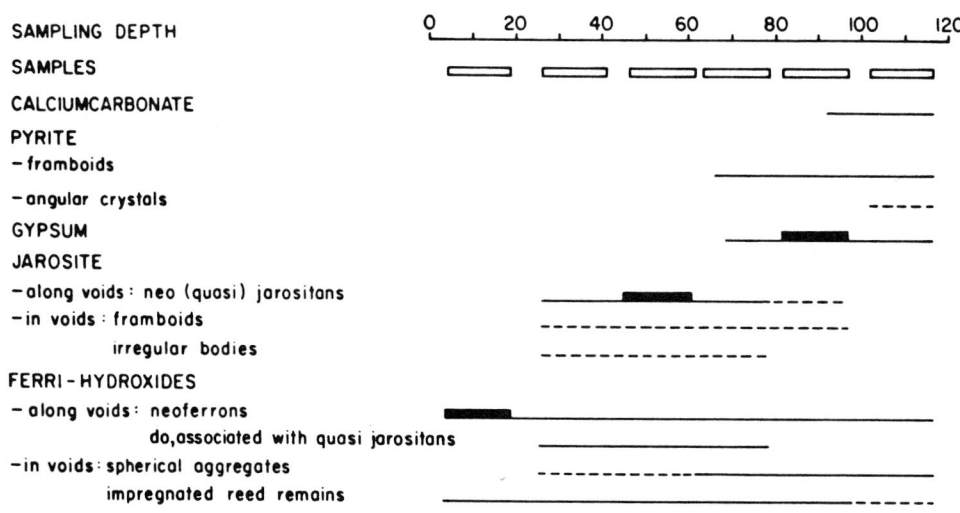

TABLE 3. Some important micromorphological observations of Profile Zuidplas I.

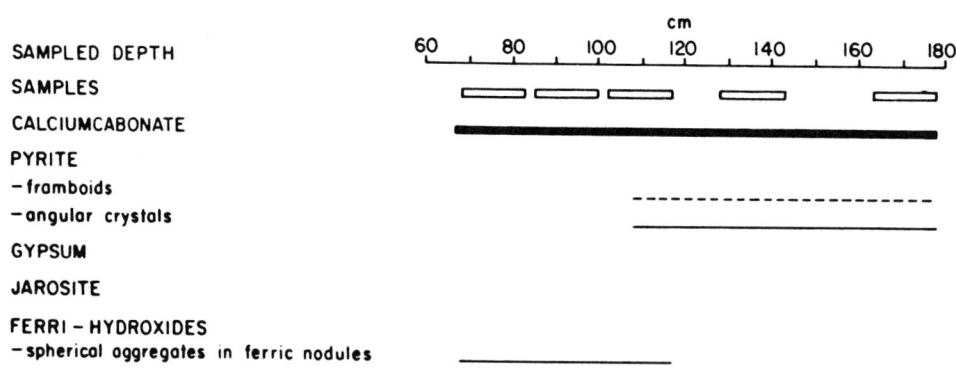

TABLE 4. Some important micromorphological observations of Profile Purmerend II.

ments and diatoms are numerous. Biogenic voids and planes occur throughout the profile decreasing in amount with depth.

A summary of the micromorphological phenomena concerning pyrite and its oxidation products is given in Table 4. Part of the pyrite is present as randomly distributed angular crystals of 5-15 microns, part of it filling diatoms. This form is predominant. Another form of pyrite are the framboids of 10-40 microns in diameter in clusters associated with reed remains. The boundary between pyritic and non-pyritic soil material lies at 98 cm below the soil surface. In the 128-143 cm layer a diffuse halo of ferri-hydroxides around pyrite accumulations was observed. Moreover in the 68-117 cm layer large, irregular ferric nodules occur with spherical isotropic centres showing the original shape and size of the pyrite accumulations (diatoms and angular pyrite; photo 11). The nuclei of these former pyrite accumulations locally still show a reflection with incident light. Gypsum is absent within the sampled depth.

Summary

The observations on the four investigated profiles reveal that a distinction should be made between oxidation of pyrite in a calcareous environment and in a non-calcareous environment. The differences refer to character and morphology of the oxidation products. In a calcareous environment the oxidation products of pyrite are amorphous ferri-hydroxides and gypsum (cf. profile Purmerend 2 and Zuidplas 1 below 92 cm).

Profile Purmerend 2 lacks gypsum for reasons which will be explained later. The amorphous ferri-hydroxides occur as clusters of spheres which have the same sizes as the pyrite framboids occurring in the surrounding groundmass (Purmerend 2). In other cases, apart from these clusters of spheres, neoferrans were observed along large channels (Zuidplas 1 below 92 cm). The clusters of spheres are surrounded by a glaebular halo to a large (Purmerend 2) or a small (Zuidplas 1) extent. Sometimes, especially in the lowest part of the oxidized zone, the ferri-hydroxide spheres still contain a nucleus of pyrite.

In a non-calcareous environment the oxidation products of

pyrite are amorphous ferri-hydroxides and jarosite (cf. Mijdrecht 2, Haarlemmermeer 1 and Zuidplas 1 above 92 cm). The attention will be focussed on these two products. Gypsum and goethite were sometimes found in small quantities. Their presence will be explained later. In non-calcareous profiles the following zonation of pyrite oxidation products is found.

Zone I. In the lower part of the oxidized zone amorphous ferri-hydroxides form the most important oxidation products, although some jarosite occurs. The amorphous ferri-hydroxides occur as neoferrans along large channels, as iron impregnated reed remains and only occasionally as spherical ferric nodules of similar sizes as the pyrite framboids occurring in the surrounding groundmass. Jarosite occurs in this zone as neo- and/or quasicutans and as jarosite framboids of similar sizes as the pyrite framboids occurring in the surrounding groundmass. The quasi-jarositans are always associated with neoferrans. The jarosite framboids only occur in voids. Sometimes (Haarlemmermeer 1: 94-109 cm) the jarosite framboids still contain a nucleus of pyrite.

Zone II. In the middle part of the oxidized zone jarosite is dominant, although amorphous ferri-hydroxides regularly occur. Jarosite occurs mainly as neo- and quasicutans along large channels, occasionally as framboids and irregularly shaped bodies in those channels. The quasi-jarositans are associated with neoferrans. The jarosite framboids in the voids tend to change into jarositic spheres stained by ferri-hydroxides with decreasing depth within this zone. Part of the irregularly shaped bodies of jarosite within this zone may be stained by ferri-hydroxides and they generally contain skeleton grains and reed remains. The amount of ferri-hydroxides is higher than in zone I. Morphologically no differences in occurrence of these compounds were observed: they occur predominantly as neoferrans and iron impregnated reed remains.

Zone III. In the upper part of the oxidized zone mainly amorphous ferri-hydroxides occur, although in the lower part of this zone some jarosite is observed. Ferri-hydroxides occur in larger amounts than in zone II. They are present as neoferrans and iron impregnated reed remains, but to a somewhat lower extent also as irregularly shaped ferric nodules with a diffuse boundary. Moreover some spheric ferric nodules were found, like those occurring in zones I and II. The morphology of the jarosite in this zone is similar to that of the jarosite in Zone II but it is strongly stained by ferri-hydroxides.

The transition between zones I and II and zones II and III is rather diffuse and tonguing than abrupt and smooth. The transition between zone I and the reduced subsoil is rather sharp and tonguing.

INTERPRETATION AND DISCUSSION

The genesis of the above mentioned phenomena will now be explained using reactions (1) through (6) in Table 5 and Figure 1. Reactions (1) - (5) are according to Van Breemen (1972). First the genesis will be discussed in a calcareous environment, afterwards the genesis in a non-calcareous environment.

TABLE 5. Chemical and microbiological reactions concerning oxidation of pyrite.

(1) $3FeS_2 + 6H^+ + 10\tfrac{1}{2}O_2 + 3H_2O \rightarrow 3Fe^{2+} + 12H^+ + 6SO_4^{2-}$
(pyrite)

(2) $12Fe^{2+} + 30H_2O + 4O_2 \rightarrow 12Fe(OH)_3 + 24H^+$

(3) $6CaCO_3 + 12H^+ \rightarrow 6Ca^{2+} + 6H_2O + 6CO_2$

(4) $6Ca^{2+} + 6SO_4^{2-} + 12H_2O \rightarrow 6CaSO_4 \cdot 2H_2O$
(gypsum)

(5) $3Fe(OH)_3 + K^+ + 2SO_4^{2-} + 3H^+ \leftrightarrow KFe_3(SO_4)_2(OH)_6 + 3H_2O$
(jarosite)

(6) $3Fe(OH)_3 \rightarrow 3FeOOH + 3H_2O$
(goethite)

When oxygen enters in a calcareous, pyrite-containing, reduced soil material (for instance when the groundwater level is lowered), the pyrite is oxidized (reaction 1). According to this reaction ferro ions, sulphate ions and hydrogen ions are formed. The ferro ions are oxidized into ferri-hydroxide and hydrogen ions (reaction 2). The oxidation reactions (1 and 2) in a calcareous environment, however, do not give rise to a lowering of the pH, since the hydrogen ions are neutralized by calcium carbonate (reaction 3). The calcium ions formed in reaction (3) are combined with the sulphate ions formed in reaction (1) to gypsum (reaction 4). Figure 1 indicates that pyrite oxidation in a calcareous environment immediately results in the formation of ferri-hydroxides. The latter are immobile at that pH level. Consequently the oxidation of pyrite results in in situ formation of ferri-hydroxides. This explains why the amorphous ferri-hydroxide segregations occur as bodies with the same size and spherical shape as the pyrite framboids from which they originate. These spherical ferri-hydroxide segregations sometimes still contain a nucleus of pyrite.

The neoferrans found in Zuidplas 1 (below 92 cm) and the glaebular halo's around the clusters of spheres of ferri-hydroxide occurring in Purmerend 2 and Zuidplas 1 (below 92 cm) are assumed to be gley phenomena. The absence of gypsum in profile Purmerend 2 (a moderately well-drained profile) is probably due to solution and removal by percolating water through this profile, which process has been active for several hundreds of years.

When oxygen enters in a non-calcareous, pyrite-containing soil material pyrite oxidation starts like in a calcareous environment, according to reaction (1). The produced hydrogen ions, however, are not neutralized by calciumcarbonate and consequently give rise to lowering of the pH. As Figure 1 indicates, basically two processes may occur.
(a) When oxygen enters relatively slowly, the pH is lowered slowly too: the ferro ions produced according to reaction (1) are oxidized into ferri-hydroxides. As the pH continues to decrease under oxidative conditions, the ferri-hydroxides are transformed into jarosite (reaction 5). The sulphate ions necessary for this reaction originate from reaction (1), the potassium ions may be derived from the soil solution and/or the

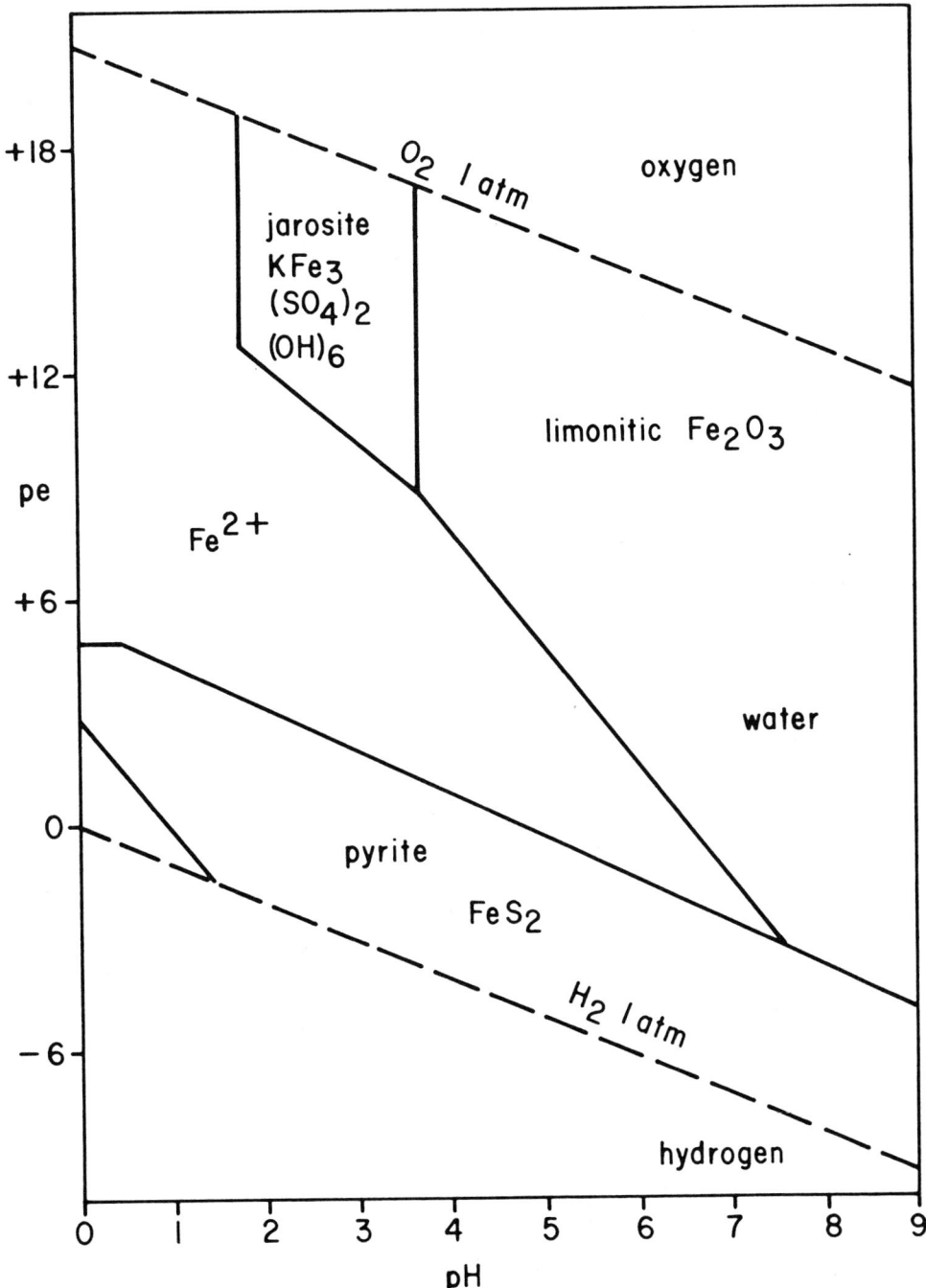

Figure 1. pe: pH diagram of pyrite, limonitic Fe_2O_3, jarosite and dissolved K^+, SO_4^{2-}, Fe^{2+} and Fe^{3+} at log $[SO_4^{2-}]$ = -2.3, log $[K^+]$ = -3.3, log $[Fe^{2+} + Fe^{3+}]$ = -5 and 1 atm total pressure. (Derived from van Breemen, 1972).

soil complex. When the hydrogen ions are removed by percolation the jarosite (reaction 5 from right to left) is hydrolized into relatively immobile ferri-hydroxides. The potassium- and sulphate ions are removed by percolation. The final ferri-hydroxides occur in other places than the original pyrite. This follows from the fact that the ferro- and ferri-compounds, present before the formation of jarosite, occur at low pH levels and consequently are mobile in solution. The ferri-hydroxides formed from jarosite, however, occur at relatively high pH levels and are relatively immobile.

(b) When oxygen enters quickly, the pH is lowered abruptly. The system is not in equilibrium and formation of jarosite from pyrite through the ferro- and ferri-stages is instantaneous. For this reason the jarosite occurs at the same place as the original pyrite. The jarosite finally is transformed into ferri-hydroxides as the hydrogen ions are removed by percolation under oxidative conditions.

The alternation of zones with dominantly ferri-hydroxides and jarosite will be explained by using the set of reactions (a) taking place when a relatively slow oxidation occurs.

In zone I the oxygen enters the reed remains and the groundmass through channels. At the wall of the channel ferro ions are oxidized into ferri-hydroxides. The following step is movement by diffusion of ferro ions (derived from the pyrite by oxidation) from the surrounding groundmass to the walls of the channel. From this process neoferrans and iron impregnated reed remains result. The process has not been active long enough in zone I to lower the pH to such a level that apprecable amounts of jarosite are formed. Ferri-hydroxides occurring as neoferrans are consequently dominant. In zone II the reactions have proceeded much further. The pyrite is nearly completely oxidized and consequently the pH is so low that ferri-hydroxides were transformed into jarosite which is dominant here. The neoferrans, as they occur in zone I, have been transformed either totally into neojarositans or partly into neoferrans/quasijarositans. The irregular bodies of jarosite, containing skeleton grains and reed remains, are assumed to have been transported mechanically as pieces of neojarositans which were formed at a higher level in the profile.

In zone III hydrogen ions are largely removed by percolating water, and the pH is consequently higher than in zone II. Whereas jarosite still occurs in zone II, jarosite in zone III is hydrolizing into ferri-hydroxides. Consequently the neojarositans are transformed either into neoferrans or neoferrans/quasijarositans. The latter can not be distinguished from the neoferrans/quasijarositans occurring in zone II, which were not yet transformed into neojarositans.

Along the most aerated voids each zone proceeds to greater depth than in the surrounding groundmass. Therefore transitions between the zones are tongueshaped. The tongues between zone I and the reduced subsoil respectively between zone II and I are formed along channels. The tongues between zone III and II are formed both along planes and channels. In these tongues phenomena as described above occur at a greater depth than in the zone where they occur dominantly. The tongues moreover show phenomena which can not be explained by using the concept of slow pyrite oxidation. In the voids in the tongues phenomena may occur which result from a rapid oxidation of pyrite and consequently rapid pH lowering. In zone I for instance jarosite framboids, sometimes with a pyrite nucleus, were found. They were formed in situ by a rapid oxidation. These jarosite framboids occur in great amounts in zone I of profile Haarlemmermeer 1 (75-109 cm). In this profile the groundwater level was lowered artificially in 1967 by some decimeters. In zone I of the profiles Haarlemmermeer 1 and Mijdrecht 2 spherical ferric nodules were found together with some gypsum. The occurrence of these phenomena may be explained by assuming that parts of the above mentioned profiles originally contained some calcium carbonate (or calcium ions on the adsorption complex). The initial stages of the pyrite oxidation consequently were the reactions 1, 2, 3 and 4 in Table 5, active in a calcareous environment.

Some goethite was found in profiles Haarlemmermeer 1, Zuidplas 1 and Mijdrecht 2 mainly in zone III. It was formed from the amorphous ferri-hydroxides by dehydration according to reaction 6. It was only found in the largest channels: the only places in these imperfectly drained soils which periodically are strongly dessicated.

ACKNOWLEDGEMENTS

We would like to thank Professor Dr. L. J. Pons for the description and sampling of the investigated profiles, and all the others that contributed, in any way, to the compilation of this article.

REFERENCES

Breemen, N. van, 1972. Soil forming processes in acid sulphate soils. Proceedings International Symposium on Acid Sulphate Soils, Wageningen, August 13-20 (in press).

Brewer, R., 1964. Fabric and mineral analysis of soils. John Wiley, London/New York/Sydney. 470 pp.

Dam, D. van and Pons, L. J., 1972. Micromorphological observations on pyrite and its pedological reaction products. Proceedings International Symposium on Acid Sulphate Soils, Wageningen, August 13-20, 1972 (in press).

Eswaran, H., 1967. Micromorphological study of a 'cat clay' soil. Pedologie 17, nr. 2, pp. 259-265.

Harmsen, G. W., 1954. Observations on formation and oxydation of pyrite in soil. Plant and Soil V, nr. 4, pp. 324-359.

Jongerius, A. and Heintzberger, G., 1963. The preparation of mammoth-sized thin sections. Soil Survey Papers, nr. 1, Soil Survey Institute, Wageningen.

Moorman, F. R., 1963. Acid Sulphate soils of the tropics. Soil Science, vol. 94, pp. 271-275.

Pons, L. J., 1964. A quantitative microscopial method of pyrite determination in soils. Proceedings Micromorphological Symposium, Arnhem. A. Jongerius, Ed., pp. 401-409.

Pons, L. J. and van de Kevie, W., 1969. Acid Sulphate soils in Thailand. Soil Survey Reports of the Land Development Department Bangkok, nr. SSR-81-65 pp.

Quispel, A., Harmsen, G. W. and Otzen, D., 1962. Contribution to the chemical and bacteriological oxydation of pyrite in soil. Plant and Soil IV, nr. 1, pp. 43-55.

Rickard, D. T., 1970. The origin of framboids. Lithos 3, pp. 269-298.

Slager, S., Jongmans, A. G. and Pons, L. J., 1970. Micromorphological of some tropical alluvial clay soils. J. Soil Sci., Vol. 21, nr. 2, pp. 233-241.

Soil Taxonomy, 1970. Selected chapters from the unedited text of the National Cooperative Soil Survey.

COLOURED ILLUSTRATIONS

For Figures 1, 4, 5, 6, 7, 9 and 11 refer to the following Plate:

Plate C1. Pyrite framboids occurring in clusters in a reed remnant; combined transmitted plane and incident light (x100).

Plate C4. Association of a neoferran and a quasijarositan; crossed polarizers (x160).

Plate C5. Well crystallized goethite in a reed remnant in a large void; crossed polarizers (x100).

Plate C6. Jarosite framboids occurring in voids; crossed polarizers (x160).

Plate C7. Jarosite framboids, locally with a nucleus of pyrite; crossed polarizers (x160).

Plate C9. Gypsum crystals, partly filling a channel, crossed polarizers (x100).

Plate C11. Aggregate of ferri-hydroxide spheres, showing shape and size of the pyrite framboids from which they originate. Some of the spheres still contain a nucleus of pyrite; plane light (x160).

[*Editors' Note:* The plate is reproduced here in black and white.]

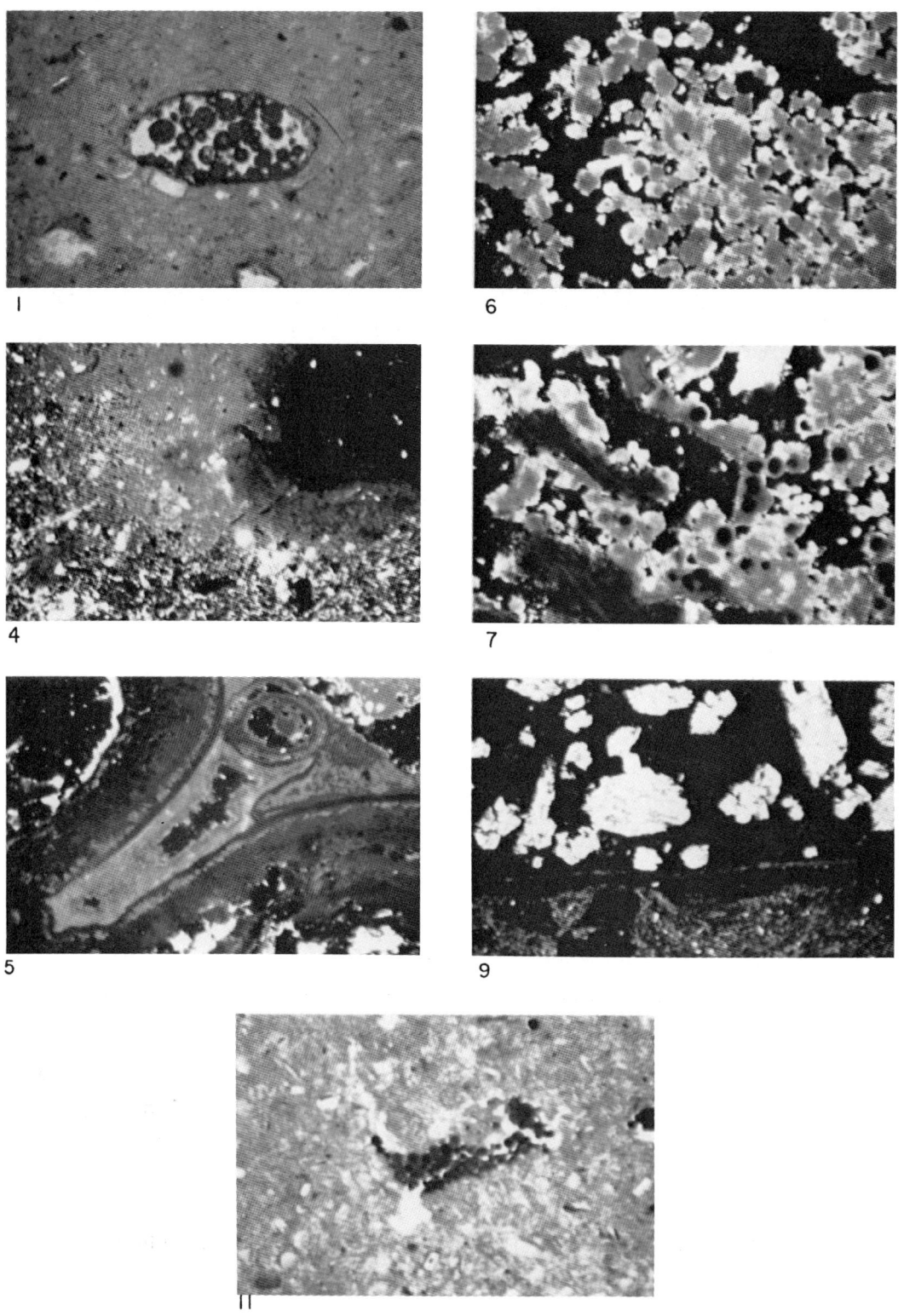

336

AUTHOR CITATION INDEX

Abrol, J. P., 296
Abtahi, A., 312
Acton, D. F., 217
Afton, C. J., 38
Albareda, J. M., 240, 269
Aleixandre, V., 240
Alexander, L. T., 20, 244, 249, 269
Alia, M. T., 269
Altemüller, H.-J., 7, 38, 177, 231, 240
American Geological Institute, 119
Anderson, R. E., 38
Aydinyan, R. K., 230
Avery, B. W., 240

Bäbler, S., 151
Babel, U., 137, 151, 152
Bal, L., 7, 100, 137, 262
Balster, C. A., 249
Banfield, C. F., 38
Baños, C., 132
Barratt, B. C., 137, 164, 165
Barzanji, S., 312
Baver, L. D., 85, 119
Beaudou, A. G., 36, 262
Beckmann, W., 7, 95, 209
Belisle, J., 38
Bellinfante, N., 132
Betekhtin, A. G., 313
Birina, A. G., 312
Bisdom, E. B. A., 7, 36, 37, 38, 312
Biswell, K. J., 8
Bloomfield, C., 20
Boersma, O., 7
Bolt, G. H., 95
Bouma, J., 7, 38
Bourbeau, G. A., 38
Bower, C. A., 296
Brasher, B. R., 248
Breeman, N. van, 334
Bresson, L. M., 36, 38
Brewer, R., 8, 20, 36, 41, 42, 95, 100, 101, 112, 119, 132, 137, 165, 182, 217, 223, 231, 240, 244, 248, 255, 262, 296, 334

Brinch-Hansen, J., 95
Brown, G., 240
Brown, R. W., 165
Bruckert, S., 36
Bulfin, M., 165
Bullock, P., 8, 36, 38, 137, 182
Buol, S. W., 244, 248
Butler, B. E., 85

Cady, J. G., 20, 36, 37, 240, 241, 244, 297
Carrol, D., 262
Chandler, R. F., 144
Chatelin, Y., 36, 262
Cline, M. G., 20, 231, 240, 244
Conry, M. J., 132
Coulson, J., 165

Dalrymple, J. B., 20, 37, 240, 255
Dam, D. van, 334
Davidson, S. E., 248
Davol, F. D., 244
De Coninck, F., 37, 101, 132
Deer, W. A., 165
Delvigne, J., 37, 38, 262
Dever, R. T., 269
Dobrovolsky, G. V., 312
Dobrovolsky, V. V., 38, 289
Ducloux, J., 7
Duncan, D. R., 248

Ehwald, E., 144
Eswaran, H., 132, 224, 312, 334

Fedoroff, N., 8, 36, 37, 137, 223
Fenton, G. R., 20
Feofarova, I. I., 231, 289, 312
FitzPatrick, E. A., 38, 119, 182
Flach, K. W., 37
Fowler, H. W., 165
Franzmeier, D. P., 248
Frei, E., 20, 95, 231, 240, 244
Fripiat, J. J., 240
Fry, W. H., 240

Author Citation Index

Gady, J. G., 269
Gamble, E. E., 241, 244
Gastuche, M. C., 240
Gedroiz, K. K., 95
Gerasimov, I. P., 230
Gerasimova, M. I., 311
Geyger, E., 8, 95, 209
Goertzen, J. V., 296
Gorbunov, N. I., 231
Greene-Kelly, R., 37
Großkopf, W., 151
Grossman, R. B., 248
Guertin, R. K., 38, 223

Haldane, A. D., 36, 240, 244
Handley, W. R. C., 151
Hanna, F. S., 312
Harmsen, G. W., 334
Hartman, F., 144
Hatch, F. H., 165
Heiberg, S. O., 144
Heintzberger, G., 334
Henstra, S., 312
Hole, F. D., 38, 244, 248
Holowaychuk, N., 244
Hoover, M. D., 144
Hopkins, L., 38
Howell, L., 38
Howie, R. A., 165
Hoyos, A., 269
Humbert, R. P., 20

Jackson, M. L., 244
Jacot, A. P., 20
Jager, A., 7, 8, 37, 182
Janse, A. R. P., 95
Johnston, J. R., 240
Jongerius, A., 7, 8, 20, 36, 37, 85, 95, 101, 119, 132, 137, 144, 165, 177, 182, 255, 334
Jongmans, A. G., 335
Jung, E., 119

Kühnelt, W., 20, 144
Katchinski, N. A., 85, 289
Kellogg, C. E., 244
Khosla, B. K., 296
Kilmer, V. J., 249
Kononowa, M. M., 152
Kowalinski, St., 37, 177, 182
Krumbein, W. C., 85
Kubiëna, W. L., 8, 11, 20, 37, 38, 86, 95, 112, 119, 132, 137, 144, 152, 165, 182, 209, 217, 223, 231, 241, 262, 269
Kulikov, A. V., 312
Kullmann, A., 95, 152

Laatsch, W., 95
Labenets, Ye. M., 231
Laruelle, J., 241
Leeper, G. W., 86
Little, W., 165
Lord, T. M., 38
Loughnan, F. C., 262
Ludwieg, F., 269
Lukashev, K. I., 112
Lund, Z. F., 297
Lundgren, H., 95
Lunt, H. A., 144

Müller, P. E., 144
Mückenhausen, E., 241
McCaleb, S. B., 20, 231, 241, 244
McKeague, J. A., 38, 223
McMillan, N. J., 86, 119, 217, 241
Mackenzie, R. C., 269
Mackney, D., 37
Matelski, R. P., 244
Maucorps, J., 101
Meunier, A., 38
Meyer, F. H., 152
Michalyna, W., 38
Michot, P., 101
Mick, A. H., 20
Miedema, R., 223
Milfred, C. J., 38
Minashina, N. G., 231, 241, 244, 289
Mitchell, J., 86, 119, 217, 241
Mitchell, W. A., 269
Mochalova, E. F., 262
Moorman, F. R., 334
Mortland, M. M., 217
Muñoz Taboadela, M., 269
Moss, H. C., 217
Murphy, C. P., 8, 38, 182
Murphy, P. W., 20

National Soil Survey Committee of Canada, 217
Nettleton, W. D., 37
Nitzsch, W., 95
Nyun, M. A., 241

Osman, A., 224
Osmond, D. A., 20, 217, 241
Otzen, D., 334

Pagé, F., 38, 223
Paneque, G., 38, 132
Parfenova, E. I., 231, 241, 262
Parsons, R. B., 249
Peterson, J. B., 20, 240, 241
Pettapiece, W. W., 38
Pettijohn, F. J., 86

Pol'skii, M. N., 289
Polynov, B. B., 231
Ponomareva, S. I., 289
Pons, L. J., 165, 334, 335
Popov, I. V., 231
Preobrazhenskiy, I. A., 312

Quispel, A., 334

Racz, Z., 177
Raeside, J. D., 20
Ramann, E., 144
Rastall, R. H., 165
Raymond, R. E., 269
Retzer, J. L., 244
Reuter, G., 8, 224
Reynders, J. J., 224
Rice, C. M., 86
Richards, L. A., 249
Rickard, D. T., 335
Righi, D., 101
Robertsen, B., 183
Robin, A. M., 101
Robinson, G. W., 86
Romans, J. C. C., 183
Romashkevich, A. I., 231, 262
Roose, E. J., 255
Ross, G. J., 38
Rozanov, A. N., 231
Ruchin, L. B., 112
Ruhe, R. V., 241, 249
Rutherford, G. K., 8, 37

St. Arnaud, R. J., 217
Sanchez Calvo, M. Del C., 240, 241, 269
Sander, B., 42, 95
Sarkisyan, S. G., 312
Saskatchewan Institute of Pedology, 217
Schaefer, G. M., 244
Scheffer, F., 269
Schlichting, E., 95
Schoonderbeek, D., 7, 37, 182
Schuffelen, A. C., 95
Sehgal, J. L., 261, 312
Sekera, F., 95
Selino, D., 36
Sherman, D. G., 244
Shoba, S. A., 312
Shrock, R. R., 86
Simonson, R. W., 244
Simpson, D. P., 165
Slager, S., 223, 335
Sleeman, J. R., 38, 95
Sloss, L. L., 85
Smith, G. D., 241
Smith, R. E., 38
Soil Conservation Service, 297

Soil Survey Staff, 1951, 165
Soil Survey Staff, 1960, 38, 165, 297
Soil Survey Staff, 1967, 165
Soil Taxonomy, 335
Sokolova, T. A., 312
Starykh, S. N., 231
Stephen, I., 20, 240, 241
Stevens, J. H., 183
Stockdill, S. M. J., 165
Stoops, G., 8, 36, 38, 101, 119, 132, 137, 224, 262, 312
Stremme, H., 231
Swanson, C. L. W., 20

Targulian, V. O., 255, 312
Tavernier, R., 241
Terzaghi, K., 95
Thiel, F., 312
Thorp, J., 241, 244
Titova, N. A., 262
Torrie, J. H., 38
Tressler, R. E., 249
Tselishcheva, L. K., 312
Turner, R. H., 8, 38, 182
Tursina, T., 8, 36, 137

U.S. Dept. of Agriculture, 86, 249

Valassis, V., 248
Valentine, K. W. G., 38, 223
Van Vliet-Lanoe, B., 8, 183
Van de Kevie, W., 334
Velde, B., 38
Veneman, P. L. M., 38
Vepraskas, M. J., 38
Verheye, W., 224
Vikulova, M. F., 230

Walker, J. L., 248
Wang, C., 38
Welte, E., 269
Whiteside, E. P., 217
Whittig, L. D., 20, 269
Wieder, M., 262
Williamson, W. O., 241, 249
Wittich, W., 152

Yaalon, D. H., 240, 262
Yarilova, E. A., 231, 241
Yarzhemskiy, Ya. Ya., 312
Yesilov, M. S., 248

Zachariae, G., 11, 152
Zakharov, S. A., 86
Zimmerman, K., 11
Zingg, Th., 86
Zussman, J., 165

SUBJECT INDEX

Numbers printed in bold type refer to pages where the terms are defined

Aggregate, 180, 184–209, 275, 304
Alfisols, 18, 63, 129, 237, 238, 242, 248
Andosols; 129
Argillan. *See* Coating
Argillicol, **154**
Aridisols, 245, 248
Arrangement. *See* Fabric

Beidellite, 225
Braunerde, 17, 163, 258, 259
Braune Vererdung, 265, 270
Braunlehm
   earthy, 265
   fabric, 3, 163, 259
   soiltype, 234, 258
   Teilplasma, 259
Brickearth, 237
Brown earths, 16, 17, 57, 59, 63, 227

Calcite, 227, 272, 273, 280–282, 322, 325
Carbonates, 259, 280, 304, 311
C/f-related distribution, **103**–111. *See also*
   Elementary fabric; Related
   distribution pattern
   chitonic, 99, **105,** 116
   enaulic, 99, **105**
   gefuric, 99, **105,** 117
   monic, 99, **104**
   porphyric, 99, **105,** 106
Chernozem, 57, 210–212, 227, 276
Chestnut brown soil, 57, 58
Cinnanon-brown soil, 225, 227
Clay
   accumulation, 225
   band, 234
   coating. *See* Coating
   cutan. *See* Coating
   fibers, 221
   film. *See* Coating
   illuvial, 28

illuviation, 236, 245. *See also* Coating
incrustation, 221, 225, 226. *See also*
   Coating
matrix, 17
migration, 17, 240
orientation, 225–230, 232–240. *See also*
   Orientation of plasma; Plasmic fabric
peptization, 48, 226, 229
pseudomorph, 5, 225, 287
scale, 221, 225
secondary, 221, 225–230
skin. *See* Coating
translocation, 21
Cleavage blocks, 180, 211, 213
Coarse material, 99, 103, 114, 122
Coating (of clay)
   artificial formation, 234, 243
   complex, 176
   compound, 176
   definition, **243, 250**
   destruction, 245–248
   distribution, 243, 287
   formation, 52, 62, 237, 242–244
   incrustation, 226
   occurrence, 19, 222, 242, 251
   synonyms, 221
Crumb, 174, 181, 198, 209, 276

Distribution pattern, 88
Domain, 220. *See also* Orientation of plasma

Efflorescence, 88
Elementary fabric, 2, 40, 44, **47**–74, 92, 98,
   103. *See also* Related distribution
   pattern
   agglomeratic, 40, 49, **64,** 65
   bleached sand, 65
   chernozem, 50, 58
   chlamydomorphic, 40, 49, **61**–63, 116
   intertextic, 40, 50, **57**–61

Subject Index

magmoidic, 40, **68**–70
mortar, 40, **70**
plectoamictic, 40, **63,** 64, 118
porphyropectic, 40, **54**–56
porphyropeptic, 40, **56,** 57
rendzina, **67**
Excrements
   disintegration, 139, 141
   evolution, 16
   fresh, 168
   occurrence, 153
   similarity, 25
Experiments, 19, 29, 245

Fabric, 40, 41, 43–46, **89**–96, 262. See also Related distribution pattern; Structure
   banded, **118,** 182, 210-**212**-217
   cleavage block, 213
   complex, 118
   elementary. See Elementary fabric. See also Related distribution pattern
   Erde, 3, 17
   formal, 92, 96
   functional, 92, 96
   isoband, 182, 210-**212**-217
   Lehm, 117
   soil. See Soil fabric
Ferran, 28
Ferri-argillan. See Coating
Field grading, **77.** See also Texture
Fine material, 99. See also Plasma
Forest soil, 162, 276, 287
Fossil formation, **80**
Fragments, 181, 209
Framboid, 316
Framework member, 99, **114**

Glauberite, 309
Gley, 27
Gleysoil, 210, 212, 213
Grain, 99, **122**
Gray-brown podzolic soil, 18, 237, 238, 239, 242
Gray earth, 57
Groundwater soil, 62
Gypsum, 283, 301–311, 313–333

Halite, 303, 309
Horizon
   albic, 121, 123
   argillic, 28, 33, 128, 236, 242–249, 251
   cambic, 290
   eluviated, 210
   illuvial, 56, 64, 128, 226, 237, 238, 280
   organic, 16. See also Humus layers; Mull; Moder
   oxic, 290
   placic, 174
   pseudosandy, 307
   spodic, 24, 25, 128
Humicol, **154**
Humiskel, **154**
Humus, 277–280. See also Organic matter
   amorphous, 135, 161, 175
   coating, 57
   dispersed, 140, 141, 175, 287
   particles, 66
   sclerotial, 162
   zoogenic, 140, 143
Humus-carbonate soil, 67
Humus form, 134, 138, 146. See also Moder; Mull; Raw humus
Humus layers
   F, 16
   H, 16, 150
   L, 16, 150

Illuviation, 286, 295
Image analysis, 31
Inceptisols, 317, 314, 321
Incrusted, 221, 225. See also Clay; Coating
Iron
   coating, 238
   concretion, 226, 228, 284
   evolution, 262–269
   ferran, 28
   neoferran, 317
   newformation, 284, 317
   pan, 174
Iwatoka, 235

Jarosite, 317

Krasnozem, 287

Laterization, 18, 235, 265, 267, 270
Laterite (lateritic) soil, 55, 287
Linear extensibility, 295
Lithiskel, **154**
Litho-relict, **80**

Matrix, 41, 99, **113**
Mechanical accumulation, 139, 251–257
Mesomorphology, 299, 300
Micro-aggregate, 140, 174, 276. See also Excrements
Micromorphology, 1
Micromorphometry, 1
Micropedology, **1,** 12, 15
Microstructure. See Structure
Mirabilite, 302, 307
Moder, 17, **139,** 140, 153, 161
   rendzina, 154, 162
   silica, 162
   swollen, 162
Mollisols, 248, 324

Subject Index

Morphoanalytical, 2, 4, 134, 258
Morphogenetic, 2, 40, 134
Muck soil, 59, 65
Mull, 17, **142,** 154, 163
Mull-like moder, 162, 163

Neoalban, 28
Neoferran, 317
Normal basic related distribution pattern, 100, **123.** *See also* Related distribution pattern
   granic, **123**
   plasmic, 100, **124**
   phyric, 100, 124
   porphyric, 100, **125**

Organic matter, 24, 134–136, 138–144, 166–176, 277–280. *See also* Humus
   accumulations of, **168**
   agglomerated, 171
   aggregate, **168**
   aggregation, 139
   amorphous, 135, 136, 161, **172,** 175
   argillicol, 154
   coating, 66, **168**
   expanded, 155
   fine material, **149**
   humicol, **154**
   humiskel, **154**
   lithiskel, **154**
   monomorphic, 136, **167**
   mullicol, 154
   polymorphic, 136, **167**
Organic residues, 147, 277
Orientation of plasmas. *See also* Plasmic fabric
   continuous, **87**
   domain, 220
   flecked, **87**
   fluidal structures, 220
   random fiber, 221–229
   scales, 221, 225
   steaks, 17
   stress, 28, 245
   striated, **87**
Ortstein, 284
Oxisols, 131, 251, 290

Peat, 161
Pedality, 76, **77.** *See also* Structure
Pedological feature, 4, 41, **80,** 82, 114, 121. *See also* specific pedological features
Pedoplasmation, 26, 260, **296**
Pedo-relicts, **80**
Pellets, 25, **168**
Phenoclast, 114
Phytolith, 287
Plant remain, 147, 171

Plasma, 2, **40, 47, 79,** 99, 107, **121,** 135, 210
   coagulated, 174
   concentration, **80**
   flocculation, 48, 59, 258
   formation, 260
   orientation. *See* Orientation of plasma; Plasmic fabric
   pectized, 98
   peptized, 51, 98, 220, 258
   separation, **80,** 221
Plasmic fabric, 4. *See also* Orientation of plasma
   asepic, 221
   formation of, 247
   lattisepic, 293
   omnisepic, 221
   striated, 223, 225, 287
   vosepic, 293, 296
Plasmic soil material, **154**
Podzol. *See* Spodosol
Podzolic, 210, 212, 213, 225, 228, 287
Polynite, 5
Pores, 122, 180
   biopores, 276
   cavities, 181, 209
   classification of, 184–209
   channels, 181
   cracks, 181, 276
   fissures, **185,** 209
   intergrain, 181
   vesicle, 181
   vugh, 181
Porepattern, 180
Porosity, 275
Peds, 81
Pseudogley, 258, 268, 270
Pseudomorphs, 247, 287, 295
Pyrite, 228, 313–333

Randomfiber clay, 221, 225, 227, 229
Ranker, 16
Raw humus, 65, **138,** 146–152, 154, 162
Red earth, 17, 56, 63, 69, 258, 270
Red-brown earth, 18
Red-yellow podzolic soil, 238
Referred orientation, 87
Related distribution pattern. *See* Elementary fabric; C/f related distribution; Normal basic related distribution pattern; Related distribution of framework members and matrix; Related distribution of plasma and skeleton grains; Specific related distribution pattern
Related distribution of plasma and skeleton grains, 88. *See also* Related distribution patterns
   agglomeratic, **88**
   granular, **88**

Subject Index

    intertextic, **88**
    porphyritic, **88**
    porphyroskelic, 114
Related distribution of framework members and matrix, 114. *See also* Related distribution pattern.
    banded, **118**
    chlamydic, **116**
    fragmic, **116**
    fragmoidic, **116**
    gefuric, **117**
    granic, 100, **114**
    granoidic, 100, **116**
    granular, 114
    plectic, 114
    porphyroskelic, **114**
Rendzina, 16, 40, 67, 162
Roterde. *See* Red earth
Rote Vererdung, 265, 270
Rotlehm, 17, 235, 258, 265, 267, 270
Rubefaction, 18, 235, 258, 265, 267, 270. *See also* Rotlehm.

Saline soil, 299
Salt
    crust, 301
    mineral, 280
    neoformation, 260
Saprolite, 26, 290
Scale, 221, 225
Sedimentary relict, **80**
Shrink-swell potential, 247
Sierozem, 225, 228
Skeletal function, 99
Skeleton grain, 2, 4, 40, **47, 79,** 107, 120, 154, 210
S-Matrix, **113,** 121, 156, 251
Soil fabric. *See also* Fabric
    concept, 17, 45, 180
    definition, 43, 76, **94, 96,** 156
    genesis, 260
Soil organism, 138
Solod, 226
Solonchak, 301
Solonetzic soil, 210, 212, 213, 226
Specific related distribution patterns, 100, 123, **128.** *See also* Related distribution patterns

    agglutinic, 100, **131**
    congelic, 100, **129**
    dermatic, 100, **128**
    intertextic, 100, **128**
    reticulic, 100, **131**
Spodosol, 18, 24, 56, 62–64, 66, 124, 129, 166, 225, 233
Streaks (birefringent), 17, 220. *See also* Plasma orientation
Structure, 41, 45, 76, 180. *See also* Fabric
    basic, **81,** 82
    complete, **82,** 84
    cracked, 181
    crumbly, 181, **205**
    cryic, 182
    elementary, 81
    fluidal, 200. *See also* Orientation of plasma
    fragmented, 181, 202
    formation, 207
    jointed, 181, 202
    massive, 155
    microstructure, 155
    pelleted, 155
    platy, 155
    pore-, **205**
    porous, 181
    primary, **81,** 82
    secondary, **81,** 82
    single grain, 155
    soil structure, **77,** 180
    spongy, 155, 181, **205**
    types, 201–207

Takyr, 226, 229
Terra fusca, 234. *See also* Braunlehm
Texture, 41, 45, 46
    soil, **77**
Thenardite, 302, 307, 309
Thin-section preparation, 15
Translocation, 250, 280, 287

Ultisols, 129, 290
Ultra-thin section, 31

Wind-erosion, 60

Yellow-gray earth, 18

About the Editors

GEORGES STOOPS obtained M.Sc. and Ph.D. degrees in geology and mineralogy from the State University of Ghent, Belgium. From 1962 to 1967, he was assistant and lecturer at Lovanium University in Kinshasa, Zaïre. In 1967 he became a lecturer, and later professor of mineralogy and micropedology, at the State University of Ghent and at the International Training Centre for Post Graduate Soil Scientists. Dr. Stoops has also been a visiting professor or scientific adviser at several foreign universities and institutes (e.g. Cameroon, Morocco, Malaysia, Tunisia).

Since 1969 he has been a member of the International Working Group on Soil Micromorphology. He is also chairman of the International Soil Science Society's Subcommission on Micromorphology.

Dr. Stoops's main research interests are soil micromorphology, the petrography of rock weathering and laterites, and the mineralogy of arid soils. He has published more than 70 papers on these topics.

HARI ESWARAN obtained his Ph.D. from the University of Ghent, Belgium, where he was a teaching assistant until 1976. In 1976 and 1977 he was visiting professor at Cornell University and in 1980 accepted the position of Program Leader of the Soil Management Support Services, Soil Conservation Service, U.S. Department of Agriculture. He has published more than 80 papers and is currently the editor of *Soil Taxonomy News*.